ANIMAL MANURE ON GRASSLAND AND FODDER CROPS. FERTILIZER OR WASTE?

Developments in Plant and Soil Sciences

1. J. Monteith and C. Webb, eds.,
   Soil Water and Nitrogen in Mediterranean-type Environments. 1981. ISBN 90-247-2406-6
2. J.C. Brogan, ed.,
   Nitrogen Losses and Surface Run-off from Landspreading of Manures. 1981. ISBN 90-247-2471-6
3. J.D. Bewley, ed.,
   Nitrogen and Carbon Metabolism. 1981. ISBN 90-247-2472-4
4. R. Brouwer, I. Gašparíková, J. Kolek and B.C. Loughman, eds.,
   Structure and Function of Plant Roots. 1981. ISBN 90-247-2510-0
5. Y.R. Dommergues and H.G. Diem, eds.,
   Microbiology of Tropical Soils and Plant Productivity. 1982. ISBN 90-247-2624-7
6. G.P. Robertson, R. Herrera and T. Rosswall, eds.,
   Nitrogen Cycling in Ecosystems of Latin America and the Caribbean. 1982. ISBN 90-247-2719-7
7. D. Atkinson et al., eds.,
   Tree Root Systems and their Mycorrhizas. 1983. ISBN 90-247-2821-5
8. M.R. Sarić and B.C. Loughman, eds.,
   Genetic Aspects of Plant Nutrition. 1983. ISBN 90-247-2822-3
9. J.R. Freney and J.R. Simpson, eds.,
   Gaseous Loss of Nitrogen from Plant-Soil Systems. 1983. ISBN 90-247-2820-7
10. United Nations Economic Commission for Europe.
    Efficient Use of Fertilizers in Agriculture. 1983. ISBN 90-247-2866-5
11. J. Tinsley and J.F. Darbyshire, eds.,
    Biological Processes and Soil Fertility. 1984. ISBN 90-247-2902-5
12. A.D.L. Akkermans, D. Baker, K. Huss-Danell and J.D. Tjepkema, eds.,
    Frankia Symbioses. 1984. ISBN 90-247-2967-X
13. W.S. Silver and E.C. Schröder, eds.,
    Practical Application of Azolla for Rice Production. 1984. ISBN 90-247-3068-6
14. P.G.L. Vlek, ed.,
    Micronutrients in Tropical Food Crop Production. 1985. ISBN 90-247-3085-6
15. T.P. Hignett, ed.,
    Fertilizer Manual. 1985. ISBN 90-247-3122-4
16. D. Vaughan and R.E. Malcolm, eds.,
    Soil Organic Matter and Biological Activity. 1985. ISBN 90-247-3154-2
17. D. Pasternak and A. San Pietro, eds.,
    Biosalinity in Action: Bioproduction with Saline Water. 1985. ISBN 90-247-3159-3
18. M. Lalonde, C. Camiré and J.O. Dawson, eds.,
    Frankia and Actinorhizal Plants. 1985. ISBN 90-247-3214-X
19. H. Lambers, J.J. Neeteson and I. Stulen, eds.,
    Fundamental, Ecological and Agricultural Aspects of Nitrogen Metabolism in Higher Plants. 1986. ISBN 90-247-3258-1
20. M.B. Jackson, ed.
    New Root Formation in Plants and Cuttings. 1986. ISBN 90-247-3260-3
21. F.A. Skinner and P. Uomala, eds.,
    Nitrogen Fixation with Non-Legumes. 1986. ISBN 90-247-3283-2
22. A. Alexander, ed.
    Foliar Fertilization. 1986. ISBN 90-247-3288-3
23. H.G. v.d. Meer, J.C. Ryden and G.C. Ennik, eds.,
    Nitrogen Fluxes in Intensive Grassland Systems. 1986. ISBN 90-247-3309-X
24. A.U. Mokwunye and P.L.G. Vlek, eds.,
    Management of Nitrogen and Phosporus Fertilizers in Sub-Saharan Africa. 1986. ISBN 90-247-3312-X
25. Y. Chen and Y. Avnimelech, eds.,
    The Role of Organic Matter in Modern Agriculture. 1986. ISBN 90-247-3360-X
26. S.K. De Datta and W.H. Patrick Jr., eds.,
    Nitrogen Economy of Flooded Rice Soils. 1986. ISBN 90-247-3361-8
27. W.H. Gabelman and B.C. Loughman, eds.,
    Genetic Aspects of Plant Mineral Nutrition. 1987. ISBN 90-247-3494-0
28. A. van Diest, ed.,
    Plant and Soil: Interfaces and Interactions. 1987. ISBN 90-247-3535-1
29. United Nations, ed.,
    The Utilization of Secondary and Trace Elements in Agriculture. 1987. ISBN 90-247-3546-7
30. H.G. v.d. Meer, R.J. Unwin, G.C. Ennik and T.A. van Dijk, eds.,
    Animal Manure on Grassland and Fodder Crops. Fertilizer or Waste? 1987. ISBN 90-247-3568-8

Animal Manure on Grassland and Fodder Crops. Fertilizer or Waste?

Proceedings of an International Symposium of the European Grassland Federation, Wageningen, The Netherlands, 31 August–3 September 1987

Edited by

H.G. VAN DER MEER
Centre for Agrobiological Research (CABO)
Wageningen, The Netherlands

R.J. UNWIN
Agricultural Development and Advisory Service (ADAS)
Woodthorne, Wolverhampton, U.K.

T.A. VAN DIJK
Institute for Soil Fertility (IB)
Haren, The Netherlands

G.C. ENNIK
Centre for Agrobiological Research (CABO)
Wageningen, The Netherlands

1987 **MARTINUS NIJHOFF PUBLISHERS**
a member of the KLUWER ACADEMIC PUBLISHERS GROUP
DORDRECHT / BOSTON / LANCASTER

Distributors

for the United States and Canada: Kluwer Academic Publishers, P.O. Box 358, Accord Station, Hingham, MA 02018-0358, USA
for the UK and Ireland: Kluwer Academic Publishers, MTP Press Limited, Falcon House, Queen Square, Lancaster LA1 1RN, UK
for all other countries: Kluwer Academic Publishers Group, Distribution Center, P.O. Box 322, 3300 AH Dordrecht, The Netherlands

Library of Congress Cataloging in Publication Data

ISBN-13: 978-94-010-8137-5 e-ISBN-13: 978-94-009-3659-1
DOI: 10.1007/978-94-009-3659-1

This Symposium was organized by
The Netherlands Society for Grassland and Fodder Crops

Organizing Committee

W.H. Prins, Haren
T.A. van Dijk, Haren
Mrs. E.J.L. Hotke-Staal, Wageningen
H. Korevaar, Lelystad
E.A. Lantinga, Wageningen
H.G. van der Meer, Wageningen
P.J.M. Snijders, Wageningen

Financial support for the Symposium was provided by:
- Agricultural University Wageningen
- Financieringsoverleg 'Mestonderzoek', Wageningen
- Netherlands Fertilizer Institute (NMI), The Hague
- Netherlands Society for Grassland and Fodder Crops (NVWV), Wageningen

Preface

On the occasion of its twenty-fifth anniversary, in 1985, the Netherlands Society for Grassland and Fodder Crops (NVWV) agreed to organize an International Symposium on a topic related to intensive grass and fodder production systems. The theme selected was "Animal manure on grassland and fodder crops: Fertilizer or waste?" This Symposium was organized under the auspices of the European Grassland Federation and held at the International Agricultural Centre in Wageningen from 31 August to 3 September 1987.

The problems connected with the disposal of animal waste have received much attention in recent years, especially in regions with intensive animal husbandry. Whereas the production of animal manure per hectare agricultural land increased strongly, the need for it decreased because of the introduction of cheap inorganic fertilizers which are easier to handle and have a more reliable effect on crop growth. As a consequence, many farmers dispose of animal manure as cheaply as possible, whilst avoiding damage to grassland and crops and paying little attention to effective use of the plants nutrients contained in the manure. Present practices of manure handling and application often lead to environmental problems. The rise in awareness of these problems renewed interest in possibilities to improve the utilization of nutrients from animal manure in crop production. Research on this topic has been stimulated in many countries during the last decade and the aim of this Symposium was to review and assess present-day knowledge. These Proceedings contain 16 invited papers dealing with different aspects of the use of animal manure on grassland and fodder crops and 29 short communications summarizing information presented on posters.

The Organizing Committee aimed to have the printed Proceedings available at the start of the Symposium. This meant that the authors had to submit the first draft of their paper well before the end of 1986. In general, they succeeded very well in doing this, leaving enough time for a good interaction between editors and authors on the contents of the papers. We would like to thank the authors for their co-operation and for their patience in dealing with our comments. We hope that authors and readers will enjoy the results of our common efforts.

For review of some invited papers and most of the short communications, we acknowledge with thanks the help of R.B. Thompson (Hurley), K. Dilz, P.A.I. Ehlert, H.G. van Faassen, A. de Jager, L.C.N. de la Lande Cremer,

G. Lebbink, J. van der Molen, J.J. Neeteson, M. van Noordwijk, W.H. Prins, G.J.G. Rauw, C.M.J. Sluijsmans and K.W. Smilde (Haren), J.H. Geurink, J.J.M.H. Ketelaars, H. van Keulen, J.H.A.M. Steenvoorden and Mrs. G.W.J. van de Ven (Wageningen).

Finally, we thank the staff of Martinus Nijhoff Publishers for their co-operation in publishing these Proceedings.

Wageningen, June 1987

H.G. van der Meer
R.J. Unwin
T.A. van Dijk
G.C. Ennik

CONTENTS

Value of animal manures: changes in perception

W P WADMAN, C M J SLUIJSMANS and
L C N DE LA LANDE CREMER
Institute for Soil Fertility, P.O. Box 30003, 9750 RA Haren, The Netherlands

Key words: animal manure, value and use, history, manure surpluses

Abstract. The paper describes the development in the production and appreciation of animal manure in Western Europe, mainly in The Netherlands, during the last 100 years.

In the past, animal manure was highly valued as a source of plant nutrients. Animal manure can still be considered as such, the more so as its intrinsic value has risen due to higher soil fertility levels and the use of mineral additives in animal nutrition. The need for these nutrients has decreased, however, since mineral fertilizers have shown to be adequate substitutes.

In certain regions of Europe the amount of animal manure produced has increased greatly as a consequence of increasing livestock density, which has occurred mainly since 1950. However, the area planted with crops that are most responsive to organic-manure application has not increased proportionally. Therefore the perceived value of the available animal manure supply has decreased further.

In order to prevent detrimental effects on the environment a number of countries have restricted the use of animal manure. This may be a third reason for a less positive view of animal manure compared with earlier days.

It can be concluded that, depending on the degree of intensification, animal manures in Western Europe have turned from a precious resource into a waste product.

Introduction

The oldest agricultural cropping system, shifting cultivation, did not depend on animal manures for crop production. In the course of time, cropping systems were developed in which the supply of nutrients to arable land was drastically increased by the use of animal manures. Manures were so important that animal herds were maintained mainly for their manure [45]. In these systems the nutrients that were applied were actually obtained from the rangeland that produced the fodder or the litter. In this way an improvement of soil fertility in one place was accompanied by a loss of fertility in another. In this form of agriculture non-leguminous crops could benefit from nitrogen (N) fixed from the air elsewhere. This input of N from biological N fixation meant that in most of these systems phosphorus (P) limited yield more strongly than N [13].

The need for manure depended on the nutrient content of the soil. Therefore, a rather exceptional situation existed around 1800 on the rela-

H.G. v.d. Meer, et al. (eds.), Animal Manure on Grassland and Fodder Crops.
 ISBN 978-94-010-8137-5.

tively rich marine clay soils in the northern Netherlands, where animal manures were already looked upon as a waste product [24].

Other ways of manuring employed were the uncovering of fertile soil layers, the relocation of soil which had been enriched with farmyard manure (FYM), the use of green manure and the application of refuse and night soil from towns [17, 45]. With these practices farmers in Western Europe succeeded in raising grain yields to about 1–2 $t\,ha^{-1}$.

After the introduction of mineral fertilizers in the mid-19th century, it still took several decades before the farmers became accustomed to the use of these new materials [41]. Stocking densities increased. The animal feed was mainly produced on the livestock farms themselves and the use of imported concentrate feeds was limited. The need for animal manure as a source of nutrients decreased, but animal manure did not become a waste product, although liquid manure was not always saved. This can be explained by assuming that it was profitable to save nutrients instead of buying them, or that animal manure was appreciated for effects which could not be expected from mineral fertilizers. The latter is illustrated by the fact that research was started in 1935 to create humus, as found in the fertile Black Soils, from FYM [24].

In periods when mineral fertilizers were scarce, animal wastes were again highly valued as a source of nutrients. During the second World War the government of The Netherlands encouraged the construction of liquid manure pits to save nutrients [1] and after the war a part of the Marshall Funds was allocated to finding methods to optimize the storage and use of manures [24].

After 1950 great changes took place in the structure of agriculture. New methods of storing animal manures were developed because of the scarcity of labour. Saving the nutrients was no longer the main argument; separate collection of the liquid and solid fractions of animal manure made way for the collection of slurry. In Central Europe this development had already taken place. Livestock densities and the use of concentrates increased.

Today the value of mineral fertilizers has been proven. The value of animal wastes is determined by their content of available nutrients for crop growth and of constituents contributing to "extra" effects on crop yield. When undesirable components are also taken into account, this value can be described as its intrinsic value. However, a potential user of animal manure considers not only the intrinsic value. To him its usefulness is determined by the extent to which animal manures can be used advantageously under his specific circumstances.

In the following chapters developments in the intrinsic value of animal manures and their use, including their environmental impact, are considered.

Developments in the amount of animal manure produced

After 1950 in Europe the number of farm animals increased (Table 1). A combination of factors contributed to this development [9, 37, 43]:

- increased forage production combined with a greatly increased use of concentrates;
- developments in disease control for large groups of animals;
- a large increase in labour productivity;
- an increase of prosperity which increased the demand for meat and dairy products;
- the foundation of the Common Market which enlarged the market outlets.

This development was not equal in all parts of Europe, so that large differences in livestock density now exist between countries (Table 2) and amongst regions (Fig. 1, [33]). With increasing livestock densities the production of animal feeds and the application of animal manures were increasingly divorced. Nowadays animal feeds are often grown elsewhere and frequently abroad. Systems of animal production which were more or less non-land-dependent made their appearance, especially in western and northern Belgium, southern, central and eastern Netherlands and in regions of the Federal Republic of Germany, Denmark and France [33]. The increase in the number of animals resulted in an increase in the production of animal manure. The development of the production of animal manure in The Netherlands is given in Table 3.

The method of storage of animal manures changed. Instead of separate storage of faeces (mixed with straw) and liquid manure, combined storage (slurry) has become the vogue, although in certain regions of Europe animal wastes are still mainly available as FYM [43].

Table 1. Number of cattle and pigs in some E.C. countries in December 1983 relative to the mean number over the period 1957–1960 (= 100). Calculated after data from Eurostat (Luxembourg)

	Number of cattle in 1983	Number of pigs in 1983
	(1957–1960 = 100)	
The Netherlands	162	417
Belgium and Luxembourg	114	324
Federal Republic of Germany	126	154
United Kingdom*	119	127
France	126	134

* mean over 1957–1959 = 100

Table 2. Total number of livestock and the livestock density in 1982 in some countries, calculated after data from [10]

	Total livestock (1000 CE*)	Livestock density (CE per ha agricultural land)
The Netherlands	5,022	2.5
Belgium and Luxembourg	2,944	2.0
Federal Republic of Germany	10,667	0.9
United Kingdom	11,600	0.6
France	16,100	0.5

* CE = Cattle Equivalent: Number of animals producing 90 kg N yr^{-1} in their manure (as defined in [39]).

Changes in the intrinsic value of animal manures

The intrinsic value of animal manures is determined by the kind, amount and efficiency of use of its components. The components playing a role are plant nutrients (present or added), other chemical constituents, water, organic matter, pathogens and weed seeds.

The total amount of nutrients produced in animal manure of housed animals in The Netherlands is shown in Fig. 2. As a comparison the amount of nutrients applied in the form of mineral fertilizers is also included [27]. The increase in livestock density (Table 1) and consequently the increase in animal manure production as given in Table 3 is reflected in these figures. As livestock density is considerably lower in other countries, the amount of nutrients in animal manure produced per ha is smaller in those countries than in The Netherlands (Table 2 and [10]).

The nutrient content of animal manures is mainly determined by the nutrient content and digestibility of the animal feed. The mineral composition of the feed is determined by its constituents, the fertility of the soil on which it was produced and the amount of additives, e.g. P, copper (Cu)

Table 3. Production of manure by housed animals (1000 t yr^{-1}) and the area of agricultural land (1000 ha) in The Netherlands from 1950 [3, 4]

		1950	1960	1970	1982
Cattle	(as slurry)*	19,662	24,383	28,050	34,102
Fattening calves	(as slurry)	5	172	964	1,253
Pigs	(as slurry)	2,643	3,918	8,873	14,737
Poultry	(as solid manure)	946	997	1,103	1,903
Agricultural land		2,235	2,339	2,187	2,000

* Production in 180 days; steers for fattening included; fattening calves excluded

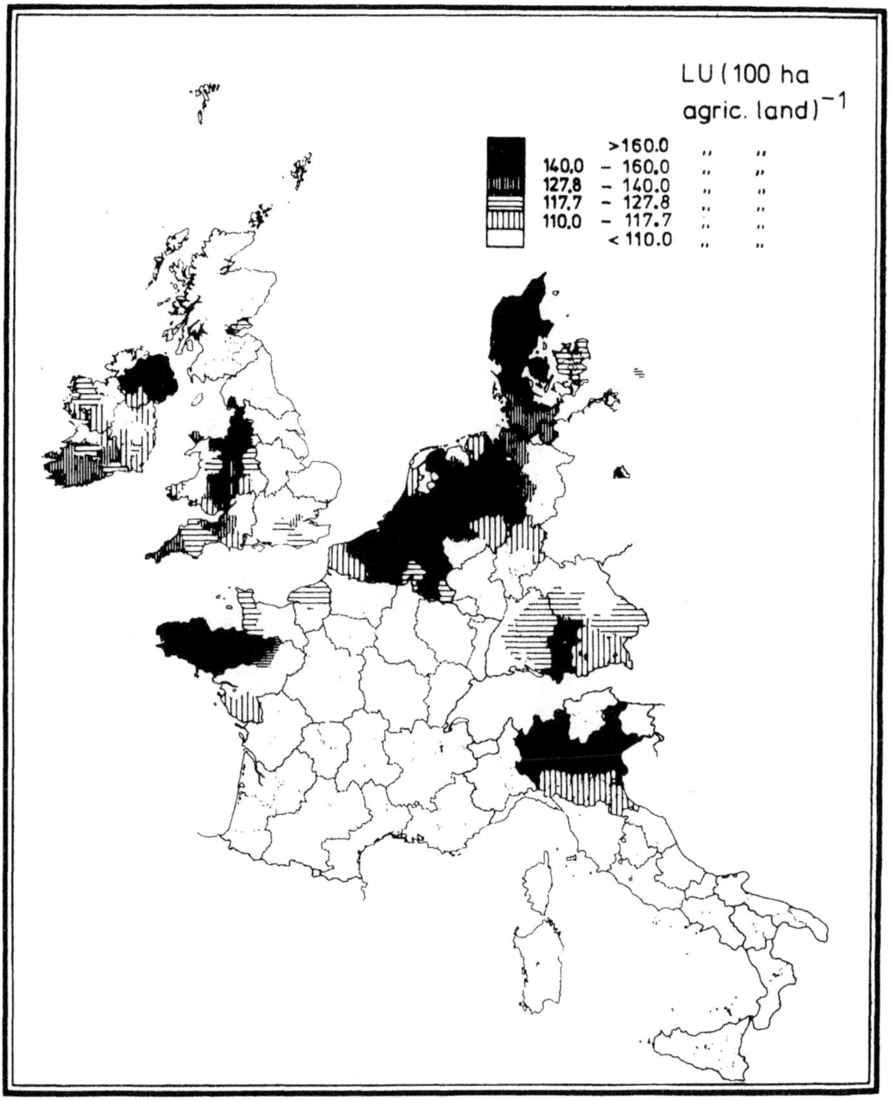

Figure 1. Livestock density in the period 1970–1974 in some regions of the EC (u: livestock unit), from [33]. The definition of LU approximates that of CE in Table 2 [33].

and zinc (Zn), which are used to improve animal performance. In general, it is assumed that improved soil fertility led to higher nutrient contents of animal manures in the early years of the 20th century [7, 24]. However, no reliable data are available to support this assumption. During the second World War the nutrient contents of animal manures decreased (Fig. 3) as a result of diminished availability of concentrates and fertilizers [1]. After the

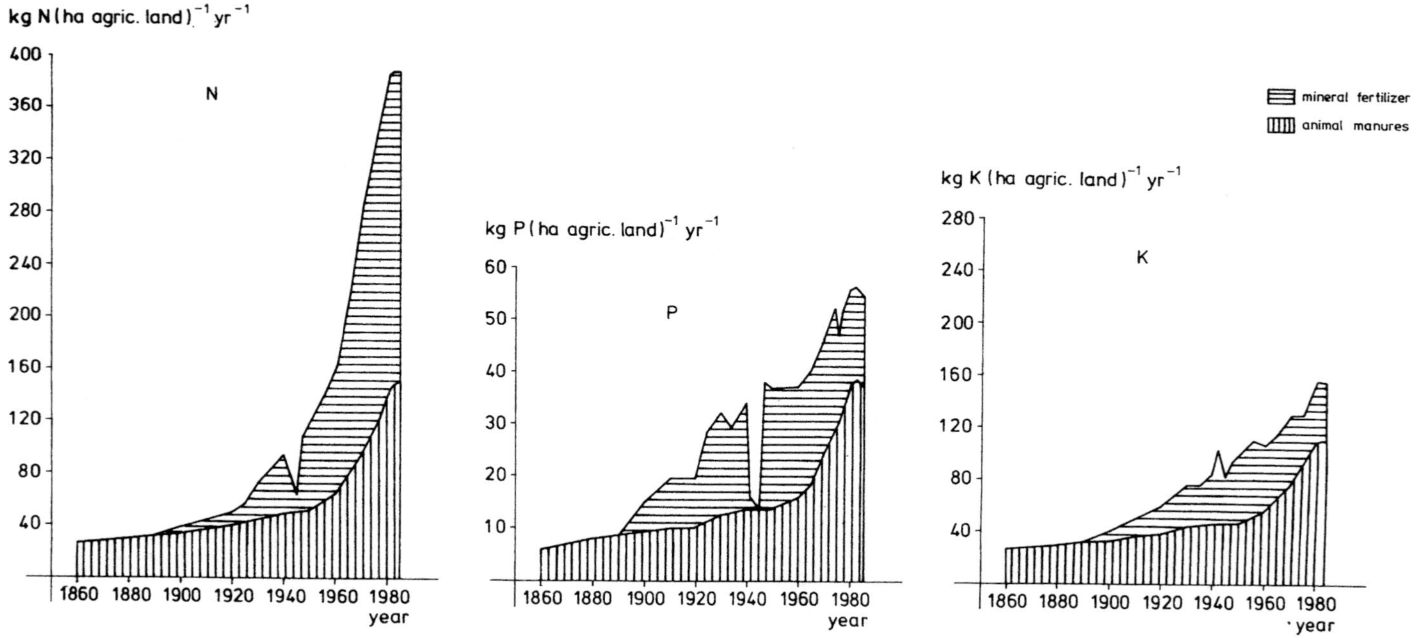

Figure 2. Changes in the amounts of nutrients applied in the manure of housed animals and in mineral fertilizers in The Netherlands (in kg per ha agricultural land per year) [27].

war, in addition to an increase in the use of concentrates and fertilizers, other factors played a role in determining the nutrient contents of animal manures. A change took place in the production of roughages: silage maize increasingly replaced mangolds, sugar beet tops and turnips. Furthermore grass for silage was harvested earlier in spring. The contents of a number of nutrients in animal manure decreased after the war. In The Netherlands the potassium (K) content of FYM started to decrease in about 1949 (Fig. 3) as a result of a more careful use of K fertilizers to prevent the occurrence of grass tetany. Recently the P content of animal feeds, and therefore also of animal manures, has decreased as a result of lower additions of P to feeds (Table 4). The Cu content of the feed for fattening pigs has been lowered several times in agreement with EC regulations and will be lowered again. This also affects its Zn content.

In general, the composition of animal manures has improved during the last 135 years. Of course the quality can in part be controlled by preventing dilution with extraneous water, but the variation in the composition of manure dry matter [27, 43] cannot be controlled because it is chiefly the result of the variability in the composition of the feed.

The plant availability of nutrients in animal manures has also changed. The release of N from animal wastes for crop growth depends primarily on the form in which N is present [40]. In its turn this depends on the kind of animal and on the method and duration of storage. Slurry from pigs and

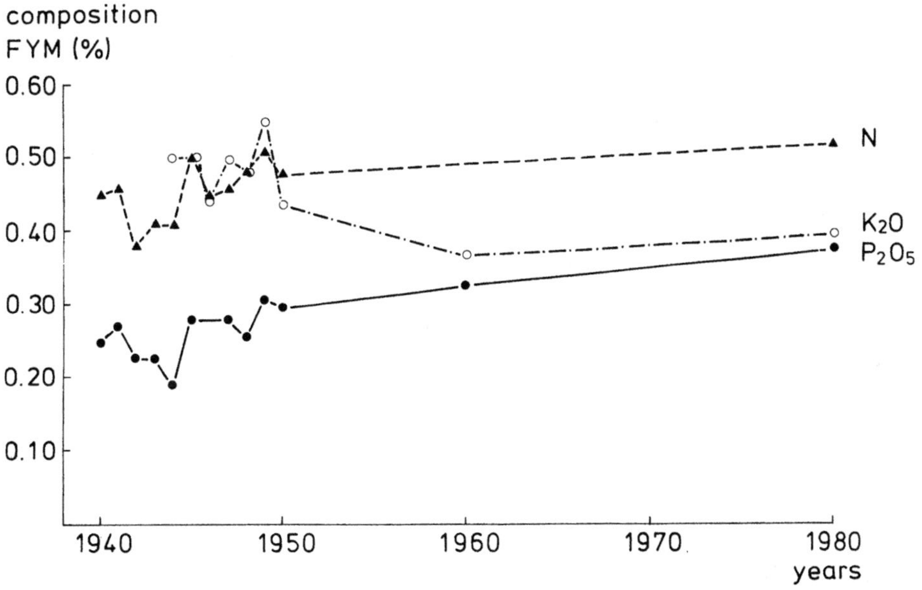

Figure 3. Nutrient contents (N, P and K) in FYM in The Netherlands from 1940.

Table 4. Phosphorus (P) content of concentrates (% of product) used in The Netherlands from 1976 [16]

Concentrates for	1976/77	1980/81
	% P	
Cattle	0.62	0.50
Pigs	0.77	0.57
Broilers	0.78	0.65
Laying hens	0.75	0.63

poultry is relatively rich in inorganic N and therefore has a higher short-term efficiency than that of dairy cattle [39]. From this point of view the greater increase in numbers of pigs and poultry compared to cattle brought about an increased average N efficiency of the total animal manure supply.

With increasing length of storage more of the organic compounds are transformed into easily decomposable substances, or are mineralized completely, resulting in a higher N efficiency. Any losses of the mineralized N due to volatilization decrease N efficiency. Depending on the length and method of storage, the N in the animal manure may have a higher or lower efficiency. However, all in all it is believed that recent developments in method and duration of storage have not led to large changes in N efficiency. The same is true for the efficiency of P and K [10].

Heavy metals in animal wastes originate from agricultural practices (for example the use of sewage sludge, pesticides, inorganic P fertilizers or phosphates in concentrates) and non-agricultural emissions. In some recent studies [6, 27] the average heavy-metal contents of animal manures have been reported. Earlier data have not been found. It may be expected that the heavy-metal contents of animal manures increased in the 20th century. In future the heavy-metal contents of manures may be reduced.

It may be assumed that the organic-matter content of animal wastes under normal conditions in The Netherlands has changed little in the course of this century.

The change in storage methods for manures may have enhanced the vitality of pathogens [31] or of seeds in manures, but this is probably of minor practical importance.

Developments in the usefulness of animal manures

Usefulness is determined by the extent to which animal manure can be used advantageously. It varies widely. Important factors are the need for plant nutrients, the expected specific effect of the animal manure and the costs of manure handling and of alternative fertilizers. The potential user may also be subject to legal restrictions. The plant nutrient requirement on a given farm depends on the balance between inputs and offtake, and the occurrence of losses. In arable farming increasing amounts of minerals are removed from the farm in marketed crops, so that a constant or even an increased import is necessary. In animal husbandry the offtake of minerals in milk and meat is small, whereas the input increases with increasing livestock density. It may lead to a situation in which the amount of nutrients in the animal manure exceeds the requirements of the crops grown. Application of the surplus may adversely affect crop yield and quality. In a recent study [28] a number of examples were given of the negative effects of excess amounts of nutrients. The elements which limit the amount of animal manure to be applied to arable crops and grassland are N and K, respectively [39]. Other components may be harmful to specific crops, for instance chloride (Cl) to potatoes for industrial starch production. On grassland surface-spread animal manure may smother the grass and adversely affect yield [34, 36]. In situations where surpluses exist, farmers tend to 'dump' animal manure on fields for silage maize which is a crop that tolerates large amounts of plant nutrients. It is clear that the value of animal manure on these farms is low or even negative. Where intensive livestock holdings are concentrated geographically, the surplus problems apply to whole regions. Sale of the surplus to arable farms within the region, or in neighbouring regions, would seem to be an obvious solution.

Since the introduction and further development of mineral fertilizers, plant nutrients are abundant in industrialized countries, and animal manures are no longer indispensable as a source of nutrients. The choice between mineral fertilizers and animal manures on the arable farm depends on their relative price, which becomes less favourable for animal manure as the distance for transport increases. Other negative aspects of animal manure are the variable efficiency of its nutrients and the higher costs of application. On the other hand, animal manure is often appreciated because of its specific effects. It has been widely reported [8, 10, 14, 19, 26, 29] that in arable farming the use of organic manures may increase crop yields to levels which can not be obtained with mineral fertilizers alone. However, the factors responsible are not fully understood. In some experiments inadequate mineral fertilization [8, 14, 29, 44] or moisture stress [19] may have played a part. Effects due to the "slow release" of N [8] could not always be shown [44]. Specific beneficial effects of animal manures have also been reported on grassland [10, 15, 30, 32], but again have not always been observed [15, 20]. There is no definite answer to the question where, when and to what extent these positive effects can be expected. Consequently it is difficult to quantify any historical development of this aspect of animal manure. In spite of these

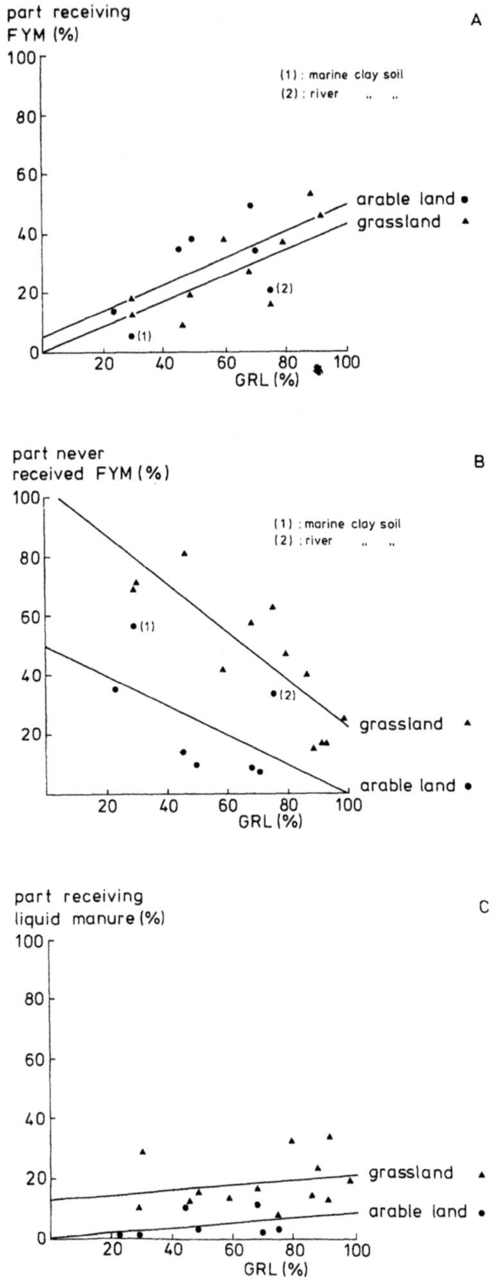

Figure 4. Proportions of arable land and grassland receiving FYM (4A), never received FYM (4B) and receiving liquid manure (4C) as a function of the grassland area (GRL) in 1950 in The Netherlands. Grassland area is expressed as the fraction of the sum of the areas of grassland and arable land.

uncertainties we assume that the preferences for animal manure in the past were based on, among other considerations, the existence of a specific effect.

Figures 4A, B and C show to what extent FYM and liquid manure were used on arable land and grassland around 1950 in different regions in The Netherlands (after data from [2, 12, 25]). In these figures the grassland area (GRL) more or less reflects manure production. It can be concluded that FYM was preferred on arable land (Figs. 4A and B). Of the arable crops, beet and potatoes were favoured as they were considered to be the most responsive. From Figs. 4A and B it can also be seen that the preference was less evident on marine and river clay soils. From Fig. 4C it can be concluded that liquid manure was preferred on grassland. Furthermore, comparing Fig. 4A with Fig. 4C, it can be concluded that the area of arable land receiving liquid manure was smaller than the area receiving FYM.

Though other factors, such as distance over which the animal manure had to be transported and the trafficability of the land had their effect, the foregoing suggests that FYM was valued because of its nutrient content and was applied on sites which needed complete fertilization. Root crops were held to be the most responsive, indicating the existence of a specific organic-manure effect particularly for these crops. Many investigations confirm that root crops [14, 26], as well as silage maize [11, 38], respond most favourably to organic manures.

It can be concluded that the total area planted to root crops and silage maize can serve as a measure of preference for organic manures. There is another reason for doing so. Root crops, if frequently grown on the same site, do not leave sufficient residues to prevent a decline in the soil organic-matter content. Changes in the area planted to root crops and maize in The Netherlands after 1950 are shown in Table 5. It can be seen that, following a decrease after 1950, the need for organic manure increased strongly again after 1970. This was mainly due to the large expansion of the area cropped to silage maize. From Table 5 (last column) it can be seen that in The

Table 5. Total area cropped to silage maize and root crops (A), production of organic matter in the manure of housed animals (B) and this production per unit area of maize and root crops (C) in The Netherlands, calculated after data from [3, 4]

Year	A (ha)	B (million t yr^{-1})	C (t ha^{-1} yr^{-1})
1950	292,567	1.57	5.4
1960	271,100	1.98	7.3
1965	277,006	2.11	7.6
1970	276,720	2.56	9.3
1974	352,200	3.00	8.5
1982	449,300	3.52	7.8

Netherlands as a whole the amount of animal manure theoretically available per ha of "responsive" crops increased after 1950 and has remained more or less stable since 1965. The increase implies a decrease in appreciation according to the law of supply and demand.

Increasing use of animal manures may lead to detrimental environmental effects. The main problems which arise are:

— pollution of groundwater and surface waters due to leaching and surface run-off of N, P, K, and organic matter [5, 22, 23, 42];
— volatilization of ammonia [18];
— accumulation of heavy metals in the topsoil [6, 35].

Up to 1967 there was no evidence that nitrate leaching affected the quality of groundwater intended for drinking water in The Netherlands [21]. However, a substantial increase in nitrate content in shallow groundwater was observed. More recently the nitrate content of deep groundwater has greatly increased at several locations [42].

In areas with manure surpluses a continued downward movement of P to the groundwater is to be expected if current high rates of animal manure application continue [5], especially on soils with a low P-fixing capacity.

In 1982 in The Netherlands the contribution of animal manure to the total nationwide emission of ammonia amounted to approximately 85% [18; citing data from The Institute for Meteorology and Oceanography in Utrecht]. This constitutes a substantial contribution to the acid rain phenomenon.

The use of most animal wastes will lead in most cases to an increased accumulation of heavy metals in soils [6]. The application of pig slurry will generally result in accumulation of Cu. A further restriction of the use of Cu in concentrates appears to be inevitable. In arable farming it is expected [6] that over the next 100 years the proposed restrictions on the use of animal wastes will suffice in most cases to keep the contents of heavy metals below the maximum acceptable limits set for these elements.

Since 1971 it has been forbidden to dispose of animal manure in The Netherlands by directly discharging it into surface waters. In 1981 the Nuisance Act was adopted in relation to the establishment of new livestock farms. Since 1984 the enlargement of pig and poultry farms has been forbidden, so as to stop further increases of manure surpluses. In 1985 milk production was restricted by EC agreements, to reduce over-production. In 1987 a limit will be set to the amount of P that may be applied to soil in the form of animal manure, and the time of application of animal manures will also be regulated to protect the quality of soil and water.

Among the countries in Europe, legislation on the use and production of animal manures varies considerably. In Great Britain, where manure surpluses do not constitute a substantial problem, the direct discharge of animal manures into surface waters is prohibited. Application to land is allowed (but not encouraged) at high rates, provided that run-off or leaching to surface waters does not result. In Belgium legislation seems to be rather tolerant: only an "abnormal" use of animal manure is prohibited. In other countries (France, Italy, Federal Republic of Germany, Sweden and Denmark) legislation goes further and restrictions are imposed on the time and/or amount of application of animal manures. Legal measures have also been taken to relate manure production in intensive holdings to the acreage of arable land (France, Sweden). In some countries the storage capacity of animal manures is regulated (Sweden).

Conclusions

The intrinsic value of animal manure, characterized by kind, amount and efficiency of use of its components, has most likely increased during this century as a consequence of improved soil fertility and the addition of certain elements to feeds to achieve better animal performance. On the other hand, such additions and some other agricultural practices might result in high contents of undesirable constituents in the animal manure, with the prospect of negative side effects. This problem has been recognized and can be controlled.

The need for nutrients in the form of animal manure in crop production has decreased due to the introduction and growing appreciation of the value of mineral fertilizers. The production of animal manure has increased enormously in Western Europe as a consequence of increased livestock densities. Therefore the value placed on animal manures has declined. Since organic manure seems to exert specific favourable effects on a number of crops, a market for it will probably remain, dependent on the price. Root crops and silage maize are the main responsive crops. The area planted to these crops has not grown in proportion to the livestock density. The prospects for selling animal manure are poor, unless sufficient evidence can be produced of specific benefits for crops other than those mentioned.

Restrictions with regard to the amount and time of application, based on expected side effects, reduce further the perception of the value of animal manures.

Gradually animal manure has turned from a precious resource into a waste material in areas of high livestock densities in Western Europe.

References

1. Anon (1949) Verslag over de landbouw in Nederland over 1942 t/m 1945. Verslagen en Mededelingen van het Ministerie van Landbouw, Visserij en Voedselvoorziening 1, 244 pp. The Hague, The Netherlands: Staatsdrukkerij en Uitgeverijbedrijf
2. Anon (1954) Verslag over de landbouw in Nederland over 1952. Verslagen en Mededelingen van het Ministerie van Landbouw, Visserij en Voedselvoorziening 1, 454 pp. The Hague, The Netherlands: Staatsdrukkerij en Uitgeverijbedrijf
3. Anon (1976) De produktie van dierlijke mest van 1950–1974. The Hague, The Netherlands: Centraal Bureau voor de Statistiek
4. Anon (1984) De produktie van dierlijke mest in 1982. The Hague, The Netherlands: Centraal Bureau voor de Statistiek
5. Breeuwsma A, Wosten J H M, Vleeshouwer J J, Slobbe A M van and Bouma J (1986) Derivation of land qualities to assess environmental problems from soil surveys. Soil Science Society of America Journal 50, 186–190
6. Breimer T and Smilde K W (1986) De effecten van organische-mestdoseringen op de zware-metaalgehalten in de bouwvoor van akkerbouwgronden. In Themadag "Organische stof in de akkerbouw". Themaboekje 7, pp 54–67. Lelystad, The Netherlands: Proefstation voor de Akkerbouw en Groenteteelt in de Vollegrond
7. Buchner A and Sturm H (1985) Gezielter düngen: intensiv-wirtschaftlich-umweltbezogen. Frankfurt, Federal Republic of Germany: DLG-Verlag
8. Cooke G W (1977) The roles of organic manure and soil organic matter in managing soils for higher crop yields. A review of the experimental evidence. Proceedings of the international Seminar 'Management in intensive agriculture' (Tokyo), pp 53–64. Tokyo, Japan: The Society of the Science of Soil and Manure
9. Daalen M van (1985) De ontwikkeling van de intensieve veehouderij. Cahiers Bio-wetenschappen en Maatschappij 10 (3), 14–23
10. Dijk T A van and Sturm H (1983) Fertiliser value of animal manures on the continent. Proceedings 220. London: The Fertiliser Society
11. Dilz K, Hag B A ten, Lammers H W and Lande Cremer L C N de la (1984) Fertilization of forage maize in the Netherlands. Netherlands Nitrogen Technical Bulletin 14, 1–25
12. Draisma M (1958) Cultivation and fertilisation of field crops as practiced in the Netherlands. Verslagen Landbouwkundige Onderzoekingen no. 64.9, 168 pp. The Hague, The Netherlands: Staatsdrukkerij en Uitgeverijbedrijf
13. Frissel M J (1977) Cycling of nutrients in agricultural ecosystems. Agro-ecosystems 4, 354 pp. Amsterdam, The Netherlands: Elsevier
14. Haan S de (1980) Einfluss von organischer Düngung auf das maximal erreichbare Ertragsniveau in langjährigen Niederländischen Feldversuchen. Landwirtschaftliche Forschung (Kongressband 1979) Sonderheft 36, 389–404
15. Hart M L't (1950) Organische stof en grasland. Landbouwkundig Tijdschrift 62, 532–542
16. Henkens Ch H (1985) De problematiek van mestoverschotten. Meststoffen 3, 9–15
17. Heuveln B van (1985) Geschiedenis van de bemesting. Cahiers Bio-wetenschappen en Maatschappij 10 (3), 7–13
18. Janssen Th W (1985) Ammoniakproblemen in Nederland. Cahiers Bio-wetenschappen en Maatschappij 10 (3), 24–33
19. Johnston A E (1986) Soil organic matter, effects on soils and crops. Soil Use and Management 2 (3), 97–105
20. Köhnlein J and Vetter H (1965) Die Wirkung des Stalldüngers auf Grünland. Zeitschrift für Acker- und Pflanzenbau 122 (1), 79–93

21. Kolenbrander G J (1972) Does leaching of fertilizers affect the quality of groundwater at the waterworks? Stikstof, Dutch Nitrogenous Fertilizer Review 15, 8–15
22. Kolenbrander G J (1981) Limits to the spreading of animal excrement on agricultural land. In Brogan J C, ed. Nitrogen losses and surface run-off, pp 443–468. The Hague, The Netherlands: Martinus Nijhoff/Dr Junk
23. Kolenbrander G J (1982) Fertilisers and pollution. Transactions of the 12th international Congress of Soil science 'Managing soil resources to meet the challenges to mankind', panel discussion papers 5. Whither soil research, pp. 248–266. New Delhi: The Indian Society of Soil Science
24. Kolenbrander G J en Lande Cremer L C N de la (1967) Stalmest en gier. Waarde en mogelijkheden. Wageningen, The Netherlands: Veenman en Zonen
25. Koopmans J (1960) Survey of productivity level. 2. Fertilization of grassland as practised in the Netherlands. Verslagen Landbouwkundige Onderzoekingen no. 66.5, 112 pp. The Hague, The Netherlands: Staatsdrukkerij en Uitgeverijbedrijf
26. Kundler P (1982) De wichtigsten Aussagen der Dauerversuche zur Erhöhung der Bodenfruchtbarkeit. Tagungsberichte 205, pp 5–16. Berlin: Akademie für Landwirtschaftswissenschaften der DDR
27. Lande Cremer L C N de la (1978) Bewertung der Wirtschaftsdünger in der Niederlande. Bericht über das internationale Kolloquium "Düngeberatung" (Gumpenstein), pp 113–117. Gumpenstein, Austria: Bundesversuchsanstalt für alpenländischen Landwirtschaft
28. Lande Cremer L C N de la (1985) Long-term effects of farm slurries application in the Netherlands. In Williams J H, Guidi G and Hermite P l', eds. Long term effects of sewage sludge and farm slurries applications. Proceedings CEC-seminar (Pisa), pp 84–90. London: Elsevier
29. Lande Cremer L C N de la (1986) Resteffecten, specifieke effecten, structuureffecten, nevenwerkingen, humus- en organische-stof-werkingen van organische bemestingen. In Themadag "Organische stof in de akkerbouw". Themaboekje 7, pp 32–43. Lelystad, The Netherlands: Proefstation voor de Akkerbouw en Groenteteelt in de Vollegrond
30. Lehne I, Raue U, Barth F J and Fischer F (1973) Einfluss gestaffelter N-Gaben in Form von organischen und mineralischen Düngern sowie Beregnung auf Ertragshöhe und Qualität von einjährigen Weidelgras und Zuckerrüben. Archiv für Acker- und Pflanzenbau und Bodenkunde 17, 935–947
31. Linton A H (1979) Report on the Dublin workshop on 'Animal and human hazards associated with the utilisation of animal effluents'. In Hawkins J C, ed. Engineering problems with effluents from livestock, pp 3–22. Brussels: Commission of the European Communities
32. Meer H G van der, Thompson R B, Snijders P J M and Geurink J H (1987) Utilization of nitrogen from injected and surface-spread cattle slurry applied to grassland (this volume)
33. Neander E, Beneker G and Uphoff P (1978) Die Ausbringung tierischer Exkremente auf landwirtschaftlich genützten Flächen in der Gemeinschaft. 2. Charakterisierung der Regionen mit intensiver Tierhaltung. A. Bericht. Mitteilungen über Landwirtschaft 48. Brussels: Kommission der Europäischen Gemeinschaften
34. Prins W H and Snijders P J M (1987) Negative effects of animal manure on grassland due to surface spreading and injection (this volume)
35. Riemsdijk W H van, Lexmond Th M, Enfield C G and Zee S E A T M van der (1987) Phosphorus and heavy metals: accumulation and consequences (this volume)
36. Schechtner G, Tunney H, Arnold G H and Keuning J A (1980) Positive and negative effects of cattle manure on grassland with special reference to high rates of application. In Prins W H and Arnold G H, eds. The role of nitrogen in intensive grassland production.

Proceedings of the international Symposium of the European Grassland Federation (Wageningen), pp 77–93. Wageningen, The Netherlands: PUDOC

37. Scheltinga H M J (1970) Bio-industrie. Hoeveelheid, hoedanigheid, variatie en behandeling der afvalstoffen. H_2O 3 (22), 569–575
38. Schröder J and Dilz K (1987) Cattle slurry and farmyard manure as fertilizers for forage maize (this volume)
39. Sluijsmans C M J, Dijk T A van, Kolenbrander G J, Lande Cremer L C N de la, Smilde K W and Werkhoven C H E (1978) The spreading of animal excrement on utilized agricultural areas of the Community. 1. Scientific basis for the limitations of quantities and criteria for rules thereon. Information on Agriculture 47. Brussels: Commission of the European Communities
40. Sluijsmans C M J and Kolenbrander G J (1977) The significance of animal manure as a source of nitrogen in soils. Proceedings of the international Seminar 'Management in intensive agriculture' (Tokyo), pp 403–411. Tokyo, Japan: The Society of the Science of Soil and Manure
41. Snelders H A M (1985) De ontvangst van Liebig's kunstmest in Nederland. Landbouwkundig Tijdschrift 62, 532–542
42. Steenvoorden J H A M, Fonck H and Oosterom H P (1986) Losses of nitrogen from intensive grassland systems by leaching and surface run-off. In Meer H G van der, Ryden J C and Ennik G C, eds. Nitrogen fluxes in intensive grassland systems. Developments in Plant and Soil Sciences, Volume 23, pp 85–97. The Hague, The Netherlands: Martinus Nijhoff Publishers
43. Vanacker L (1981) Dierlijke mest: afval of grondstof? Ph.D. thesis. Gent, Belgium: Rijksuniversiteit
44. Wadman W P (1987) Effect of organic manure on crop yield in long-term field experiments. INTECOL-symposium, 4–6 June 1986, Ekenäs. Submitted for publication in the Intecol Bulletin
45. Wit C T de (1972) Food production: past, present and future. Stikstof, Dutch Nitrogenous Fertilizer Review 15, 68–80

Effects of herbage composition and supplement feeding on the excretion of nitrogen in dung and urine by grazing dairy cows

A M VAN VUUREN[a] and J A C MEIJS[b]

[a] *Institute for Livestock Feeding and Nutrition Research (IVVO), P.O. Box 160, 8200 AD Lelystad, The Netherlands*

[b] *CLO-Institute for Animal Nutrition 'De Schothorst', Meerkoetenweg 26, 8218 NA Lelystad, The Netherlands*

Key words: nutrition, dairy cows, nitrogen excretion, herbage, supplementation

Abstract. Theoretically a 600 kg cow producing 25 kg milk per day can utilize dietary nitrogen (N) with a maximum efficiency of 40–45%. The actual efficiency of N utilization in cows grazing intensively managed pastures, is 15 to 25%. When young leafy grass is the sole feed this figure cannot be altered substantially without reducing animal performance. Supplementing the diet of grazing dairy cows with a low-protein, high-energy feed (special concentrates or maize silage) increases the efficiency of N utilization, mainly because it lowers N intake. First results using maize silage indicate an improvement in N utilization of 25 to 30%.

Introduction

Intensification of pasture management by higher nitrogen (N) input and better grazing management has led to higher N concentrations in herbage. Van der Meer [12] estimated that on farms using abundant N fertilizer only 16% of the N input is removed in milk and liveweight. Important causes for such a low utilization of N are an inefficient uptake by herbage of N in faeces, urine and slurry and low efficiency of N utilization by ruminants.

Knowledge of the limiting factors in the utilization of N from herbage by dairy cows, both for grassland managers and nutritionists, is an important step to improve the efficiency of N utilization, thus preventing or diminishing the emission of N to the environment.

In this paper we will focus on the utilization of N by grazing dairy cows and indicate possibilities for nutritional improvement.

Nitrogen metabolism of dairy cows

Compartmentation of the proximal part of the digestive tract of ruminants slows the transit of feed, thus enabling a predigestion of slowly degradable

H.G. v.d. Meer, et al. (eds.), Animal Manure on Grassland and Fodder Crops.
 ISBN 978-94-010-8137-5.

feed rich in fibre. The presence of cellulase producing micro-organisms in the forestomachs gives ruminants the ability to utilize energy from cellulose. Delayed passage of the feed, and microbial digestion in the forestomachs also affect the protein metabolism of ruminants.

New concepts that enable the prediction of the amount of absorbable amino acids reaching the intestines are being developed in order to take this redistribution of N into account. Such systems distinguish between two sources of amino acids entering the small intestines. One source is the non-degraded feed protein that escapes microbial digestion. The proportion of this "bypass protein" depends mainly on the rate of degradation and on the retention time of feed in the forestomachs. The second source of amino acids is microbial protein, synthesized in the forestomachs.

Microbial protein synthesis depends on the supply of (i) precursors such as peptides, amino acids, ammonia, carbon skeletons, phosphorus, and sulphur, (ii) other essential nutrients and (iii) energy [7]. Under optimal conditions the amount of protein synthesis is related to the amount of energy extracted from the feed. Demeyer and Tamminga [7] estimate a synthesis of 35 g microbial N (220 g crude protein) per kg organic matter digested in the forestomachs under ideal circumstances. However, the efficiency of microbial protein synthesis on fresh herbage seems to be somewhat higher, but reasons for this are still not clear [4]. On the other hand, we also have to bear in mind, that a significant amount of the organic matter from herbage digested in the forestomachs is protein, hence limiting the amount of energy available for the growth of the microbes [21].

Not all the protein entering the small intestines will be absorbed: intestinal digestion of different proteins may vary considerably [10]. Also, the absorbed amino acids will not be completely recovered in animal protein. Part of the N is lost during metabolism.

Minimum loss of nitrogen by dairy cows

Maintenance

In a recent review Owens [19] concluded that a 600 kg cow fed at maintenance level loses 52 to 82 g N day^{-1} (Table 1). Several factors attribute to this minimum N loss. A small proportion is lost with skin and hair. A second fraction is endogeneous urinary N (EUN), originating from N containing metabolites (e.g. amino acids, nucleic acids) expelled during metabolic processes of body tissues. A third route of N excretion is faecal output. Metabolic faecal N (MFN) consists of the residues of secreted enzymes, sloughed intestinal cells and of the gastro-intestinal biota.

Table 1. Minimum N losses ($g\,N\,day^{-1}$) of dairy cows [19]

A: MAINTENANCE (600 kg liveweight)				
		skin, hair	2	
		endogeneous urinary N	10–15	
		metabolic faecal N	40–65	
				67
B: MILK PRODUCTION (N content: 5.2 $g\,kg^{-1}$)				
	per kg:	endogeneous urinary N:		
		– absorbed nucleic acids	0.8	
		– metabolized during synthesis	0.9	
		metabolic faecal N	2.4	
			4.1	
	per 25 kg:			103
TOTAL				170

Production

The production of milk protein is also coupled with inherent losses of N, originating from processes within the gastro-intestinal tract and intermediary metabolism. The production of 1 kg milk requires 460 NEL (Net Energy for Lactation, in VEM per kg dry matter (DM); 1 VEM = 6.9 kJ, [6]), under normal conditions equivalent to approximately 250 g organic matter fermented in the rumen (FOM). As stated by Owens [19] urinary N excretion increases with about 3 g per kg·FOM, resulting from extra microbial nucleic acids, synthesized in the rumen and subsequently absorbed from the intestinal tract.

Another urinary N loss is due to the fact that the efficiency of N utilization is less than 100%. If the composition of the amino acids absorbed from the intestines is ideal, e.g. matches potential needs, the efficiency of amino acid utilization is considered to be approximately 85% [18]. Also extra faecal N output has to be taken into account. Various methods for estimating MFN are used. Considering a feed containing 900 g organic matter and 700 g digestible organic matter per kg DM, MFN is estimated as 4.3 (US-NRC system [20]), 4.6 (French-PDI system [22]) or 4.9 [5] g per kg DM ingested. Under normal conditions 460 NEL requires a DM intake of approximately 0.5 kg.

Thus N loss of a 600 kg cow, producing 25 kg milk day^{-1} and fed on energy and protein balance, calculated with the assumptions described above, will be at least 170 g N day^{-1} (Table 1). In this ideal situation at a production of 130 g milk N day^{-1} the efficiency of utilization of dietary N is 43%. Under practical conditions efficiencies will be lower, because of a lower N digestibility and a less ideal composition of absorbed amino acids.

Nitrogen losses by grazing dairy cows

As indicated by Van der Meer [13] intensification of grassland production and N fertilization has led to a decrease of the efficiency of N utilization. Cows weighing 550 kg and producing 20 kg milk day^{-1} utilized 21% of the ingested N on a farm where no N fertilizer was applied. On farms with an intensive N fertilization (Nitrogen Pilot Farms) only 16% of ingested N is retained in milk and liveweight (Table 2), which is far from the theoretical maximum efficiency of 40–45%.

Balance trials have been performed by Van der Honing *et al.* [9] using dairy cows fed fresh herbage indoors. The efficiency of utilization of N in these experiments was 22 to 25% (Table 3), which is higher than that calculated for the Nitrogen Pilot Farms. The reason for this difference seems mainly a different N intake.

The efficiency of N utilization in grazing dairy cows can be estimated from experiments of Meijs [14, 15, 16]. Weight gain of the spring-calving animals was small during these trials and therefore we assumed N retention to be negligible. On average about 78% of the N ingested with the herbage was excreted with faeces and urine (Table 4). This result agrees with data from indoor-experiments (74–78%), and makes it clear that the utilization of grass N on well fertilized pastures is generally less than the theoretical maximum. Efficiency of N utilization in the experiments of Meijs was

Table 2. Estimated utilization of N of unfertilized and heavily fertilized herbage by dairy cows, weighing 550 kg and producing 20 kg fat corrected milk per day [13]

N rate (kg ha^{-1} yr^{-1})	N in herbage DM (g kg^{-1})	N intake (g day^{-1})	N excretion (g day^{-1})		
			milk	faeces	urine
0	29.6	506 (100%)	106 (21%)	117	283
383	44.0	647 (100%)	106 (16%)	111	430

Table 3. Nitrogen balance of cows fed fresh herbage and 1 kg concentrates per day (basic data from [9])

Institute	N in herbage DM (g kg^{-1})	N intake (g day^{-1})	N excretion (g day^{-1})			N retention by animal (g day^{-1})
			milk	faeces	urine	
Anim. Physiol., Wageningen	37.2	521 (100%)	88 (17%)	126 (24%)	279 (54%)	28 (5%)
IVVO, Lelystad	38.8	460 (100%)	107 (23%)	117 (25%)	227 (49%)	9 (2%)

Table 4. Nitrogen intake and N excretion in the milk of grazing cows supplemented with a maximum of 1 kg DM day^{-1} [14, 15, 16]

Year	N in herbage DM ($g\,kg^{-1}$)	N intake ($g\,day^{-1}$)	N excretion in milk ($g\,day^{-1}$)	Efficiency of N utilization (%)
1978	36	521	122	23.4
1979	38	575	127	22.1
1981	38	560	126	22.5
1982	38	572	107	18.7
1983	35	525	123	23.4
Mean	37	551	121	22.0

somewhat higher than on the Nitrogen Pilot Farms, because of a lower herbage N content, and a higher N excretion with the milk.

Reduction of N losses by grazing animals can be pursued by two approaches:

- change the factors which determine the efficiency of N utilization without alteration of the diet (same amount of fresh herbage), or
- reduce the fraction of herbage in the diet by feeding supplements.

Reduction of nitrogen losses with herbage as sole feed

One obvious way to improve N utilization is by increasing milk protein production. However, there are indications that milk production of grazing dairy cows not receiving supplementation is limited to 20–25 kg day^{-1} by the restricted intake of energy [14] and by an insufficient supply of absorbable amino acids in the small intestine [4]. The inadequate supply of amino acids in the small intestine would arise from the excessive degradation of protein from fresh forages (Tables 5 and 6). This excessive loss of protein N in the forestomachs can not be compensated for by microbial synthesis.

In fact, improvement of N utilization can be achieved either by decreasing N intake or by increasing protein supply to the small intestine possibly resulting in a higher milk protein production. With herbage as the sole feed, the intake of N can only be decreased by lowering the crude protein (CP) content of the herbage. Crude protein levels in temperate grass species vary widely and may range from 70 to 150 g per kg DM in sub-optimal situations up to 300 g per kg DM in highly productive regions, where pastures are heavily fertilized [11]. Lowering N fertilization will decrease CP intake, mainly because of a decreased CP content in the herbage. Lower N fertilization may also decrease the rate of CP degradation in the forestomachs (Table 5), resulting in a higher proportion of feed protein escaping from

Table 5. Effect of N fertilization on the crude protein (CP) content in herbage, the rate of disappearance of CP from nylon bags incubated in the forestomachs (k_d) and estimated supply of digestible feed CP entering the small intestine (DPI). (S. Tamminga, personal communication)

N rate ($kg\ ha^{-1}\ yr^{-1}$)	CP in herbage DM ($g\ kg^{-1}$)	k_d ($\%\ hr^{-1}$)	DPI in herbage DM ($g\ kg^{-1}$)
0	203	8.6	44
250	261	9.7	52
400	288	10.3	57
700	332	10.9	58
maize silage	94	1.4	14

rumen degradation. However, the absolute supply of feed amino acids to the small intestine will not be altered, because of the simultaneously occurring lower CP intake.

A lower CP intake may also be achieved by feeding more mature grass. As grasses mature CP content decreases, as does the rate of protein degradation in the forestomachs (Table 6). However, digestibility and hence NEL content of more mature herbage is substantially lower. Thus energy intake with this herbage will be less, and consequently milk production will decline.

Another possibility to decrease N content is fractionation of grass into a protein-rich juice and a fibrous pressed residue. However, the NEL/CP ratio in the pressed residue is only changed moderately compared to the original material [24]. Furthermore this method is expensive and energy consuming.

Several research groups have tried to increase protein supply to the small intestine by inhibiting protein degradation in the forestomachs. One approach has been the inclusion of tannin containing legumes like sainfoin (*Onobrychis viciifolia* [4]) or *Lotus pedunculatus* [2]. Proteolysis of tannin-protein complexes is reduced or takes place more slowly. The tannin-protein

Table 6. Effect of stage of herbage maturity on crude protein (CP) content, rate of disappearance of CP from nylon bags incubated in the forestomachs (k_d) and estimated supply of digestible feed CP entering the small intestine (DPI). (S. Tamminga, personal communication)

Weeks after mowing	CP in herbage DM ($g\ kg^{-1}$)	k_d ($\%\ hr^{-1}$)	DPI in herbage DM ($g\ kg^{-1}$)
3	269	12.2	41
4	275	10.1	43
5	256	11.0	39
6	213	9.9	32
7	156	9.6	27
8	200	8.6	32

bonds are cleaved under the acidic conditions in the abomasum and thus do not interfere with the proteases excreted by the host-animal. However, in some cases the effects of tannins on animal performance have not been very promising [1].

Influencing proteolysis in the forestomachs has also been attempted by using formaldehyde treatment to affect the molecular structure of proteins or Monensin to affect the metabolism of microbes. However, the effects have been found variable [3].

There is little data on comparisons between different herbage species. Differences have been reported, but they may be partly overruled by variations occurring during the growing season [4].

Reduction of nitrogen losses by supplement feeding

Cows grazing on well fertilized pastures consume 50–100% more protein than they require. The easiest way to reduce N consumption is to replace part of the N-rich herbage by roughage or concentrates with a low protein content. During the summer of 1985 a stall-feeding experiment was done at Lelystad in which herbage and a combination of herbage and maize silage were compared. The aim of this experiment was to compare both rations at the same level of net energy intake, and therefore feed supply was restricted. Herbage contained 39 g N per kg DM, maize silage 17 g N per kg DM.

The results of the N balance measurements are shown in Table 7. The N in the faeces and urine was decreased by 28% when 49% of herbage DM was replaced by maize silage. This effect was mainly due to the reduced intake of N with this treatment. However, despite the lower protein intake, the protein content of the milk was increased significantly ($P < 0.05$) by maize silage. Most probably the microbial protein synthesis in the rumen was improved by the supply of energy, mainly carbohydrates, from maize silage.

Another way of reducing the N content of the diet is by supplementing herbage with concentrates. Several authors have reported a substantial

Table 7. Nitrogen balance of cows fed fresh herbage or a combination of herbage and maize silage (51/49 on DM basis)

Ration	N intake ($g\,day^{-1}$)	N excretion ($g\,day^{-1}$)			Efficiency (%)
		milk	faeces	urine	
Herbage	626	107	158	361	17
	(100%)	(100%)	(100%)		(100%)
Herbage + maize silage	494	118	178	198	24
	(79%)	(110%)	(72%)		(141%)

decrease of the ammonia concentration in the rumen when supplementing N-rich forages [8, 23]. The effect on N excretion will be less than with supplementation of maize silage, because of the higher N content in concentrates. Moreover, the quantities of concentrate which may be fed in addition to herbage are limited, due to possible disturbances of rumen fermentation [23]. However, when carbohydrates in the concentrate consist of highly but slowly digestible structural polysaccharides, such as soya bean hulls and palm kernel expeller, the negative effects on herbage intake and milk fat may be smaller [17].

Conclusion

Possibilities for an improvement of the efficiency of N utilization in ruminants with grass as the sole feed are limited. In most cases such measures are coupled with a decline of animal performance due to a decrease of NEL intake. A better response seems possible if grazing cows are supplemented with good quality concentrates or forages with a relatively high energy and low protein content. Preliminary experiments gave promising results. However, the consequences of such feeding systems on farm management and economics are not yet clear.

References

1. Barry T N and Duncan S J (1984) The role of condensed tannins in the nutritional value of *Lotus pedunculatus* for sheep. 1. Voluntary intake. British Journal of Nutrition 51, 485–491
2. Barry T N and Manley T R (1984) The role of condensed tannins in the nutritional value of *Lotus pedunculatus* for sheep. 2. Quantitative digestion of carbohydrates and proteins. British Journal of Nutrition 51, 493–504
3. Beever D E, Losada H R, Gale D L and Spooner M C (1984) The effect of Monensin addition or formaldehyde application on the digestion of fresh forage by growing cattle. Animal Production 38, 538
4. Beever D E and Siddons R C (1986) Digestion and metabolism in the grazing ruminant. In Milligan L P, Grovum W L and Dobson A, eds. Control of digestion and metabolism in ruminants, pp 479–497. Englewood Cliffs: Prentice-Hall
5. Boekholt H A (1977) Nitrogen metabolism of the lactating cow and the role of gluconeogenesis from amino acids. Doctoral thesis, Agricultural University, Wageningen
6. CVB Manual (1977) Manual for the calculation of the nutritive value of roughages. Lelystad, The Netherlands: Advisory Centre for Animal Feeding
7. Demeyer D I and Tamminga S (1987) Microbial protein yield and its prediction. In Alderman G and Jarrige R, eds. Feed evaluation and protein requirements systems for ruminants. Report (in press). Luxembourg: Commission of the European Communities
8. Griffiths T W (1982) Effects of supplementation of grass silages on nitrogen retention in growing heifers. In Griffiths T W and Maguire M F, eds. Forage protein conservation and utilization, pp 219–223. Luxembourg: Commission of the European Communities

9. Honing Y van der, Wieman B J, Donselaar B van and Steg A (1982) Digestibility and utilization of energy of grass from pasture by dairy cows and sheep. Report IVVO No. 150. Lelystad, The Netherlands: Institute for Livestock Feeding and Nutrition Research
10. Hvelplund T (1987) Amino acid content of feed and microbial protein and their intestinal digestibility. In Alderman G and Jarrige R, eds. Feed evaluation and protein requirements systems for ruminants. Report (in press). Luxembourg: Commission of the European Communities
11. Lyttleton J W (1973) Proteins and nucleic acids. In Butler G W and Bailey R W, eds. Chemistry and biochemistry of herbage, pp 63–103. London: Academic Press
12. Meer H G van der (1982) Effective use of nitrogen on grassland farms. In Corrall A J, ed. Efficient grassland farming. Proceedings of the 9th Meeting of the European Grassland Federation (Reading), pp 61–68. Occasional Symposium No. 14 of the British Grassland Society. Hurley, UK: The British Grassland Society
13. Meer H G van der (1982) Efficiënt gebruik van stikstof uit kunstmest en organische mest op het weidebedrijf. Gebundelde Verslagen No. 23, pp 67–80. Wageningen: Nederlandse Vereniging voor Weide- en Voederbouw
14. Meijs J A C (1981) Herbage intake by grazing dairy cows. Doctoral thesis, Agricultural University, Wageningen
15. Meijs J A C (1983) The influence of concentrate supplementation on milk production by grazing dairy cows. 2. Report of the experiment on herbage intake in 1982. Report IVVO No. 149. Lelystad, The Netherlands: Institute for Livestock Feeding and Nutrition Research
16. Meijs J A C (1985) The effect of concentrate supplementation on herbage intake and milk production by grazing dairy cows. 4. Report of the grazing experiment in 1984. Report IVVO No. 167. Lelystad, The Netherlands: Institute for Livestock Feeding and Nutrition Research
17. Meijs J A C (1986) Concentrate supplementation of grazing dairy cows. 2. Effect of concentrate composition on herbage intake and milk production. Grass and Forage Science 41, 229–235
18. Oldham J D (1987) Efficiencies of amino acid utilization. In Alderman G and Jarrige R, eds. Feed evaluation and protein requirements systems for ruminants. Report (in press). Luxembourg: Commission of the European Communities
19. Owens F N (1987) Maintenance protein requirements. In Alderman G and Jarrige R, eds. Feed evaluation and protein requirements systems for ruminants. Report (in press). Luxembourg: Commission of the European Communities
20. Owens F N (1987) The US-NRC approach to ruminant nitrogen usage. In Alderman G and Jarrige R, eds. Feed evaluation and protein requirements systems for ruminants. Report (in press). Luxembourg: Commission of the European Communities
21. Tamminga S (1982) Nitrogen and amino acid metabolism in dairy cows. Doctoral thesis, Agricultural University, Wageningen
22. Vérité R (1987) Present situation of protein evaluation for ruminants in France: the PDI system. In Alderman G and Jarrige R, eds. Feed evaluation and protein requirements systems for ruminants. Report (in press). Luxembourg: Commission of the European Communities
23. Vuuren A M van, Koelen C J van der and Vroons-de Bruin J (1986) Influence of level and composition of concentrate supplements on rumen fermentation patterns of grazing dairy cows. Netherlands Journal of Agricultural Science 34, 457–467
24. Wieringa G W, Meer H G van der and Geurink J H (1980) Production and utilization of energy and protein in grass in relation to nitrogen rate, green crop fractionation and dairy feeding systems. In Prins W H and Arnold G H, eds. The role of nitrogen in intensive grassland production, pp 103–111. Wageningen: PUDOC

Manure as a source of nitrogen and phosphorus in soils

H G VAN FAASSEN and H VAN DIJK
Institute for Soil Fertility, P.O. Box 30003, 9750 RA Haren, The Netherlands

Key words: nitrogen, phosphorus, manure, chemical composition, soil, fertilizer value, mineralization-immobilization-turnover

Abstract. Nitrogen (N) and phosphorus (P) are present in manure as various inorganic and organic compounds. The distribution of N and P between these compounds depends on the origin of the manure (the species, age and production level of the animal and the corresponding composition of their diets), on the housing, manure collection and manure storage systems and on any possible manure treatment. During storage of slurry the composition moves towards 40–70% of total N being present as ammonium-N (NH_4^+-N) and, more slowly, towards 80–90% of total P being present as calcium phosphates.

After field application all components of the manure, including NH_4^+-N that is not immediately lost due to NH_3 volatilization and inorganic phosphates enter the mineralization-immobilization-turnover (MIT) of the soil. The availability of N for plant uptake, *i.e.* the outcome of MIT for N, cannot be simply derived from the distribution of mineral and organic N in the manure. Quantitatively, N losses may have a larger impact on the actual N availability than differences in net mineralization rates of the organic fraction.

When differences in "fertilizer value" of manure phosphate are found, it is unlikely that they are due to the presence of organic phosphates in manure.

Introduction

Good fertilization practice has more than ever to provide a well balanced nutrient supply to the crop during its growth, and to minimize emissions to the environment. For manure this means that we have to know its direct contribution to the amount of plant-available mineral nutrients as well as the fate of its organic nitrogen (N) and phosphorus (P) in mineralization-immobilization-turnover (MIT). This paper reviews the assessment of manure as a source of N and P for crops.

Chemical composition of manures

Manure is here defined as the faecal wastes and urinary excretions of livestock and poultry. Manure is often diluted with water ("slurry") or it may be dried and may contain some waste feed, bedding or litter. At least three major factors influence the chemical composition of manures:

H.G. v.d. Meer, et al. (eds.), Animal Manure on Grassland and Fodder Crops.
 ISBN 978-94-010-8137-5.

- the species and age of animal;
- the composition of diets (corresponding with the production level of the animal);
- the housing system and the manure collection, storage and treatment systems.

Faeces and urine are waste products of the feed and fodder metabolized by the animal. The digestibility of feed components is shown in Table 1. Similar methods of analysis have been applied to determine faeces components (Table 2). The dry matter (DM) in feed that is soluble in neutral detergent is largely digested by the animal while this fraction when found in faeces is primarily of metabolic origin. The faeces compositions in Table 2 reflect the differences in feed composition and digestive capabilities of farm animals. The higher N content for poultry is due to the common route of faeces and urine. Urine is not included in the figures for swine and ruminants. The high fibre contents of ruminant faeces reflect lower digestibility

Table 1. Division of forage organic matter by a system of analysis using detergents; source [20]

Fraction	Components	Nutritional availability	
		Ruminant	Nonruminant
Category A			
Cell contents (soluble in neutral detergent)	Lipids	Virtually complete	Highly available
	Sugars, organic acids, and water-soluble matter		
	Starch		
	Nonprotein nitrogen		
	Soluble protein		
	Pectin		
Category B			
Cell-wall constituents (fibre insoluble in neutral detergent)	Attached protein	Complete	High
Soluble in acid detergent	Hemicellulose	Partial	Very low
Insoluble in acid detergent (acid-detergent fibre)	Cellulose	Partial	Very low
	Lignin	Indigestible	Indigestible
	Lignified nitrogen compounds	Indigestible	Indigestible
	Heat-damaged protein	Indigestible	Indigestible
	Keratin	Indigestible	Indigestible
	Silica	Indigestible	Indigestible

Table 2. Typical chemical compositions of faeces for the major species of farm animals; source [18]

Faeces source	Neutral detergent soluble	Nitrogen	Hemi-cellulose	Cellu-lose	Lignin	Ash
	(% of dry matter)					
Broilers (caged)	69	6.5	16	11	4	22
Laying hens (caged)	65	6.2	17	15	3	28
Swine (growing & fattening)	60	3.0	20	15	5	17
Beef cattle (fattening)	53	3.0	22	17	8	7
Dairy cattle (lactating)	41	2.6	21	25	13	9
Dairy heifers (all forage-fed)	32	2.0	20	28	20	12
Sheep (all forage-fed)	45	2.5	15	28	15	13

of their diets. Even within a given class of livestock considerable variation in manure composition should be expected. Young animals are fed more easily digestible feed because of their lower digestive capabilities; their manure can be rich in N (e.g. veal calf manure) and will contain less fibre. The distribution of N between faeces and urine, based on figures for the United States, is shown in Table 3, together with the major N components. Variations in manure composition reported from different countries may result from any of the factors mentioned.

In view of its fate in soil, manure N may be considered to be present in three fractions:

– inorganic N (ammonium and incidentally nitrate) and rapidly mineralizable N, for instance from urea and uric acid;

Table 3. Distribution and forms of N in fresh faeces and urine of farm animals; adapted from [19]

Species	% of total N		Nitrogen form (% of total N)				
	Faeces	Urine	Amino acid (protein)	Urea	Ammonium-N	Uric acid	Other
Beef cattle	50	50*	20	35	0.5	–	44
Dairy cattle	60	40*	23	28	< 0.5	–	49
Sheep	50	50*	21	34	< 1.5	–	43
Swine	33	67	27	51	–	–	22
Poultry	25	75	27	4	8	61	1

* Will be much higher on N fertilized grass; extra N in the ration is excreted in urine.

- easily decomposable organic N compounds with a low C/N ratio (e.g. proteins and amino acids);
- "resistant", slowly mineralizable organic matter with a high C/N ratio (e.g. lignocellulosic fibre complex).

Manure also contains easily decomposable N-free organic compounds (e.g. fats, fatty acids and simple carbohydrates).

The P content of animal feed varies widely. In contrast to N which is present almost completely in organic forms (mainly protein), P in feeds is partly inorganic, and mineral phosphate is often added. In pig feed, for example, inorganic phosphate may vary from 10 to 50% of total phosphate. As a result, the form and content of phosphate in fresh manure varies widely. Data from Peperzak *et al.* [16] are illustrative. The inorganic P content may range from 0.3 to 2.4% of DM, the organic P content from less than 0.1 to 1%. In fresh chicken manure organic phosphate may amount to 60% of total phosphate, in pig manure to 40%. Faeces contribute most to the total as well as to the organic phosphate. In urine, P is mainly present in the form of inorganic phosphates [10]. Using another fractionation procedure, Gerritse [8] found that the main organic phosphates in fresh pig slurry were inositol hexaphosphate (phytin) and adenosine triphosphate.

Transformations during storage and treatment

Nitrogen compounds

When faeces and urine are deposited, their composition immediately starts to change. Urea and uric acid from urine are rapidly hydrolysed to ammonia (NH_3) and carbon dioxide (CO_2) by enzymes from faecal bacteria. With further ammonification during storage, the composition moves towards 40–70% of total N being present as ammonium-N (NH_4^+-N), depending on manure type and NH_3 loss. Gross overall processes during anaerobic storage include the degradation of high molecular weight compounds like plant fibre and microbial and metabolic protein to low molecular weight compounds that are often volatile and malodorous. Spoelstra [21] showed that 24% of the plant fibre and 43% of the crude protein of swine manure were decomposed during 70 days anaerobic storage. Acetic, propionic, and butyric acids and small amounts of other lower fatty acids and CO_2 were the main products formed; in addition some methane (CH_4) was formed. Van Velsen [24] showed that about 50% of the organic matter from liquid piggery wastes could be biodegraded to CH_4 and CO_2 by adapted microbes that tolerated the high NH_4^+-N concentrations of this type of manure. The organic matter which had been decomposed consisted mostly of hemicellulose, cellulose and soluble compounds.

Ammonification and CO_2 evolution lead to an increase in pH of manure, whereas fatty acid production may decrease pH. Loss of NH_3 due to volatilization during storage depends on its partial pressure (governed by NH_4^+-N concentration, pH and temperature) and on the method of filling (top or bottom), wind speed, agitation and crust formation. "Deep-pit" storage of poultry manure leads to high losses of N. The drying is so slow that the uric acid can be degraded almost completely. Rapid drying can considerably reduce this loss. Aerobic storage and treatment can prevent odour nuisance but can lead to N loss due to NH_3 volatilization or nitrification-denitrification [3, 4, 5, 14, 22, 23]. The presence of nitrate in manure is indicative of aerobic storage or treatment.

Phosphorus compounds

The P cycle in pig slurry was investigated [10] by adding radioactive $^{32}PO_4$-ions and measuring the increase of radioactivity in various fractions over the course of time. In all fractions the specific activity approached the value calculated for complete distribution of ^{32}P among all phosphates. It thus appears that all P in manure must be part of a microbial cycle. From a simplified model (Fig. 1) microbial turnover times for P in pig slurry were calculated to be in the order of 10–20 weeks.

The distribution of P in different fractions of pig slurry after 1 to 6 months of storage is given in Table 4 [9]. It was further found (R.G. Gerritse, unpublished) that the relative phosphate composition in other manures is similar. Organic P constituted 5–15% of total P after storage for two months or after anaerobic digestion, irrespective of the initial distribution between organic and inorganic forms. Thus, as a result of the microbial P cycle a similar equilibrium is reached, though the equilibration time may vary depending on the nature of the organic P compounds.

The inorganic phosphates in manure are mainly present in the form of crystals or precipitates of calcium phosphates. Upon separation of manure into a liquid and a solid fraction most of the phosphates will be present in the solid fraction.

Transformations of manure nitrogen in the soil

Organic compounds from manure are used by soil organisms as a source of energy and nutrients to build up their cells. Therefore gross mineralization

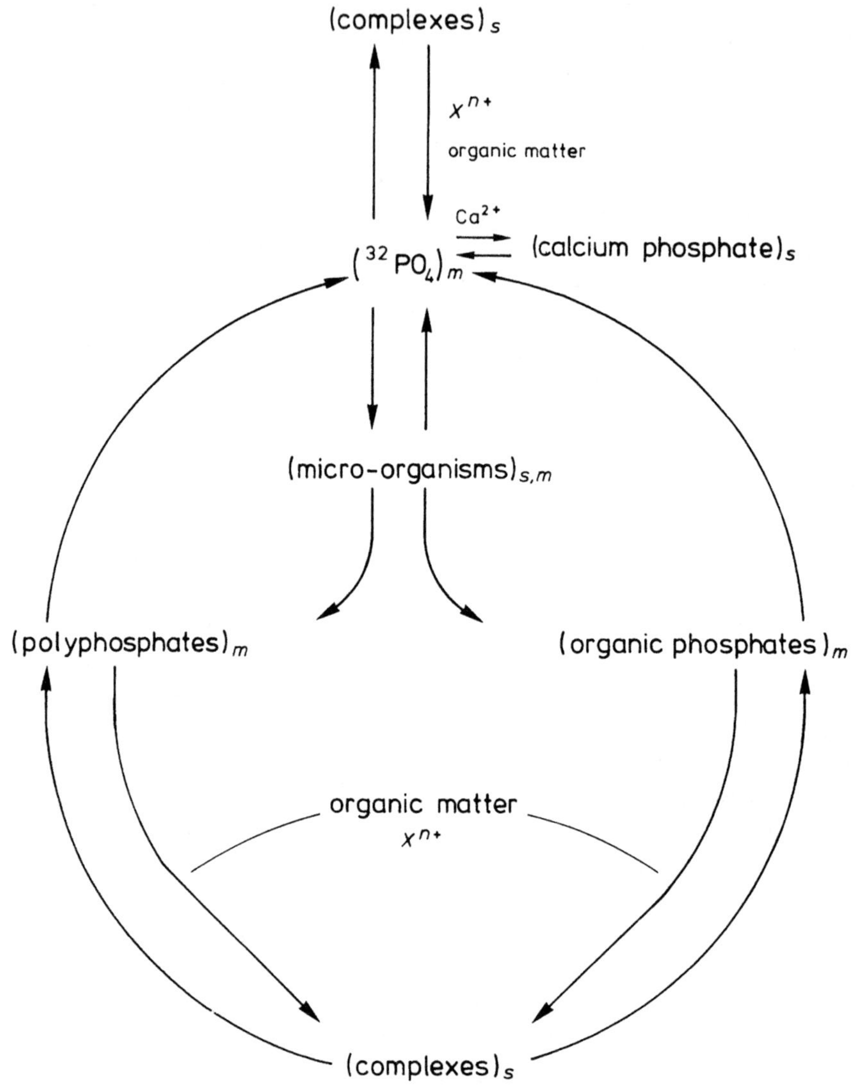

Figure 1. Model for the P cycle in pig slurry, according to which labeled phosphate ($^{32}PO_4^{3-}$) is distributed (adapted from [10]). s = solids; m = mobile in solution; x^{n+} = polyvalent cations (Ca^{2+}, Cu^{2+}, . . .).

of organic compounds always goes together with gross immobilization of N in the soil biomass (bacteria, fungi, soil fauna). Easily degradable organic compounds are thus rapidly transformed into CO_2, H_2O and cells. Depending on the N content, or C/N ratio, of these compounds and of the cells formed, inorganic N will be liberated or immobilized. So-called "resistant"

Table 4. Distribution of P among organic and inorganic P compounds in stored pig slurries; source [9]. Ten slurries were analysed. Dry matter content ranged from 5 to 10% and time of storage from 1 to 6 months

	P content (% of DM)	Average % of total P
Total inorganic P	1.5–2	85
Total organic P	0.2–0.3	15
Inorganic P in solution	0.01–0.2	5
P in inositol hexaphosphates (IHP)	0.01–0.1	2
P in microorganisms	0.02–0.04	1.5
P in adenosine-5-triphosphate (ATP)	0.000–0.1	1
Organic P in solution	0.01–0.03	1

organic compounds from manure are only slowly and less effectively attacked by soil organisms for energy or nutrient supply. This resistant organic matter has a much higher C/N ratio and longer turnover time than easily decomposable (labile) organic matter and by immobilization of N it will form a substantial pool of stabilized soil organic N (Fig. 2).

Simple and reliable methods for the direct determination of easily decomposable and resistant organic matter fractions from manure are not available. Even if such methods were at our disposal, the availability of N for a crop cannot be accurately deduced solely from the distribution of N over these fractions [1, 17].

The efficiency of manure N for plant growth generally depends on the outcome of two groups of processes:

- The extent of N losses into i) the air due to NH_3 volatilization and to denitrification, ii) groundwater due to leaching of nitrate, iii) surface waters due to leaching or run-off.
- The rate of N supply to the growing crop's requirements through MIT [2, 11] after application of manure.

Nitrogen losses from manures will be dealt with in a separate contribution [12]. Here only the role of copper (Cu) in chemo-denitrification is mentioned. It has been shown in laboratory experiments that the rate of nitrate reduction under neutral to alkaline conditions is sharply increased in the presence of Cu. With pig slurry providing the soil with large amounts of Cu, chemo-denitrification may contribute substantially to N loss under certain conditions (Hofman and Van Cleemput, unpublished results).

Model calculations

Model calculations on the transformation of manure N in soil can be made with the conceptual scheme shown in Fig. 3.

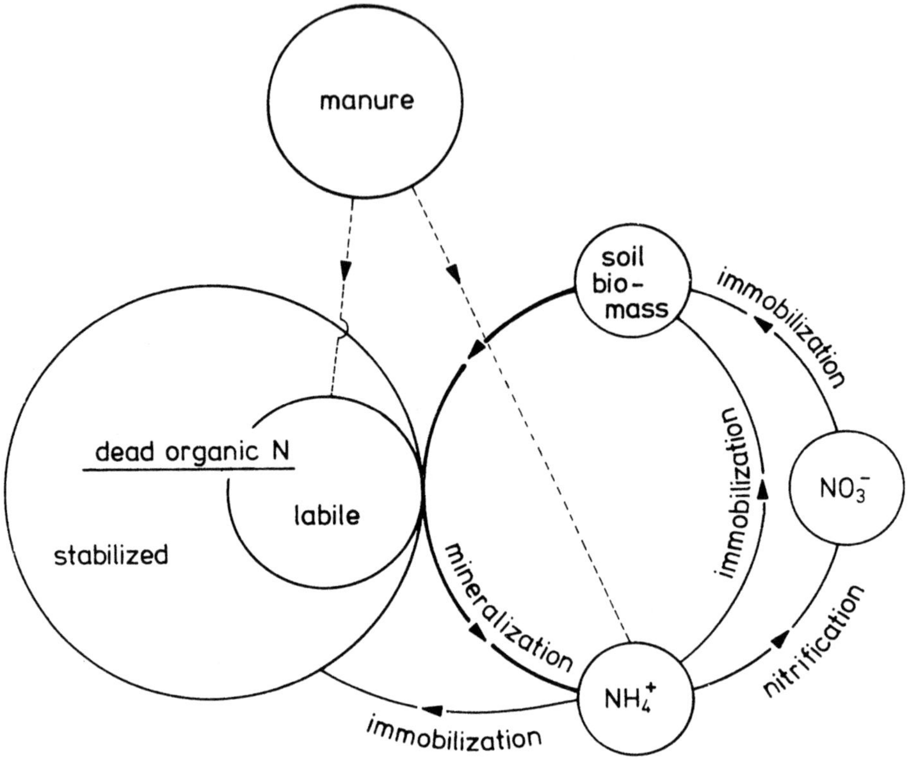

Figure 2. Model for the mineralization-immobilization-turnover in the soil (adapted from [11]).

The biodegradation of various organic matter fractions from manure is assumed to occur at different (first order) rates resulting in certain amounts of soil biomass, depending on the growth efficiencies used. Not only the growth but also the death of the soil biomass must be calculated as well as the production of transformation products. Similar calculations were performed for the turnover in soil of crop residues [13].

In that case the model calculations could be compared with the decrease in ^{14}C from ^{14}C-labeled crop residues applied to the soil. By connecting C and N turnover as shown in Fig. 3, it is possible to calculate the fate of manure N in the course of time. Whereas it is still rather difficult to simulate N turnover in soil throughout the year, this type of modelling can be useful to describe the contribution of long-term, repeated applications of manure to the N supply of crops [7].

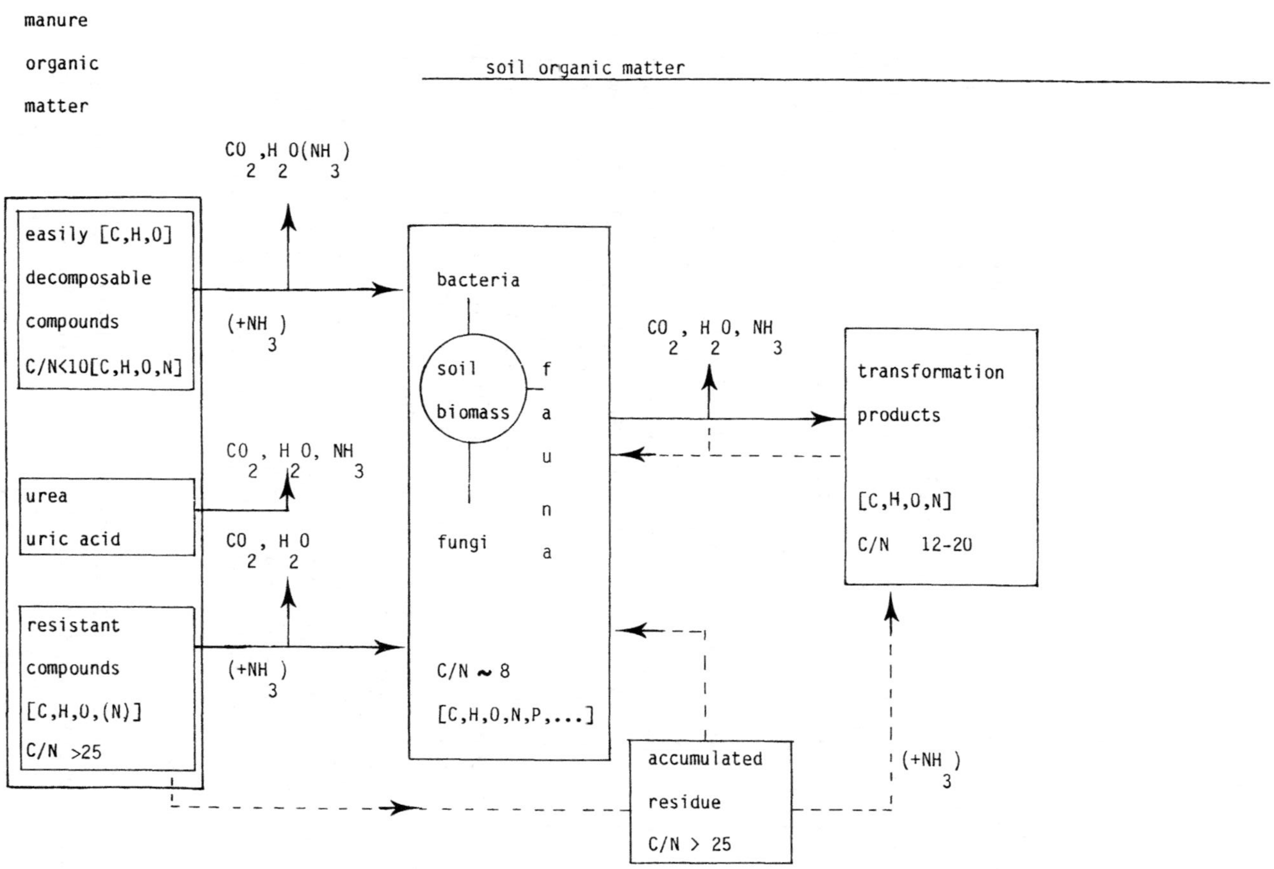

Figure 3. Conceptual scheme for calculating MIT of manure in soil.

Mineralization-immobilization-turnover (MIT)

Information on the MIT of different manures was obtained in two pot trials. The purpose was to derive potential N mineralization rates on the basis of chemical characterization of the manure.

Experimental method

In the spring of 1978 a pot trial (IB 6299) was started with cow faeces, pig faeces and chicken manures, each from 3 sources. The faeces and two of the chicken manures were freshly voided. The materials were mixed with two soils at rates corresponding to 8 and 16 t DM ha^{-1}, with or without addition of ground wheat straw at rates corresponding to 2.7 and 10.8 t DM ha^{-1}. The soils were an acid sandy soil (podzol) from Haren with 3.4% organic matter and a pH (H_2O) of 5.1, and a calcareous sandy loam soil (fluvisol) from Marknesse with 2.4% organic matter and a pH (H_2O) of 8.1. As controls some pots were filled with soils to which $(NH_4)_2SO_4$ with or without straw was added. The pots were buried in the field and were thus subject to field soil temperatures. Loss of N due to leaching was prevented.

In some pots in trial IB 6299 all the inorganic N was initially immobilized. Because of this and because farmers mostly apply liquid manure, a second pot trial was started in the spring of 1979 (IB 6328). In the latter trial anaerobically stored faeces were used with added urea to ensure that inorganic N would not be limiting mineralization of the manures in the soil. In this trial the chicken manures were mixed with sawdust instead of straw. The same soil types were used as in trial IB 6299. However, the sandy soil was taken from another part of the field and contained about 6% organic matter and had a pH (H_2O) of 6.2.

In both trials inorganic N contents of the soils were determined periodically for 1.5 years. In trial IB 6299 additional analyses were made for total N, hydrolysable N fractions and organic matter contents of the soils; pH was also measured. The faeces and manures were analysed in detail.

Results and discussion

Net mineralization of added organic N was dependent on the type of manure (e.g. animal species) as well as on soil type. Generally speaking the fresh chicken manures were mineralized fastest, followed by the pig faeces and the cow faeces. Of the organic N in the undried chicken manures, 70 to 100% was mineralized during the first 6 months of incubation (spring to winter). The N mineralization rate of the dried deep-pit chicken manure was

Table 5. Net N mineralization as a percentage of added organic N from animal manures. + U = with added urea (not considered as added organic N), − U = no urea added

	First year (spring to winter) (6 months)						Second year (12 months)					
	sandy soil (podzol)			sandy loam soil (fluvisol)			sandy soil (podzol)			sandy loam soil (fluvisol)		
	6299	6328 −U	6328 +U	6299	6328 −U	6328 +U	6299	6328 −U	6328 +U	6299	6328 −U	6328 +U
Veal calf faeces	38	39	37	28	16	20	12	13	15	7	8	10
Fattening bull faeces	22	47	49	6	8	22	13	17	7	10	11	0
Dairy cattle faeces	25	36	40	22	20	22	14	15	16	8	10	7
average cattle:	28	41	42	19	15	21	13	15	13	8	10	6
Farrowing house faeces	42	39	49	35	40	20	28	16	5	10	5	22
Fattening swine faeces	30	47	45	20	42	47	14	19	18	15	9	0
Sow faeces	48	29	38	39	18	18	17	15	15	11	11	13
average pigs:	40	38	44	31	33	28	20	17	13	12	8	12
average cattle and pigs:		38			25			15			9	
Broiler manure	95	71	–	93	68	–	7	8		7	7	–
Laying hen manure	70	98	–	67	84	–	12	4	–	13	−4	
Deep-pit manure	45	–	–	28	–	–	8	–	–	9	–	

Table 6. Characteristics of the animal manures and straw admixed to a sandy soil (podzol) and a calcareous sandy loam (fluvisol) in pot experiments (6299 and 6328) to determine net mineralization rates of N

		Dry matter	Contents, % of dry matter								C_{ox}
			organic matter	N_t	NH_4^+-N	uric acid-N	'amide-N'	α-amino-N	P_t	P_{inorg}	N_{org}
Veal calf faeces											
6299	r_I	18.2	81.9	3.81	0.47		0.31	1.29	0.64	0.39	11.8
6328	R_I	17.5	83.6	3.27	0.51		0.48	1.61	0.79	0.20	15.7
Fattening bull faeces											
6299	r_{II}	16.8	79.9	2.66	0.33		0.24	0.89	0.79	0.55	15.7
6328	R_{II}	14.6	82.5	2.71	0.27		0.52	1.02	0.86	0.60	16.8
Dairy cattle faeces											
6299	r_{III}	18.8	84.4	2.84	0.34		0.25	0.97	0.60	0.38	16.1
6328	R_{III}	14.2	80.5	3.81	0.67		0.54	1.04	1.13	0.84	14.3
Farrowing house faeces											
6299	v_I	25.4	79.4	3.90	0.64		0.28	0.85	1.85	0.90	11.2
6328	V_I	26.6	80.9	4.01	0.87		0.54	1.72	2.20	1.97	14.3
Fattening swine faeces											
6299	v_{II}	30.4	83.4	3.05	0.42		0.24	1.09	2.08	1.32	14.7
6328	V_{II}	24.9	77.5	3.87	1.13		0.61	1.47	2.36	1.72	11.0
Sow faeces											
6299	v_{III}	26.4	85.8	4.52	0.67		0.38	1.90	1.88	1.30	10.5
6328	V_{III}	30.2	80.0	2.69	0.29		0.45	1.12	1.62	0.80	15.6
Broiler manure											
6299	k_I	29.6	79.9	6.46	1.17	4.5	0.43	1.11	2.06	0.68	5.8
6328	K_I	54.4	84.3	5.5	0.37	2.8	0.68	1.17	0.90	0.25	7.7
Laying hen manure											
6299	k_{II}	23.0	75.7	6.21	0.71	3.9	0.28	0.69	1.91	0.71	5.4
6328	K_{II}	23.2	73.8	6.1	1.19	3.1	0.94	1.21	1.88	0.98	6.7

Deep-pit manure											
6299	k_{III}	81.0	47.6	3.56*	0.59	0.2	0.53	0.93	3.74	1.53	8.8
Wheat straw meal											
6299		94.0	93.1	0.58	0.01		0.06		0.10	0.075	80
6328		90.7	89.7	0.56	0.01			0.21	0.06	0.05	80

r or R = cow faeces; v or V = pig faeces; k or K = chicken manure; t = total
C_{ox} = 3/8 × chemical oxygen demand
$N_{org} = N_t - (NH_4^+\text{-N})$
* 0.44% NO_3^--N included

much lower; its uric acid had already been mineralized during storage to ammonium and nitrate, and the inorganic N thus formed must have been largely lost due to volatilization of NH_3 and perhaps denitrification of nitrate. In the sandy soil on average 38% of the added organic N from the pig and cow faeces was mineralized during the first 6 months of incubation, and in the following year on average another 15%. The corresponding figures for the sandy loam soil were 25% and 9%. The individual data are given in Table 5.

Apart from N in uric acid in the chicken manures, which was very rapidly mineralized, no relationship was found between the chemical composition of the manures (Table 6) and their N mineralization rates in either of the soil types. The N fractions hydrolysable with 6 *N* HCl in the manured soils were not related to net N mineralization rates during incubation. At most there was some increase of α-amino-N in the acid hydrolysate, reflecting immobilization of N in microbial biomass. Characterization of organic matter from manures and soils with pyrolysis mass spectrometry gave distinct "fingerprints" for the different manures and soil types, but no relationship with N mineralization rates could be found.

A linear relationship was found between the ratio C/N-organic of manures and of mixtures with straw, and the percentage net N mineralization of these materials after 6 months incubation with the soils (Fig. 4). This relationship is strongly dominated by the effect of the added straw (C/N-organic ratio of about 80 initially). It is often assumed that organic matter with C/N-organic < 25 always shows a net mineralization of N. Here it was found, especially with the cow faeces in the sandy loam soil, that net N immobilization occurred for 1 to 5 months even when C/N-organic was $\leqslant 16$. The difference in the composition of the soil microfloras may explain the stronger N immobilization in the sandy loam soil than in the sandy soil. In the calcareous sandy loam soil bacteria with a C/N of about 4 predominate; in the sandy soil fungi with a C/N of about 10 are predominant.

As soon as net N mineralization started, it proceeded almost linearly with time (Fig. 5) for a few months before leveling off. In the winter period, with temperatures around 0 °C, N mineralization stopped.

In the sandy soil in some cases strong acidification occurred, caused by nitrification of ammonium. This may have affected N mineralization in the first trial (IB 6299). In the second trial (IB 6328) this was prevented by timely addition of lime ($CaCO_3$).

The addition of urea to the manure and manure straw mixtures (IB 6328) had no clear effect on the rates of net N mineralization. Urea is very rapidly hydrolysed to ammonium, so that it should be considered as an addition of inorganic N and it therefore changed neither the C/N-organic ratio of the

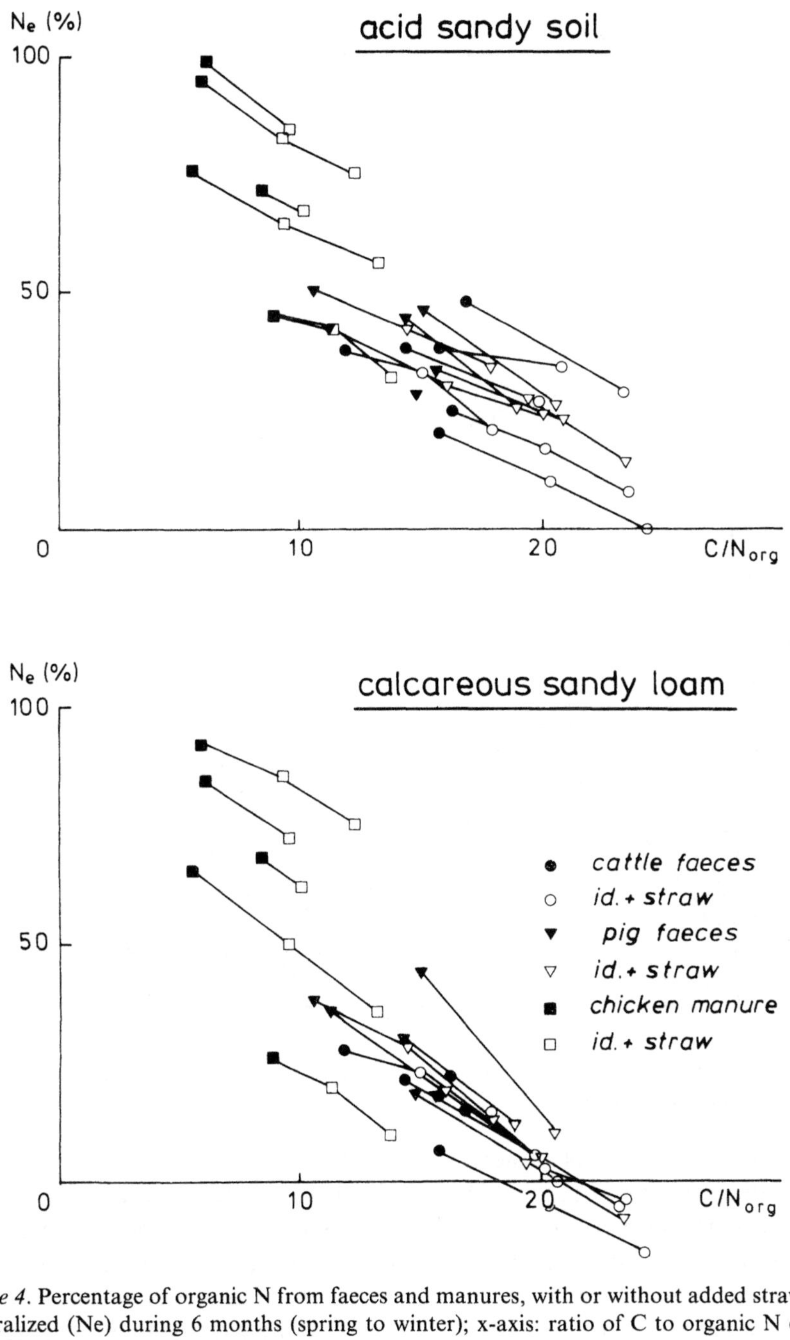

Figure 4. Percentage of organic N from faeces and manures, with or without added straw, net mineralized (Ne) during 6 months (spring to winter); x-axis: ratio of C to organic N of the materials.

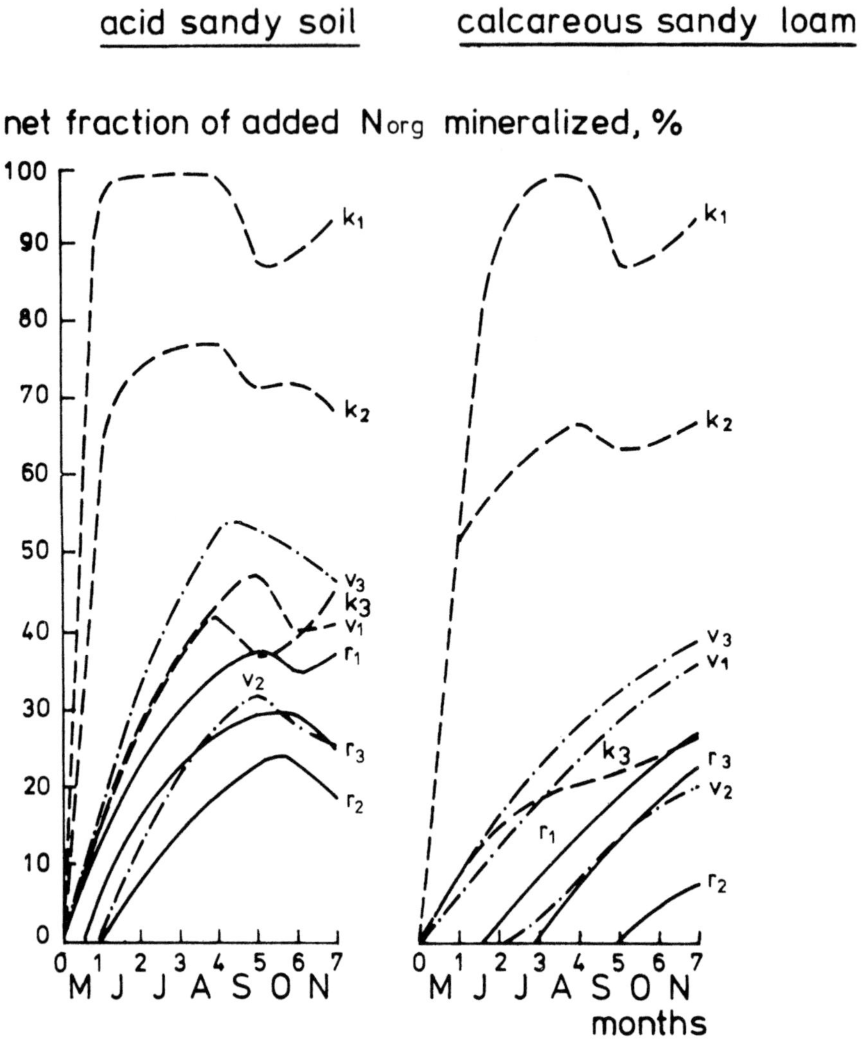

Figure 5. Net N mineralization of organic N from animal manures in two soils. For explanation of symbols see Table 6.

manures nor the net N mineralization rates. Urea addition prevented N immobilization from becoming limited by an absolute shortage of inorganic N as occurred in a few cases in IB 6299. Nitrogen immobilization during decomposition of straw-manure mixtures persisted for a long period and was of the same order of magnitude as in the treatments with straw and $(NH_4)_2SO_4$. It was not possible to follow the course of the C mineralization of manure and straw in the two soils by analysing for chromic acid-oxidizable organic matter.

Conclusions on nitrogen

The results of the pot experiments cannot be simply transferred to field conditions. For instance, no leaching of mineral N could occur and gaseous N losses due to NH_3 volatilization and denitrification were generally small. As vegetation was absent, mineral N accumulation and resulting acidification may have inhibited further mineralization. Also, the absence of living plant roots deprived the soil microflora of a source of stimulation. The pot experiments, however, clearly demonstrated the wide range of N mineralization rates that can be expected in the field, depending on the kind of manure (animal species), the storage conditions of the manure and, particularly also on the soil type (Fig. 4). All these factors should be taken into account to estimate to what extent N in manure is available to the crop in the course of time. Since organic N in manures is often less than 50% of total N, the variations in mineralization rates result in variations of inorganic N availability of only 5% of total N. In comparison, N losses may have a larger impact on the actual N availability than, for example, the origin of the manure [12]. The amount of N supplied from soil organic N increases upon long-term application of manure. This build-up makes it easier to calculate N supply to the crop [7, 15, 17].

Transformations of manure phosphates in the soil

In both soil types in the pot experiments mentioned above, mineralization of organic P from the added manures was found to take place. It appeared to be followed by immobilization of inorganic P. Because of the rather limited increase of soil total-P content due to the added manures and the greater variation in the analytical data, the results of P mineralization are less reliable than those for N mineralization. Moreover, because in both soil types more than 50% of the total-P content was present as inorganic P, it was much more difficult to detect an increase of inorganic P than to follow the sharp increases of inorganic N. Incubation experiments at 2, 20 and 30 °C [9], showed that after one year organic P from pig slurry added to a sandy soil was almost completely mineralized at 30 °C, but only about 40% at 20 °C and 20% at 2 °C.

In an experiment with undisturbed columns of a sandy soil with pig slurry mixed into the top 10 cm and subjected to a "rainfall" of 1 cm day^{-1} the net mineralization of organic P after one year at about 18 °C was only 15%. The gross mineralization (turnover) is always larger and may even lead to a net synthesis of organic phosphate as was observed when only the centrifugate of pig slurry was added to the soil columns. Different organic phosphates will not be decomposed at the same rate. Phosphates derived from nucleic acid and phytates can be persistent. Phytate-P is often the main constituent of organic phosphate in soil. In view of the fact that usually 80–90% of total phosphate in stored manure is inorganic, it cannot be expected that any differences between the "fertilizer value" of manure phosphate and of fertilizer phosphate will be due to the presence of organic phosphates in the manure.

References

1. Amberger A, Vilsmeier K and Gutser R (1982) Stickstofffraktionen verschiedener Güllen und deren Wirkung im Pflanzenversuch. Zeitschrift für Pflanzenernährung und Bodenkunde 145, 325–336
2. Dijk H van (1982) Some notes on the importance of mineralization and immobilization of nitrogen in making fertilizer recommendations. Colloquium Humus-Azote (Reims), pp 151–160. ISSS Association Française pour l'étude du sol, INRA
3. Faassen H G van (1979) Prospects for re-feeding aerobically treated liquid pig manure. In Hawkins J C, ed. Engineering problems with effluents from livestock. Proceedings of a CEC seminar (Cambridge, UK), pp 233–253. Luxembourg: Bureau Official Publications European Community
4. Faassen H G van and Dijk H van (1975) Oxidation-nitrification and denitrification of veal calf manure. In Managing livestock wastes. Proceedings of the third international Symposium on Livestock wastes, pp 314–317. St. Joseph, Michigan: ASAE
5. Faassen H G van and Dijk H van (1979) Nitrogen conversions during the composting of manure/straw mixtures. In Grossbard E, ed. Straw decay and its effects on disposal and utilization, pp 113–120. Chichester, UK: Wiley & Sons
6. Faassen H G van and Dijk H van (1985) Mineralisatiesnelheid van organische stikstof (en fosfaat) uit mest in de grond. With a summary: Mineralization rates of organic nitrogen (and phosphate) from animal manures in soil. Instituut voor Bodemvruchtbaarheid, Rapport 7–85, 90 pp. Haren, The Netherlands: IB
7. Faassen H G van and Smilde K W (1985) Organic matter and nitrogen turnover in soils. In Kang B T and Heide J van der, eds. Nitrogen management in farming systems in humid and subhumid tropics, pp 39–55. IB/IITA
8. Gerritse R G (1978) Assessment of a procedure for fractionating organic phosphates in soil and organic materials using gel filtration and HPLC. Journal of the Science of Food and Agriculture 29, 577–586
9. Gerritse R G (1981) Mobility of phosphorus from pig slurry in soils. In Hucker T W G and Catroux G, eds. Phosphorus in sewage sludge and animal waste slurries, pp 347–369. Dordrecht, The Netherlands: Reidel
10. Gerritse R G and Zugec I (1977) The phosphorus cycle in pig slurry measured from $^{32}PO_4$ distribution rates. Journal of Agricultural Science (Cambridge) 88, 101–109
11. Jansson S L and Persson J (1982) Mineralization and immobilization of soil nitrogen. In Stevenson F J, ed. Nitrogen in agricultural soils. Agronomy no. 22, 229–252
12. Jarvis S C, Sherwood M and Steenvoorden J H A M (1987) Nitrogen losses from animal manures: from grazed pastures and from applied slurry (this volume)
13. Jenkinson D S and Rayner J H (1977) The turnover of soil organic matter in some of the Rothamsted classical experiments. Soil Science 123, 298–305

14 Kirchmann H (1985) Losses, plant uptake and utilisation of manure nitrogen during a production cycle. Acta Agriculturae Scandinavica: Supplement 24, 77 pp.
15. Lande Cremer L C N de la (1985) Long-term effects of farm slurries applications in the Netherlands. In Williams J H, Guidi G and Hermite P l', eds. Long-term effects of sewage sludge and farm slurries applications. Proceedings CEC Seminar (Pisa), pp 84–90. London: Elsevier Applied Science Publishers
16. Peperzak P, Caldwell A G, Hunziker R R and Black C A (1959) Phosphorus fractions in manures. Soil Science 87, 293–302
17. Sluijsmans C M J and Kolenbrander G J (1977) The significance of animal manure as a source of N in soils. Proceedings of the international Seminar on Soil environment and fertility management in intensive agriculture, pp 403–411. Tokyo, Japan: The Society of the Science of Soil and Manure
18. Smith L W (1973) Nutritive evaluations of animal manures. In Inglett G E, ed. Symposium: Processing agricultural and municipal wastes, pp 55–74. Westport. USA: Avi Publishing Co
19. Smith L W (1973) Recycling animal wastes as protein sources. In Alternative sources of protein for animal production, pp 147–173. Washington, USA: National Academy of Sciences
20. Soest P J van (1966) Nonnutritive residues: A system for the replacement of crude fibre. Journal of the Association of Agricultural Chemists 49, 546–551
21. Spoelstra S F (1978) Microbial aspects of the formation of malodorous compounds in anaerobically stored piggery wastes. Thesis Agricultural University Wageningen, 91 pp.
22. Turner D O (1975) On-the-farm-determination of animal waste disposal rates for crop production. In Managing livestock wastes. Proceedings of the third international Symposium on Livestock wastes, pp 587–590. St. Joseph, Michigan: ASAE
23. Vanderholm D H (1975) Nutrient losses from livestock waste during storage, treatment and handling. In Managing livestock wastes. Proceedings of the third international Symposium on Livestock wastes, pp 282–285. St. Joseph, Michigan: ASAE
24. Velsen A F M van (1981) Anaerobic digestion of piggery wastes. Thesis Agricultural University Wageningen, 103 pp.

Utilization of nitrogen from injected and surface-spread cattle slurry applied to grassland

H G VAN DER MEER[a], R B THOMPSON[b], P J M SNIJDERS[c] and J H GEURINK[a]

[a]*Centre for Agrobiological Research (CABO), P.O. Box 14, 6700 AA Wageningen, The Netherlands*

[b]*Institute for Grassland and Animal Production, Hurley, Maidenhead, Berkshire SL6 5LR, United Kingdom*

[c]*Research Station for Cattle, Sheep and Horse Husbandry (PR), Runderweg 6, 8219 PK Lelystad, The Netherlands*

Key words: slurry, animal wastes, injection, nitrogen, fertilizer, grassland, nutrient utilization, nitrogen losses, specific effect

Abstract. The utilization of nitrogen (N) from cattle slurry following injection into grassland, was assessed in 4 field experiments at different sites for 3–5 years. In two experiments, surface application of slurry was included; all slurry applications were made in the spring. In each experiment three rates of slurry application were used, the maximum rates were 40 t ha^{-1} yr^{-1} and 80 or 90 t ha^{-1} yr^{-1} for the surface-spread and injected treatments, respectively. In all experiments 0, 200, 400 or 600 kg fertilizer N ha^{-1} yr^{-1} was applied, both in combination with the slurry applications or alone.

Averaged over all experiments the mean apparent recoveries of N (ANRs) from surface-spread slurry, injected slurry and the lowest rate of fertilizer N were 23%, 49% and 90%, respectively. The corresponding mean apparent efficiencies of N (ANEs) were 6.0, 11.5 and 27.4 kg dry matter (DM) per kg N applied. In the experiments enabling direct comparison with surface spreading, injection consistently resulted in higher ANR and ANE values; the average increases were 2.3 fold and 1.9 fold, respectively. Injection had a relatively greater effect on N uptake than on DM yield; causes and consequences of this are discussed.

The ANRs and ANEs for each method of slurry application differed between sites and years, but were unaffected by slurry application rate. A lag phase of several weeks was observed in N uptake and DM yield responses following the spring injection of slurry. The largest responses occurred in the second of 5–7 cuts; the lag phase was more apparent in DM yield than in N uptake.

Analysis of the responses to combined applications of slurry and mineral fertilizer N, revealed that effects of slurry and fertilizer N on both N uptake and DM yield were additive up to optimal levels of N supply. At higher rates of N supply, fertilizer N could not fully replace the effect of slurry on N uptake and DM yield, indicating a specific effect of slurry.

Herbage from two experiments was examined for nitrate content. Herbage from the plots receiving injected slurry at rates of 80–90 t ha^{-1} without fertilizer N contained excessive nitrate in the second cut. For the other treatments, accumulation of nitrate in herbage occurred when total N uptake exceeded 400–475 kg N ha^{-1} yr^{-1}. These results suggest that herbage nitrate contents should be considered when determining (i) maximum application rates for injected slurry when used without fertilizer N and (ii) optimal fertilizer N rates to be used with injected slurry.

H.G. v.d. Meer, et al. (eds.), Animal Manure on Grassland and Fodder Crops.
 ISBN 978-94-010-8137-5.

The high recovery of available slurry N indicated that injection of slurry in spring considerably reduced the losses of N associated with both surface spreading and applications made well before the start of the growing season.

Introduction

In many European countries most of the collected excrement of housed livestock is handled as slurry [2, 26, 27]. The rapid introduction of the slurry system has been a consequence of the intensification of livestock farming. Its main advantage is the saving of labour; it also reduces the quantity of bedding material required.

Dairy cattle produce approximately 10 t slurry per cattle unit during the 6 months when cattle are housed. This quantity contains on average 45 kg nitrogen (N), 8 kg phosphorus (P), and 45 kg potassium (K) [1]. Intensive grassland farms, which often have more than 3 cattle units per ha, are therefore required to dispose of large amounts of plant nutrients in slurry. The availability of relatively cheap inorganic fertilizers, which are easy to handle, has reduced the interest of farmers in using animal manures as a source of nutrients [33]. Although there is no reliable statistical information about slurry application practices, it appears that most livestock farmers dispose of slurry as cheaply as possible whilst avoiding damage to grassland and crops. At present, most dairy farms in The Netherlands have a storage capacity for slurry which is less than the amount produced in 3 months. This means that landspreading in the winter period is necessary. Furthermore, farmers try to minimize the negative effects of surface spreading of slurry to grassland in the growing season; these are smothering and scorching of the sward, rejection of herbage by grazing animals, and spreading of diseases and parasites [20, 23].

As a consequence of the limited storage capacity for slurry and the problems involved in surface spreading on grassland during the growing season, a large proportion of cattle slurry is applied throughout the winter to arable land destined for the production of silage maize and to the driest grassland fields.

The repeated applications to some fields, and not to others, and the practice of surface spreading well before the growing season are major factors responsible for the poor utilization of nutrients in slurry. The limited recycling of nutrients in collected manures is a major cause of the low overall efficiency of N used on intensively managed grassland farms [15, 21]. It may also explain the decrease of the P and K status of grassland soils which has been recently observed in The Netherlands [11]. The current practices of slurry application contribute to pollution of the natural environment by volatilization of ammonia, run-off of slurry and leaching of nitrate. The need to reduce these negative effects on the environment, and the increasingly strict legislation being introduced require application techniques that enable increased utilization of the nutrients contained in slurries.

Research on slurry injection into grassland started in The Netherlands in 1972. The main objective was to reduce offensive odours associated with surface spreading. Moreover it was thought that injection of slurry would improve the utilization of slurry N and reduce smothering and scorching of the sward and fouling of the herbage [12]. Preliminary experiments were carried out to study the optimal design of the injection equipment, considering also tine spacing and depth of injection. It was concluded that a distance of 50 cm between injection tines and an injection depth of 15–20 cm caused little damage to the sward and gave good responses to injected slurry [34]. However, the commercial slurry injector used in these preliminary experiments had a number of shortcomings that hindered accurate assessment of the fertilizer value of the injected slurry. Consequently, a special slurry application unit was constructed which enabled accurate measurement of the volume of slurry applied by injection and surface spreading and an even distribution to the injection tines [24]. The injection unit has 4 tines, each with an 18 cm wide duckfoot sweep at the base; disc coulters are located in front of the tines and press wheels behind them. Slurry is pumped into the slits created by the tines through flexible pipes ending behind the duckfoot sweeps.

This paper presents the results of 4 long-term field experiments carried out to study the utilization of N from cattle slurry applied to grassland by injection and surface spreading in spring. Preliminary results of this study have already been reported [12, 13].

Experimental methods

The effects of injection and surface application of different rates of cattle slurry on N uptake and dry matter (DM) production by grass, were studied in long-term experiments on permanent grassland at Den Ham, Ruurlo, Hemrik and Friens. The experiments at Den Ham, Ruurlo and Hemrik were situated on sandy soils with a watertable varying from less than 25 cm below soil surface during wet periods in winter to about 120 cm below soil surface during dry periods in the growing season. The experiment at Friens was on a clay soil with a slightly lower watertable. In the experiments conducted at Den Ham and Ruurlo, slurry was applied by surface spreading and injection; at Hemrik and Friens only injection was used. The slurry treatments and the years in which the experiments were conducted are summarized in Table 1. Injection of 0 t ha^{-1} yr^{-1} (Inj0) refers to cutting the sod with the injection unit without the application of slurry. The rates of slurry presented in Table 1 were target rates. The actual rates applied were recorded by the slurry application unit. They generally differed less than ten per cent from the target rates and are used in the analysis of results.

Slurry was applied with the Schepan MMM slurry application unit [24] each spring when the T sum had reached 200 °C [9] and the experimental field was sufficiently dry to support the tractor and slurry application equipment. Generally, this was between 15 March and 20 April. The applied cattle slurries had a fairly similar composition; the average (17 different slurries) contents on a fresh weight basis were 10.3% dry matter (range: 7.7–11.7), 0.51% total N (range: 0.46–0.60), 0.08% P (range: 0.04–0.11), 0.50% K (range: 0.41–0.63), 0.19% Ca (range: 0.13–0.27) and 0.07% Mg (range: 0.05–0.09).

In each experiment, the experimental treatments referred to in Table 1 were combined with four levels of mineral fertilizer N: N0, N1, N2 and N3, for which the respective target rates were 0, 200, 400 and 600 kg N ha^{-1} yr^{-1}. The annual rates actually applied were determined by the number of cuts taken. In the experiment at Hemrik each cut received a fixed rate of fertilizer N, *viz.* 0, 40, 80 and 120 kg N ha^{-1} on N0, N1, N2 and N3, respectively. In the other experiments 25, 20, 20, 15, 10 and 10% of the target annual rates was given to the first six cuts and 10% to each subsequent cut. The spring application of fertilizer N (for the first cut) was carried out immediately after the application of slurry. It was aimed to keep P, K and Mg in optimal supply by frequent uniform dressings of inorganic fertilizer to the whole experimental field; application rates were based on the results of regular soil analyses. Detailed information on the manuring practices is given elsewhere [28].

For each experiment a split-plot design was used. The slurry treatments

Table 1. Rates of surface-applied and injected cattle slurry in the experiments at Den Ham, Ruurlo, Hemrik and Friens. The codes for the experimental treatments will be used in the text: SA = surface application; Inj = injection

Site and duration of the experiment		Surface application (t ha^{-1} yr^{-1})				Injection (t ha^{-1} yr^{-1})			
		Control	SA1	SA2	SA3	Inj0	Inj1	Inj2	Inj3
Den Ham,	1979–83	0	10	20	40	0	20	40	80
Ruurlo,	1980–84	0	10	20	40	0	20	40	80
Hemrik,	1978–81	0	–	–	–	0	30	60	90
Friens,	1981–83	0	–	–	–	–	30	60	90

and Controls were the main plots, the mineral fertilizer treatments the sub-plots. Each experiment had either 3 or 4 replicate plots per treatment. The slurry and fertilizer treatments were repeated each year on the same plots.

The plots of each experiment were cut on the same day. Normally, cuts were made when the yields of the most productive treatments were estimated to be between 2.5 and 3 t DM per ha. Herbage yields were determined by harvesting areas of 5.4 m^2 from each 12.5 or 15 m^2 plot.

Responses to slurry and mineral fertilizer applications

The results obtained in the experiments conducted at the 4 sites were similar with regard to the responses of both DM production and N uptake (total N in harvested herbage) to the treatments used [28]. As an example, herbage DM yield and N uptake for the different treatments in the experiment at Ruurlo are presented in Table 2. These results show that, compared to surface application, injection substantially improved the effect of slurry on N uptake and DM yield of grass. This difference was clearly evident on the N0 plots, but was increasingly masked as the rate of fertilizer N increased.

In the absence of slurry and fertilizer N, cutting the sod by the injection unit (Inj0) resulted in a small increase in annual DM yield and annual N uptake (Table 2). This also occurred in the other two experiments where the cutting treatment was used. Averaged over 14 experiment-years, the cutting treatment increased DM yield in the absence of fertilizer N by 364 kg ha^{-1} yr^{-1} and N uptake by 15 kg ha^{-1} yr^{-1}. When the Inj0 treatment

Table 2. Effects of surface-applied or injected cattle slurry and fertilizer N on DM yield and N uptake by grass in the experiment at Ruurlo. Figures are averages of 5 years. The average rates of fertilizer N are given between brackets. Codes are explained in Table 1. LSDs (Least Significant Differences; P = 0.05) are 0.527 t DM ha^{-1} yr^{-1} and 19 kg N ha^{-1} yr^{-1}

Slurry treatment	Slurry N applied (kg ha^{-1} yr^{-1})	Herbage DM (t ha^{-1} yr^{-1}) and N yield (kg ha^{-1} yr^{-1}) at:							
		N0 (0)		N1 (193)		N2 (386)		N3 (579)	
		DM	N	DM	N	DM	N	DM	N
Control	0	5.68	141	11.36	325	12.74	464	13.42	542
SA1	51	6.67	163	11.75	349	13.25	480	13.57	554
SA2	99	6.56	166	11.98	352	13.22	494	13.64	555
SA3	185	7.24	186	11.94	358	13.58	500	13.76	557
Inj0	0	5.92	147	11.19	329	12.89	472	12.82	508
Inj1	106	7.32	196	11.87	373	13.08	490	13.64	560
Inj2	195	8.46	245	12.56	426	14.09	537	13.79	582
Inj3	378	10.88	349	13.04	480	13.85	562	13.40	572

was combined with the two highest rates of fertilizer N (N2 and N3), the cutting effect of the injector on N uptake and DM yield tended to be negative [28]. Generally, the cutting treatment slightly reduced N uptake and DM yield in the first cut at each level of fertilizer N. In subsequent cuts, N uptake and DM yields were increased where the supply of N was sub-optimal [28]. It seems that in N deficient situations the cutting effect of the injector achieved a slight increase in growth, presumably by promoting the mineralization of soil organic N, which more than compensated for the initial retardation of growth resulting from sward damage. Where the supply of N was adequate, such a compensation was not possible and the negative effect of the injector was predominant.

Assessment of responses to slurry and fertilizer nitrogen

The efficiency of use of N from slurry or mineral fertilizer can be expressed in two ways: (i) as the apparent recovery of N (ANR) which is the increase of the amount of N contained in the harvested herbage, expressed as a percentage of that applied in slurry or fertilizer, and (ii) as the apparent efficiency of N (ANE) which is the increase of harvested herbage DM per kg N applied in slurry or fertilizer. The ANR and ANE values are therefore calculated from the respective differences in N uptake and DM yield between the slurry or fertilizer treated plots and the untreated (Control) plots.

In the following text, ANRs from slurry relate to treatments in which no fertilizer N was used, and ANRs from fertilizer relate to treatments in which no slurry was applied. The calculations of ANR values for injected slurry treatments used the Control plots as references, therefore any cutting effect of the injector upon N uptake is included. The ANE values were calculated similarly for the relevant slurry and fertilizer treatments.

Apparent recovery of nitrogen from slurry and fertilizer

The average ANRs for each rate of surface-applied and injected slurry at each site are presented in Table 3. The values are the averages for the periods over which the experiments were conducted. For comparison, similarly determined ANRs from the lowest rates of fertilizer N (N1 and N2) are included.

From the data presented in Table 3, it is evident that the ANR values obtained in each experiment for both slurry application methods were not affected by the rate of application. The only striking exception was the high ANR value for the lowest rate of surface-spread slurry (SA1) at Ruurlo. This was probably caused by the experimental error involved with the

Table 3. Apparent recoveries of N (ANRs) from different rates of surface-applied or injected cattle slurry in the absence of fertilizer N, and from fertilizer N in the absence of slurry in 4 field experiments. The figures are averages over 3 (Friens), 4 (Hemrik) or 5 years (Den Ham and Ruurlo). Codes are explained in Table 1

Code N source	Average ANRs (in % of N applied) in the experiment at:			
	Den Ham	Ruurlo	Hemrik	Friens
SA1	18	43		
SA2	18	25		
SA3	22	24		
Inj1	56	52	43	39
Inj2	56	53	39	47
Inj3	49	55	46	46
N1	81	95	91	96
N2	73	84	85	93

determination of this value (a small response assessed from multiple cuts). The observed similar ANR values for the different rates of slurry indicate a linear relationship at each site, between the rate of slurry N and N yield in the harvested herbage for each application method even up to the highest rate of slurry N applied (Inj3), *viz.* 355, 378, 475 and 485 kg N $ha^{-1} yr^{-1}$ at Den Ham, Ruurlo, Hemrik and Friens, respectively. Consequently, ANR at each site can also be calculated by linear regression of N yield in the harvested herbage (N_Y) on the rate of slurry N applied (N_S), according to the equation: $N_Y = N_C + ANR * N_S$, where N_C is the N yield in the harvested herbage on the Control plots. This enables determination of annual ANR values for each application method in each experiment, which simplifies comparisons between application methods, and between sites and years. The ANR values determined in this manner for the slurry application methods used in the 4 experiments, are presented in Table 4. The ANR values from the N1 treatments, which produced similar DM yields and N uptakes as the highest rate of injected slurry (Table 2; [28]), are included in Table 4 for comparison.

The ANR values from slurry were substantially improved by injection (Tables 3 and 4). However, the ANRs from injected slurry were clearly less than those from fertilizer N, which were very high in these experiments as has often been observed in recent N-manuring experiments on grassland in The Netherlands [16]. The high ANR values for N1 and N2 may imply that the ANRs from slurry obtained in these experiments are also relatively high. The ANRs from surface-applied slurry are slightly higher than those obtained in other studies. In recent trials in the UK and Ireland, average ANR values for cattle slurry surface-spread on grassland in spring, varied from

Table 4. Annual apparent recoveries of N (ANRs) from surface-applied and injected cattle slurry, and from N1 fertilizer treatments in each experiment. ANRs from slurry were calculated by linear regression of N yield on the rate of slurry N applied, using the results of the Control plots and of the three slurry rates for each slurry application method. Averages for each experiment and the relative values compared to the N1 treatments are included

Experiment	Source of N	ANRs (in % of N applied) in experimental year:					Average	ANR relative to N1
		1	2	3	4	5		
		(1979)						
Den Ham	surf. appl. slurry	14	25	29	25	22	23	28
	injected slurry	53	47	56	46	42	49	60
	N1	83	74	97	74	77	81	100
		(1980)						
Ruurlo	surf. appl. slurry	9	22	28	28	26	23	24
	injected slurry	61	59	61	44	50	55	58
	N1	98	96	92	91	98	95	100
		(1978)						
Hemrik	injected slurry	57	49	44	29		45	49
	N1	100	109	61	100		92	100
		(1981)						
Friens	injected slurry	45	54	42			47	49
	N1	101	99	84			95	100

about 15% to 26% [18, 30, 32]. Higher values have been obtained from diluted slurry or urine and also in conditions where slurry application increased mineralization of organic N in the soil or legume content of the sward through the supply of other nutrients [23, 27]. Information on ANRs from injected slurry on grassland is very limited. At the Animal and Grassland Research Institute (AGRI) at Hurley, ANR values of 33 and 36% were found for cattle slurry injected in December and April, respectively [29].

The ANR values from surface-spread and injected slurry and N1 fertilizer were generally consistent between years and between sites (Table 4). ANRs from injected slurry were slightly higher at Den Ham and Ruurlo than at Hemrik and Friens. No explanation could be found for this difference. Studies at AGRI have shown that denitrification can cause significant losses of N from injected slurry, especially during wet periods [29] (R.B. Thompson, unpublished). This may explain the low ANRs from injected slurry in 1983 (last experimental year at Den Ham and Friens and fourth experimental year at Ruurlo), when very wet weather prevailed in April and May.

No relationship was observed between the quantity of rain within 3 days after slurry application (0–12.3 mm) and ANR from surface-spread slurry.

A similar conclusion was drawn from a series of experiments in South West England, where increased rainfall within 5 days after slurry spreading (0–48.6 mm) did not appear to improve the response to slurry N [32]. Relationships between ANR and total rainfall within several days after surface spreading of slurry are unlikely due to the very high rate of N loss through volatilization of ammonia immediately after surface applications of slurry [6, 29]. The low ANR values from surface-applied slurry in the first experimental year at both Den Ham and Ruurlo, from fertilizer in the third year and from injected slurry in the last year at Hemrik cannot be explained from the available data.

Repeated annual applications of cattle slurry for 3–5 years, did not noticeably increase annual ANRs (Table 4). This indicated that the residual effect of slurry N after the year of application was very small and that that part of slurry N which was not available in the first year became available at a very slow rate. In other studies residual N effects from cattle slurry have also been small [14, 25, 27].

In the experiments allowing direct comparison of injection with surface spreading (Den Ham and Ruurlo), the average ANRs from injected and surface-spread slurry were 59 and 26%, respectively, of that from the N1 fertilizer treatment (Table 4). For the 4 experiments, the average ANR from injected slurry was 54% of that from the N1 treatment.

Apparent efficiency of nitrogen from slurry and fertilizer

As with the ANR values in Table 3, the apparent efficiencies of N (ANEs) for the different rates of slurry applied by either surface spreading or injection were not affected by the rate of application [28]. This means that linear relationships existed at each site between the rate of slurry N and herbage DM yield up to the highest rates of slurry for each application method. Therefore, ANE can be calculated by linear regression of herbage DM yield (Y) on the rate of slurry N applied (N_S), according to the equation: $Y = C + ANE * N_S$, where C is the DM yield from the Control plots. The annual ANE values determined in this way for the slurry application methods used in the 4 experiments, are presented in Table 5. ANEs from the N1 treatments are included for comparison.

In general, ANE values in the experiment at Den Ham were lower than those in the other experiments. This probably reflects the comparatively high DM yields on the untreated plots and the relatively low maximum DM yields at Den Ham [28]. In the other experiments, the differences between maximum DM yield and DM yield on the untreated plots were larger, enabling greater initial responses to applied N.

Table 5. Annual apparent efficiencies of N (ANEs) from surface-applied and injected cattle slurry, and from N1 fertilizer treatments in each experiment. ANEs from slurry were calculated by linear regression of DM yield on the rate of slurry N applied, using the results of the Control plots and of the three slurry rates for each application method. Averages for each experiment and the relative values compared to the N1 treatments are included. The relative values can be considered as "efficiency indices" [3]

Experiment	Source of N	ANEs (kg DM per kg N) in experimental year:					Average	ANE relative to N1
		1	2	3	4	5		
		(1979)						
Den Ham	surf. appl. slurry	3.1	4.2	6.4	4.8	3.3	4.4	22
	injected slurry	10.9	9.6	10.8	8.7	7.7	9.5	47
	N1	22.2	19.1	22.7	15.4	21.6	20.2	100
		(1980)						
Ruurlo	surf. appl. slurry	3.7	7.1	8.4	10.5	7.7	7.5	26
	injected slurry	15.9	12.8	12.3	12.8	14.9	13.7	47
	N1	34.6	27.6	24.8	27.5	31.1	29.1	100
		(1978)						
Hemrik	injected slurry	12.5	14.1	12.2	6.7		11.4	35
	N1	33.8	39.5	26.9	29.4		32.4	100
		(1981)						
Friens	injected slurry	10.3	13.2	11.3			11.6	39
	N1	30.9	30.5	27.8			29.7	100

Injection was always more effective than surface application in promoting herbage DM yield, although the differences were variable (Table 5). In the experiments at Den Ham and Ruurlo, the average ANEs from injected and surface-spread slurry were 47 and 24%, respectively, of the average ANE from the N1 fertilizer treatment. Averaged over the 4 experiments, ANE from injected slurry was 42% of that from the N1 fertilizer treatment.

It should be noted that when expressed relative to fertilizer values, the ANR of injected slurry was clearly higher than the ANE value (cf. relative values in Tables 4 and 5). This indicates that N taken up by the sward from injected slurry was used less efficiently for DM production than that from fertilizer.

Distribution throughout the growing season of the effects of spring-applied slurry on nitrogen uptake and dry matter yield

The effect of spring-injected cattle slurry on N uptake and DM yield appeared to have a characteristic distribution throughout the growing season. The distribution patterns were similar in the 4 experiments [28]. The distributions over the individual cuts of the effect of different rates of injected and surface-spread slurry on N uptake in the experiment at Ruurlo are presented in Fig. 1. The equivalent distributions of the effect of slurry on DM yield at the same site are presented in Fig. 2. These results are averages of 5 experimental years from plots receiving no fertilizer N.

A relatively small part of the N uptake from spring-injected slurry took place in the first cut; the major response occurred in the second cut (Fig. 1). The effect on N uptake in later cuts was also significant, particularly at the highest rate of slurry application. On average, 20–30% of the annual effect of injected slurry on N uptake took place in the first cut, 35–45% in the second cut, 15–25% in the third cut and 10–25% in combined later cuts. However, the pattern of distribution over the individual cuts differed considerably from year to year. This appeared to be related to the cutting management applied. Heavy first cuts generally contained a larger part of the total annual response than light cuts.

The distribution of the DM yield response to injected slurry over the cuts taken (Fig. 2) was very similar to that of N uptake (Fig. 1), the major difference being the smaller contribution of the first cut in the total yield response (on average 10–20%), which was compensated for by a slightly larger response in the third and later cuts.

The observed distribution throughout the growing season of the effects of spring-injected slurry on N uptake and DM yield, indicated a limited uptake of slurry N in the first month after injection. Uptake of N in the first cut from fertilizer was only slightly reduced by the cutting effect of the injector

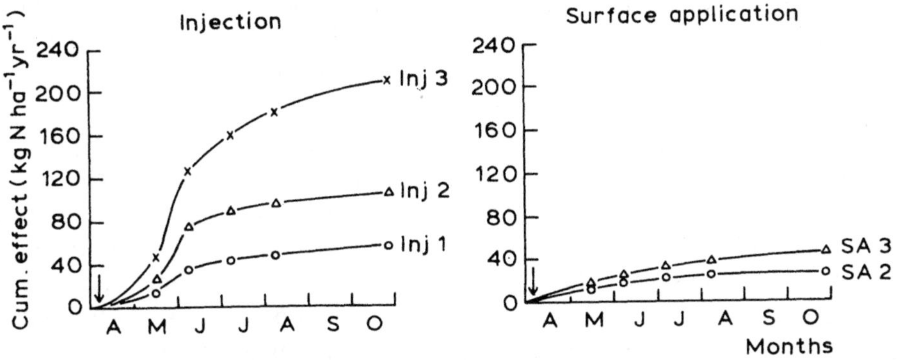

Figure 1. Distribution throughout the growing season of the effect on N uptake of different rates of slurry applied by either injection or surface spreading at Ruurlo. The average effect (1980–1984) of each treatment on the yield of N in the harvested herbage (*i.e.* the difference in herbage N yield between the treated plots and the untreated plots) in the first, second, third, fourth and combined later cuts are plotted cumulatively against the average harvest dates of the first, second, third, fourth, and last cut, respectively. Arrows indicate the average date of slurry application.

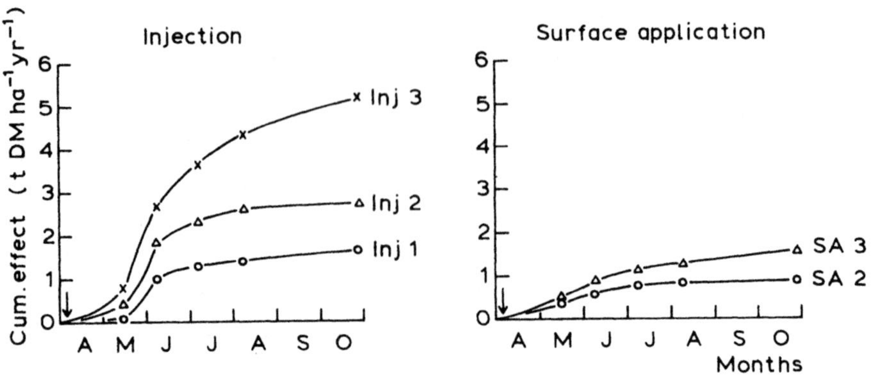

Figure 2. Distribution throughout the growing season of the effect on DM yield of different rates of slurry applied by either injection or surface spreading at Ruurlo. Effects on individual cuts from each treatment are calculated and plotted as in Fig. 1.

(Inj0). In a pot experiment (Mitscherlich pots) where cattle slurry was injected into a well developed sward of *Lolium perenne* in May, 55–60% of the total N uptake and DM yield responses were observed in the first cut, 4 weeks after injection [6]. These results suggest that the delayed response of N uptake and DM yield to spring-injected slurry in the field is caused by (i) the distance between injection tines and the depth of slurry placement (50 cm and 15–20 cm, respectively in these experiments), (ii) the limited reach and activity of grass roots in March and April, and (iii) the slow recuperation of damaged grass roots in spring. More experimental work on the influence of cutting management, time of injection and weather conditions after injection on the distribution throughout the growing season of N uptake and DM yield responses to injected slurry appears to be necessary.

Figs. 1 and 2 show that surface spreading of slurry resulted in comparatively more even distributions of the N uptake and DM yield responses. In this experiment (Ruurlo), 30–40% of the annual total of both responses was observed in the first cut. However, at Den Ham it was approximately 60%. The observed difference between the two experiments may have resulted from the slightly heavier first cuts taken at Den Ham. The experimental error involved with small total responses assessed from multiple cuts, may also hamper an accurate estimation of the distribution over the individual cuts.

Responses to combined applications of slurry and fertilizer nitrogen

The preceding section has described the response of both N uptake in the harvested herbage and herbage DM yield following slurry applications, without mineral fertilizer N, to grassland. In order to develop effective manuring guidelines it is important to assess whether the responses to N from slurry and chemical fertilizer are additive. The "three quadrants diagram" [4, 15, 16], which shows the relationships between the supply of fertilizer N, total uptake of N in the harvested herbage and herbage DM yield, provides a clear and effective method for this assessment. Examples from the experiments at Ruurlo, Hemrik and Den Ham are presented in Figs. 3 and 4. The values given are averages for the duration of each experiment. In quadrants II of these Figures, the response of DM yield to fertilizer N is shown for the plots without slurry and for those with injected (Fig. 3) or surface-spread slurry (Fig. 4). This response has two components: (i) the effect of fertilizer N on N uptake in the harvested herbage (quadrants IV), and (ii) the effect of N uptake on DM yield (quadrants I).

Injected slurry and fertilizer nitrogen

The contribution of injected slurry to N uptake when combined with a given rate of fertilizer N, is illustrated in quadrant IV of Fig. 3 as the difference in N uptake (the horizontal distance) between the curves or symbols relating N uptake to the rate of fertilizer N for the Inj2 or Inj3 treatments, and the treatment without slurry (Control). At increasing rates of fertilizer N, the effect of injected slurry on N uptake decreased. This is in agreement with the widely observed decrease in the apparent recovery of N that occurs with increasing rates of N application [15, 17].

Comparisons between the effects of injected slurry and fertilizer N on N uptake can be made by considering the vertical distance between the response curves in quadrant IV of Fig. 3. The vertical distance between the curves represents the effect of the injected slurry on N uptake in the slurry/-fertilizer treatment, expressed as the rate of mineral fertilizer N required to obtain the same response. For example, at Ruurlo (Fig. 3a) the effect on N uptake of 80 t injected slurry $ha^{-1} yr^{-1}$ (378 kg N $ha^{-1} yr^{-1}$) without added fertilizer N was equivalent to the effect of 215 kg fertilizer N $ha^{-1} yr^{-1}$. From these data it can be calculated that at this rate of injected slurry at Ruurlo, the effect of N from slurry on N uptake was 57% of the effect of N from fertilizer. This value can also be derived from the corresponding ANRs in Table 3.

The vertical distance between the curves in quadrant IV of Fig. 3a is constant up to N uptakes of approximately 450–500 kg $ha^{-1} yr^{-1}$ and increases at higher uptakes of N. This was also consistently observed in the other experiments (Fig. 3b) [28]. The constant vertical distance between the curves indicates that within a wide range of N uptakes (150–500 kg N $ha^{-1} yr^{-1}$) the effect of a given rate of injected slurry on N

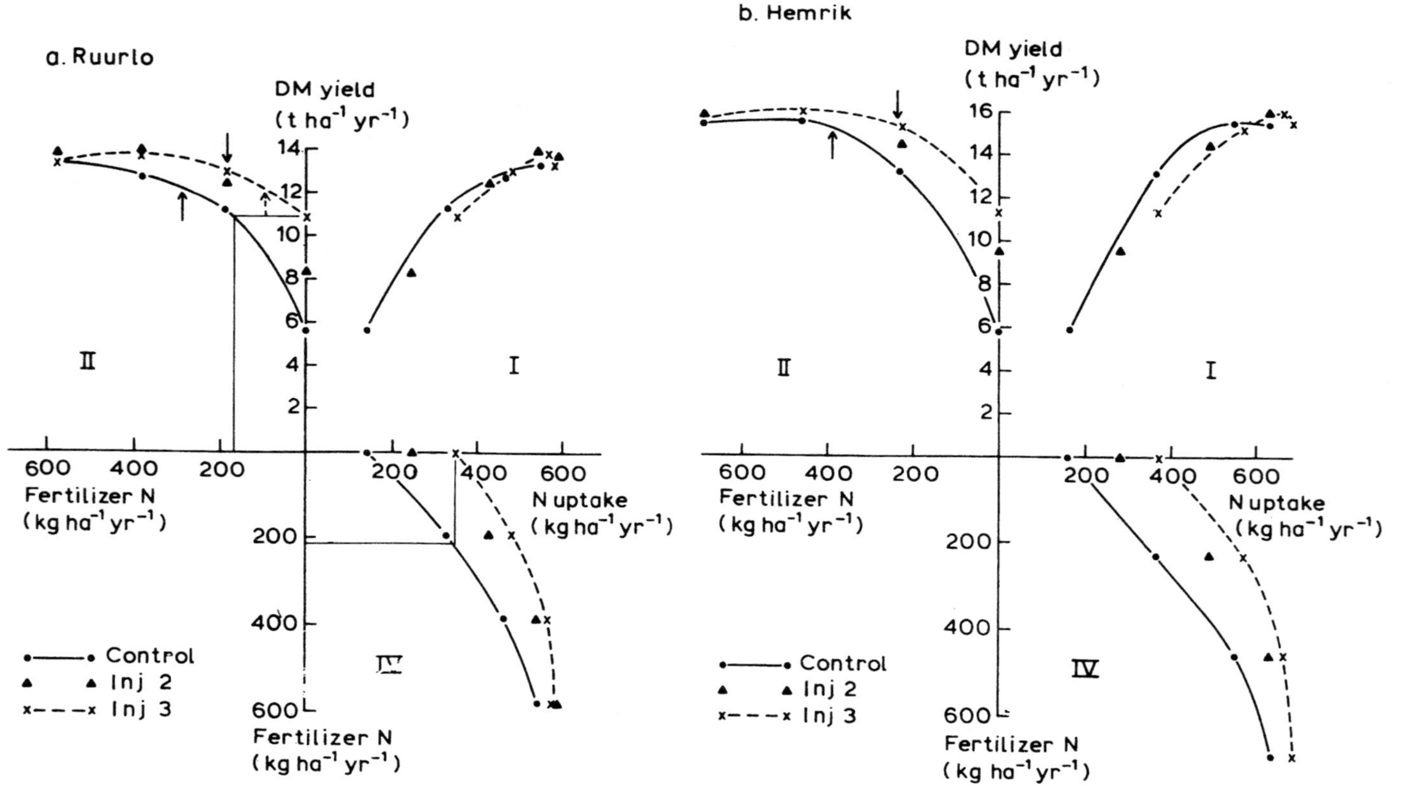

Figure 3. The effect of the rate of fertilizer N on N uptake and DM yield in the absence of slurry (Control) and combined with 2 rates of injected cattle slurry (Inj2 and Inj3). The data are averages of 5 (Ruurlo, Fig. 3a) and 4 years (Hemrik, Fig. 3b).

uptake was equivalent to a fixed rate of fertilizer N. Therefore, for N uptakes between 150 and 500 kg ha^{-1} yr^{-1} the effects on N uptake of injected slurry and mineral fertilizer N were additive. For uptakes of N higher than 500 kg ha^{-1} yr^{-1}, an increasing amount of fertilizer N was needed to replace the effect of injected slurry (Inj2 or Inj3). At very high uptakes of N, the effect of injected slurry on N uptake could not be replaced by fertilizer N. These two observations indicate a specific or additional effect of injected cattle slurry. Specific effects of animal manures have often been observed in studies with animal manures on arable crops [3].

The effects of injected slurry and fertilizer N on DM yield can be examined in a similar manner (quadrant II of Fig. 3). In this case the vertical distance between the response curves indicates the effect of injected slurry from the combined slurry/fertilizer treatments on DM yield. The horizontal distance between the curves represents the amount of fertilizer N which gives the same effect on DM yield as slurry in the combined injected slurry/fertilizer treatment. The example given in Fig. 3a shows that at Ruurlo the effect of 80 t injected slurry ha^{-1} yr^{-1} (378 kg N ha^{-1} yr^{-1}) on DM yield was equivalent to the effect of 175 kg fertilizer N ha^{-1} yr^{-1}. This means that the effect of N from Inj3 on DM yield was 46% of the effect of fertilizer N (47% in Table 5).

The horizontal distance between the response curves in quadrant II of Fig. 3 is constant for the lower rates of fertilizer N. However, at rates of fertilizer N that produce near-maximum DM yields, the horizontal distance between the curves increases, indicating that an increasing amount of fertilizer N was required to replace the effect of a given rate of injected slurry. The maximum DM yields obtained from combinations of injected slurry and fertilizer N, were slightly higher than those on plots receiving only fertilizer N. The preceding indicates that at rates of N inducing a high growth response, *i.e.* on the lower parts of the response curve, the effects of N from injected slurry and fertilizer on DM yield were additive. At higher levels of N (from combinations of injected slurry and fertilizer N), the specific effect of slurry upon DM yield was expressed, which means that the effect of injected slurry on DM yield could not be fully replaced by fertilizer N. Both the additive effects of injected slurry and fertilizer N on DM yield on the lower parts of the response curve, and the expression of the specific effect of injected slurry on DM yield at high rates of fertilizer N, were consistently observed in each of the 4 experiments [28].

As has previously been discussed, injected slurry was more effective in promoting N uptake than in promoting DM yield when these responses were compared to those obtained with fertilizer N. This is apparent in quadrant I of Fig. 3, which presents the relationship between N uptake and DM yield for the mineral fertilizer treatment and for the combined injected slurry and fertilizer treatments. In this quadrant, it is evident that the amount of N absorbed by the sward from Inj3 (without fertilizer) gave a lower DM yield than the same amount of N absorbed from fertilizer on the plots without slurry (Control). This difference was approximately 0.8 t DM ha^{-1} yr^{-1} at Ruurlo (Fig. 3a), 2.0 t DM ha^{-1} yr^{-1} at Hemrik (Fig. 3b) and 1.0 t DM ha^{-1} yr^{-1} at Den Ham and Friens [28].

The comparatively lower effectiveness of N absorbed from injected slurry in promoting DM production, probably resulted from the uneven distribution of slurry and the large single application at the start of the growing season. The uneven distribution leads to excessive uptake of N by part of the sward and to N deficiency in another part. The single large application is likely to result in some luxury uptake early in the growing season and N deficiency later.

The uneven distribution of slurry N is a consequence of the distance of 50 cm between the injection tines. Non-uniform responses of grass growth and colour to injected slurry were often observed in these experiments. A decreasing concentration of N in the herbage with increasing distance from the injection slit was also noted [28]. Non-uniform crop responses to injected sewage sludge were recorded in a recent study in the United Kingdom [7, 8]. In this UK study the non-uniformity was more pronounced on a clay soil than on a loamy sand. The observed non-uniformity of growth responses to injected slurries and sludges suggests that reduction of the distance between injection tines will improve the growth responses to slurry N. However, information in the literature on the optimal spacing is conflicting. In Ireland it was found that reduction of the distance between injection tines from 60 to 30 cm improved the growth responses to slurry N [31]. In the previously mentioned UK study the optimal spacing was 65 cm for autumn injection into all soils and for spring injection into sandy soils, whereas it was less for spring injection into heavier clay soils. In the UK study [7] the optimal distance was the best compromise between minimizing the damage caused by the injection tines (which decreased with increasing distance) and the non-uniformity of crop response (which increased with increasing distance). There is a need for more research on optimal spacings of injector tines. It is suggested that future work should consider factors such as injector design, physical distribution of the injected slurry and the influence of different soil types and conditions. Growth responses need to be considered both in terms of distribution throughout the growing season and total annual yields.

It is possible that the addition of fertilizer N could partly correct the negative consequences of the uneven physical distribution and the single large application of injected slurry on the effectiveness of slurry N in producing DM. In the 4 experiments discussed in this paper, the application rates of fertilizer N for the individual cuts were not adjusted to account for the distribution throughout the growing season of the effect of injected slurry on DM yield. Such an adjustment may slightly improve the response of DM yield to N uptake on plots receiving both injected slurry and fertilizer N.

Quadrant I of Fig. 3 shows that at very high uptakes of N (approximately 400–600 kg ha^{-1} yr^{-1}), the response of DM yield to additional N uptake was larger on plots with injected slurry than on the mineral fertilizer plots without slurry (Control). As a consequence, the difference in DM yield between the two curves at a given N uptake decreases, and the curves cross. The crossing over occurred at N uptakes of approximately 500 and 600 kg ha^{-1} yr^{-1} at Ruurlo (Fig. 3a) and Hemrik (Fig. 3b), respectively. The slurry apparently caused a small improvement in the capacity of the sward to produce DM (specific effect of slurry), which brings about improved utilization of absorbed N at high levels of N uptake (quadrant I of Fig. 3), and a small increase in the maximum uptake of N (quadrant IV). The nature of the specific effect of slurry in these experiments is not known. The supply of other nutrients possibly played a part. Chemical analyses of herbage from the plots without slurry indicated that on a few occasions K might have been in short supply. Slurry may also have improved the water holding capacity of these sandy soils by adding organic matter.

The response of DM yield to N uptake on the Inj2 + mineral fertilizer plots was intermediate between that of the mineral fertilizer only (Control) and the Inj3 + mineral fertilizer treatments (quadrant I of Fig. 3). This confirms that the differences between these response curves are related to the effectiveness of injected slurry N rather than to any direct effects of the injector.

The differences between the response curves in quadrant I of Fig. 3 indicate that at a given DM yield the N contents in the harvested herbage differed. At less than maximum DM yields they were higher on the plots receiving injected slurry.

Surface-spread slurry and fertilizer nitrogen

The effects of fertilizer N on N uptake and DM yield on the SA3 plots (surface application of 40 t slurry ha^{-1} yr^{-1}) are compared with those on the plots without slurry (Fig. 4). From a similar analysis to that applied to Fig. 3, it can be concluded that the effects of N from surface-spread slurry and inorganic fertilizer on N uptake (quadrant IV) and DM yield (quadrant II) were additive over the lower range of applied N. At higher rates of N application an increasing amount of fertilizer N was required to replace the effects of slurry on N uptake and DM yield. This indicated that surface spread cattle slurry also has a specific effect on grassland. The specific effect

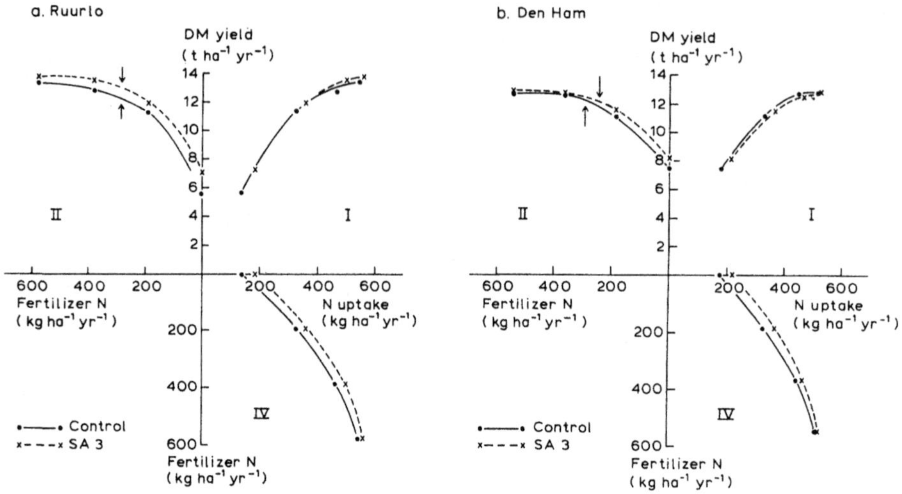

Figure 4. The effect of the rate of fertilizer N on N uptake and DM yield in the absence of slurry (Control) and combined with 1 rate of surface-spread cattle slurry (SA3). The data are averages of 5 years in both experiments.

at Ruurlo (Fig. 4a) appeared to be larger than that at Den Ham (Fig. 4b). However, as has already been mentioned the experimental error involved with the assessment of small effects in a multiple-cut system may hamper an accurate comparison of such small effects.

At Ruurlo and Den Ham, the response of DM yield to N uptake for the SA3 + fertilizer N treatment was almost equal to that for the mineral fertilizer only treatment (Fig. 4, quadrant I). At lower levels of N uptake, N absorbed from surface-applied slurry was slightly less effective for DM production than N absorbed from inorganic fertilizer. This was more evident at Den Ham than at Ruurlo (Fig. 4b). However, these differences in effectiveness were much smaller than the differences between plots receiving injected slurry (Inj2, Inj3) + mineral fertilizer and those receiving only fertilizer N (Control) (Fig. 3). At high levels of N uptake surface-spread slurry had a small positive effect (specific effect) on the response of DM yield to N uptake. The specific effect of surface-spread slurry (Fig. 4, quadrants I) appeared smaller than that of injected slurry (Fig. 3, quadrants I).

Optimal rates of fertilizer nitrogen for combined slurry and fertilizer nitrogen applications

Arrows in quadrant II of Figs. 3 and 4 indicate the optimal annual rates of fertilizer N supply assuming an economic marginal N effect of 7.5 kg DM per kg fertilizer N [19]. The optimal rates of fertilizer N and the corresponding DM yields and N uptakes for the treatments without slurry (Control), and for the highest rates of slurry (Inj3 and SA3) in the 4 experiments are presented in Table 6. The N uptake values for Ruurlo and Hemrik have been estimated from quadrant I of Figs. 3 and 4. Results from the experiments at Den Ham and Friens are derived from data published elsewhere [28].

At each site, the Inj3 and SA3 treatments reduced the optimal rate of fertilizer N, consistently increased the corresponding N uptakes, and in most cases increased the corresponding DM yields (Table 6). The small increases in DM yields at the optimal supply of N, generally observed on the slurry treated plots, are a consequence of the specific effect of slurry. The increase in the corresponding N uptakes reflects the specific effect of slurry and the higher N contents in the harvested herbage, the latter particularly on the injected slurry plots.

The data in Table 6 can be used to compare the effect on DM yield of N from slurry (injected or surface-spread) with that of fertilizer N when the total supply of N is optimal. The effect of the slurry N can be expressed as the rate of fertilizer N required to obtain an equivalent DM yield. It is the difference between the optimal rates of fertilizer N required for the combined slurry/fertilizer treatment and for the treatment without slurry (Control), added to the amount of fertilizer N required to produce the difference in optimal DM yields between the two treatments. The latter amount can be determined in quadrant II of Figs. 3 and 4 as the difference in fertilizer N requirement between the optimal yield levels without (broken arrow in Fig.

Table 6. Optimal rates of fertilizer N, and corresponding DM yields and N uptakes for the treatments without slurry (Control) and the combined slurry/fertilizer treatments, at the highest rates of slurry application (Inj3, SA3) in the 4 experiments. Optimal rates of fertilizer N are based on a marginal N effect of 7.5 kg DM per kg fertilizer N. Data are averages over the duration of each experiment

Experiment	Slurry treatment		Optimal rate of fertilizer N ($kg\,ha^{-1}\,yr^{-1}$)	DM yield ($t\,ha^{-1}\,yr^{-1}$)	N uptake ($kg\,ha^{-1}\,yr^{-1}$)
	code	slurry N ($kg\,ha^{-1}\,yr^{-1}$)			
Ruurlo	Control	0	290	12.3	395
	Inj3	378	185	13.1	480
	SA3	185	285	12.9	425
Den Ham	Control	0	295	12.2	395
	Inj3	355	165	12.5	465
	SA3	171	245	12.1	405
Hemrik	Control	0	390	15.1	480
	Inj3	475	240	15.3	580
Friens	Control	0	400	13.5	485
	Inj3	485	220	13.6	530

3a) and with (unbroken arrow in Fig. 3a) slurry application. Alternatively it can be estimated from the difference in DM yield in Table 6, considering that at this level of N supply the response of DM yield to applied fertilizer N is slightly over 7.5 kg DM per kg N (e.g. 8 kg DM per kg N). For example, from the data in Table 6 it can be calculated that the effect of 90 t injected slurry $ha^{-1} yr^{-1}$ (475 kg N $ha^{-1} yr^{-1}$) on DM yield at Hemrik was equivalent to that from (390 − 240) + (200:8) = 175 kg fertilizer N $ha^{-1} yr^{-1}$. Therefore, at Hemrik the effectiveness of N from injected slurry in promoting DM yield at the optimal supply of N was 37% (175:4.75) of that of fertilizer N. Similarly, values of 49, 48 and 40% can be calculated for N from injected slurry in the experiments at Ruurlo, Den Ham and Friens, respectively. These "efficiency indices" of N from injected slurry at the optimal supply of N are almost identical to the relative ANE values for the injected slurry treatments in Table 5. This indicates that the effects of N from injected slurry and fertilizer N on DM yield were additive up to the economically optimal supply of N.

The "efficiency indices" for N from surface-spread slurry at the optimal supply of N, which can be derived from the data in Table 6, are 40 and 22% for the experiments at Ruurlo and Den Ham, respectively.

Optimal rates of fertilizer N have also been determined for the lower rates of slurry applied in the 4 experiments [28]. In general the results in each experiment were intermediate between those of the highest rate of slurry for the same application method and the treatment without slurry presented in Table 6. Some deviating values were found, especially for the slurry treatments having only a small effect on DM yield. This may be caused by the experimental error associated with the assessment of small effects, which may also explain the unexpectedly high "efficiency index" (40%) of N from surface-spread slurry at optimal supply of N in the experiment at Ruurlo.

At the optimal supply of fertilizer N, N uptake on the treatments with injected slurry (Inj3) is much higher than on the treatments without slurry (Control) (Table 6). Analyses of herbage quality in the experiments at Ruurlo and Hemrik revealed a strong relationship between N uptake and nitrate accumulation in the herbage [28]. For both experiments the relationship was similar for all treatments with the consistent exception of Inj3 without fertilizer N. At Ruurlo, nitrate accumulation in the herbage increased strongly for N uptakes over 400 kg $ha^{-1} yr^{-1}$, and at Hemrik for N uptakes over 475 kg $ha^{-1} yr^{-1}$. Although slightly higher N uptakes appeared to be acceptable, for the treatments with injected slurry at the (assumed) economically optimal rate of fertilizer N (Table 6), there was a risk of excessive nitrate in the herbage, particularly if the herbage was to be harvested for conservation as hay or pre-wilted silage [5]. Therefore, it is suggested that care be taken when determining optimal rates of fertilizer N application, particularly following injection of slurry. For the following discussion it is assumed that the optimal rate of fertilizer N (without slurry additions) is known from experience, and that the "savings of fertilizer N" relate to situations in which optimal rates of fertilizer N (without slurry) would otherwise be used.

The possible savings of fertilizer N after injection of slurry can be derived by two different approaches, either by using the data for combined slurry/-fertilizer treatments in Table 6 or using the data for single treatments in Table 5. From Table 6 the N saving can be calculated as the difference between the optimal rate of fertilizer N required for a combined slurry/fertilizer treatment and for the treatment without slurry (Control) in the same experiment. From Table 5 it can be determined as the rate of fertilizer N giving an equivalent increase of DM yield, which is calculated by multiplying the relative ANE by the rate of slurry N applied. For example, at Ruurlo the saving of fertilizer N using Table 6 data is 290 − 185 = 105 kg N $ha^{-1} yr^{-1}$, and using data from Table 5 it is 378 * 0.47 = 178 kg N $ha^{-1} yr^{-1}$. The N savings derived from Table 6 are smaller than those from Table 5. This is caused by the specific effect of slurry which enhances the productive capacity of the sward and the DM yield at optimal supply of N (Table 6). Some additional fertilizer N is required to obtain this increase in DM yield, as has been indicated by the horizontal distance between the broken and the unbroken arrow on the corresponding response curve in quadrant II of Fig. 3a for the results at Ruurlo. Given that there is a risk of nitrate accumulation in herbage following the use of economically optimal rates of fertilizer N on the injected slurry treatments, it is suggested that relative ANE values (or efficiency indices of slurry N) be used to calculate the saving of fertilizer N after the injection of slurry. A consequence of this is that the value of the specific effect of slurry will be very small or negligible.

Nitrate contents in the harvested herbage on the plots receiving 80 and 90 t injected slurry $ha^{-1} yr^{-1}$ (Inj3) at Ruurlo and Hemrik, respectively, without fertilizer N, were higher than suggested by the general relationship between N uptake and nitrate accumulation in herbage, observed for the other treatments. For the Inj3 without fertilizer N treatments at Ruurlo and Hemrik, approximately 10% of the cuts contained herbage with nitrate contents over 0.17% NO_3^--N on a DM basis, which is considered the critical concentration in hay and pre-wilted silage with respect to animal health [5]. This indicates that a single application by injection of 378 or 475 kg slurry N ha^{-1} (Ruurlo and Hemrik, respectively) was too high. Excessive nitrate contents in the herbage did not occur after the injection of 40 or 60 t cattle slurry ha^{-1} (Inj2) in treatments without the combined application of fertilizer N.

Discussion and conclusions

Experience in these 4 field experiments indicated that spring injection of slurry in grassland generally presented no operational problems. This was also reported from other studies [7, 8, 31]. The moisture content of the soil proved to be very important in relation to damage to the sod. Injection should only be carried out when the soil is sufficiently dry to support the tractor and injection equipment, and to prevent wheel slip. When the soil was too dry, considerable damage to the sod was observed. Drying out of the surface layer of the soil after injection, resulted in the death of grass along the edge of the injection slit. When injection is conducted at the start of the growing season, insufficient soil moisture will seldom be a problem. However, injection later in the season may require irrigation prior to and/or after application.

In the experiments allowing direct comparison of surface spreading and injection (Den Ham and Ruurlo), the mean apparent recoveries of N (ANRs) from surface-spread slurry, injected slurry and the lowest rate of mineral fertilizer (N1) were 23%, 52% and 88%, respectively (Table 4). Injection improved ANR from slurry because it almost completely prevents losses of N by volatilization of ammonia as has been shown by direct measurements of these losses at Hurley [22, 29]. This means that average losses of N by ammonia volatilization from surface-spread slurry at Den Ham and Ruurlo may have been approxmately 30% of the amount of slurry N applied. Losses of ammonia of this magnitude have been directly measured [10, 29].

In the mentioned study at Hurley [29], it was observed that N losses from slurry by denitrification were larger after injection than after surface spreading. These losses were relatively small when cattle slurry (262 kg N ha^{-1}). was applied in April, *viz.* 2% and 7% of N in surface-spread and injected slurry, respectively. Denitrification losses were considerably higher (12% and 21% after surface-spreading and injection, respectively) when slurry had been applied in December at a rate of 248 kg N ha^{-1} [29]. Variations in ANR values for injected slurry (Table 4) may partly be explained by differences in denitrification losses, which may arise from different soil conditions (aeration, temperature) after injection.

In the experiment at Ruurlo, losses of N by leaching of nitrate were measured from some of the experimental treatments. A selection of results is presented by Jarvis *et al.* [10]. They show that leaching of nitrate on the slurry treated plots without fertilizer N differed only slightly from leaching losses on the untreated Control plots (approximately 20 kg N ha^{-1} yr^{-1} lost from the highest rate of injected slurry *versus* 10 kg N ha^{-1} yr^{-1} from the Control plots). From these treatments at Ruurlo no more than 2.5% of applied slurry N was lost by leaching of nitrate. At the economically optimal rates of N supply (Table 6), leaching losses were approximately 25 kg N $ha^{-1} yr^{-1}$ both on the mineral fertilizer treatments and on the combined slurry/fertilizer treatments (Fig. 3 in [10]).

Consideration of the ANR values of slurry and fertilizer presented in this paper, of direct measurements of gaseous N losses and ANR values of slurry at Hurley [29], and of leaching losses at Ruurlo [10], indicates that losses of N from spring-injected slurry in the experiments reported here were likely to be small. The difference between the ANR values for injected slurry and those for the lowest rate of fertilizer N (Table 4) provides an estimate of the proportion of slurry N remaining in the soil at the end of the experiment. This was probably slightly more than 30% of applied slurry N. This is close to the value concluded from studies with long-term heavy applications of cattle slurry to silage maize [25]. This organic N is likely to become available for crops on the long term. From this it can be concluded that a large part of the N from injected cattle slurry can be utilized for crop growth, particularly when the residual long-term effect is also taken into account.

Spring injection of cattle slurry into grassland appeared to be an attractive technique from the point of view of utilization of N and reduction of N losses. More information is required about the effects of slurry injection on the utilization of other elements and on the mineral composition of herbage. Furthermore, the economic feasibility of this application method requires attention.

References

1. Anon (1985) Dierlijke mest. Vlugschrift voor de Landbouw nr. 406. The Hague: Ministerie van Landbouw en Visserij
2. Burg P F J van, Hart M L 't and Thomas H (1980) Nitrogen and grassland – past and present situation in the Netherlands. In Prins W H and Arnold G H, eds. The role of nitrogen in intensive grassland production. Proceedings of an international Symposium of the European Grassland Federation (Wageningen), pp 15–33. Wageningen, The Netherlands: PUDOC
3. Dijk T A van and Sturm H (1983) Fertiliser value of animal manures on the continent. Proceedings No. 220. London: The Fertiliser Society
4. Frankena H J and Wit C T de (1958) Stikstofbemesting, stikstofopname en grasgroei in het voorjaar. Landbouwkundig Tijdschrift 70, 465–472
5. Geurink J H, Malestein A, Kemp A, Korzeniowski A and Klooster A Th van 't (1982) Nitrate poisoning in cattle. 7. Prevention. Netherlands Journal of Agricultural Science 30, 105–113
6. Gijsman A and Meer H G van der (1987) Effects of anaerobic digestion and application techniques on the utilization of nitrogen from cattle slurry in a pot experiment with perennial ryegrass (in preparation)
7. Godwin R J, Warner N L and Hall J E (1985) Soil injection of liquid wastes. Proceedings of the 5th international Symposium on Agricultural Wastes. St. Joseph, Michigan: American Society of Agricultural Engineers

8. Hall J E (1986) Soil injection research in the UK. In Dam Kofoed A, Williams J H and Hermite P l', eds. Efficient land use of sludge and manure. Proceedings CEC seminar (Askov), pp 78–89. London: Elsevier Applied Science Publishers
9. Jagtenberg W D (1970) Predicting the best times to apply N to grassland in spring. Journal of the British Grassland Society 24, 266–271
10. Jarvis S C, Sherwood M and Steenvoorden J H A M (1987) Nitrogen losses from animal manures: from grazed pastures and from applied slurry (this volume)
11. Keuning J A (1984) Fosfaat- en kalibemesting op het grasland is vele jaren een vergeten zaak. Gebundelde Verslagen Nr. 25, pp 84–92. Wageningen, The Netherlands: Nederlandse Vereniging voor Weide- en Voederbouw
12. Lande Cremer L C N de la (1986) Dutch experience with slurry injection. In Dam Kofoed A, Williams J H and Hermite P l', eds. Efficient land use of sludge and manure. Proceedings CEC seminar (Askov), pp 99–104. London: Elsevier Applied Science Publishers
13. Luten W, Geurink J H and Woldring J J (1982) Yield response and nitrate accumulation of herbage by injection of cattle slurry in grassland. In Corrall A J, ed. Efficient grassland farming. Proceedings of the 9th General Meeting of the European Grassland Federation (Reading), pp 185–191. Occasional Symposium No. 14 of the British Grassland Society. Hurley, UK: The British Grassland Society
14. McAllister J S V (1981) Responses to slurry-N in Northern Ireland. In Brogan J C, ed. Nitrogen losses and surface run-off from landspreading of manures. Developments in Plant and Soil Sciences, Volume 2, pp 389–393. The Hague: Martinus Nijhoff/Dr Junk
15. Meer H G van der (1982) Effective use of nitrogen on grassland farms. In Corrall A J, ed. Efficient grassland farming. Proceedings of the 9th General Meeting of the European Grassland Federation (Reading), pp 61–68. Occasional Symposium No. 14 of the British Grassland Society. Hurley, UK: The British Grassland Society
16. Meer H G van der and Uum-van Lohuyzen M G van (1986) The relationship between inputs and outputs of nitrogen in intensive grassland systems. In Meer H G van der, Ryden J C and Ennik G C, eds. Nitrogen fluxes in intensive grassland systems. Developments in Plant and Soil Sciences, Volume 23, pp 1–18. The Hague: Martinus Nijhoff Publishers
17. Morrison J, Jackson M V and Sparrow P E (1980) The response of perennial ryegrass to fertilizer nitrogen in relation to climate and soil: report of the joint ADAS/GRI grassland manuring trial – GM 20. Grassland Research Institute Technical Report No. 27. Hurley, UK: The Animal and Grassland Research Institute
18. Pain B F, Smith K A and Dyer C J (1986) Factors affecting the response of cut grass to the nitrogen content of dairy cow slurry. Agricultural Wastes 17, 189–202
19. Prins W H, Burg P F J van and Wieling H (1980) The seasonal response of grassland to nitrogen at different intensities of nitrogen fertilization, with special reference to methods of response measurements. In Prins W H and Arnold G H, eds. The role of nitrogen in intensive grassland production. Proceedings of an international Symposium of the European Grassland Federation (Wageningen), pp 35–49. Wageningen, The Netherlands: PUDOC
20. Prins W H and Snijders P J M (1987) Negative effects of animal manure on grassland due to surface spreading and injection (this volume)
21. Ryden J C (1984) The flow of nitrogen in grassland. Proceedings No. 229. London: The Fertiliser Society
22. Ryden J C and Lockyer D R (1985) Fate of nitrogen following land application of slurry. Final Annual Report 1984–85, pp 19–21. Hurley, UK: The Grassland Research Institute
23. Schechtner G, Tunney H, Arnold G H and Keuning J A (1980) Positive and negative effects of cattle manure on grassland with special reference to high rates of application. In Prins W H and Arnold G H, eds. The role of nitrogen in intensive grassland production.

Proceedings of an international Symposium of the European Grassland Federation (Wageningen), pp 77–93. Wageningen, The Netherlands: PUDOC
24. Scheepers J H (1978) Beschrijving en handleiding proefveld-drijfmestdoseermachine (Schepan MMM) met bijbehorende apparatuur. Rapport 1978, no. 31. Haren, The Netherlands: Institute for Soil Fertility
25. Schröder J and Dilz K (1987) Cattle slurry and farmyard manure as fertilizers for forage maize (this volume)
26. Simán G, Dam Kofoed A, Larsen K E, Kemppainen E and Steineck S (1987) Utilization of nitrogen from slurry applied to fodder crops (this volume)
27. Smith K A and Unwin R J (1983) Fertiliser value of organic manures in the UK. Proceedings No. 221. London, UK: The Fertiliser Society
28. Snijders P J M, Woldring J J, Geurink J H and Meer H G van der (1987) Stikstofwerking van geïnjecteerde runderdrijfmest op grasland. PR Rapport Nr. 103. Lelystad, The Netherlands: PR
29. Thompson R B, Ryden J C and Lockyer D R (1987) Fate of nitrogen in cattle slurry following surface application or injection to grassland. Journal of Soil Science (submitted for publication)
30. Tunney H (1981) Nitrogen uptake by crop – grassland. In Brogan J C, ed. Nitrogen losses and surface run-off from landspreading of manures. Developments in Plant and Soil Sciences, Volume 2, pp 161–166. The Hague, The Netherlands: Martinus Nijhoff/Dr Junk
31. Tunney H and Molloy S P (1986) Comparison of grass production with soil injected and surface spread cattle slurry. In Dam Kofoed A, Williams J H and Hermite P l', eds. Efficient land use of sludge and manure. Proceedings CEC seminar (Askov), pp 90–98. London: Elsevier Applied Science Publishers
32. Unwin R J, Pain B F and Whinham W N (1986) The effect of rate and time of application of nitrogen in cow slurry on grass cut for silage. Agricultural Wastes 15, 253–268
33. Wadman W P, Sluijsmans C M J and Lande Cremer L C N de la (1987) Value of animal manures: changes in perception (this volume)
34. Woldring J J (1977) Injecteren van drijfmest in blijvend grasland. PR Intern Rapport nr. 76. Lelystad, The Netherlands: PR

The influence of different processing methods for slurry upon its fertiliser value on grassland

H VETTER[a], G STEFFENS[a] and R SCHRÖPEL[b]
[a] *Landwirtschaftliche Untersuchungs- und Forschungsanstalt der Landwirtschaftskammer Weser-Ems, Mars-la-Tour-Straße 4, 2900 Oldenburg, Bundesrepublik Deutschland*
[b] *Staatliche Lehr- und Versuchsanstalt für Tierhaltung und Grünlandwirtschaft Spitalhof, Spitalhofstraße 9, 8960 Kempten/Allgäu, Bundesrepublik Deutschland*

Key words: slurry, aeration, anaerobic digestion, separation, dilution, yields, crop quality

Abstract. The most important slurry processing methods are aeration, anaerobic digestion, separation and dilution of slurry. Depending on the kind and intensity of treatment, slurry properties and the effect of slurry applications upon crop growth can change in a wide range of ways. Positive as well as negative yield effects can occur. The fertilising efficiencies of different slurry processing methods can be summarised as follows:

- aeration: more often negative than positive;
- anaerobic digestion, separation: slightly positive;
- dilution: generally positive.

In general appropriate slurry application is more important than slurry processing.

Introduction

Processing of slurry has different aims; e.g. to kill pathogenic agents, to decrease odours, to facilitate homogenisation and the emptying of slurry pits, to improve slurry distribution and last but not least to increase the fertilising value. The following points are of interest concerning the influence of different slurry processing methods upon fertiliser value:

- the influence upon the nutrient content of the slurry;
- the influence upon crop yield;
- the influence upon crop quality, sward composition and fodder intake when grazing of slurried fields.

The most important methods for processing slurry are aeration, anaerobic digestion, separation and dilution. Slurry additives will be discussed in another paper [19].

H.G. v.d. Meer, et al. (eds.), Animal Manure on Grassland and Fodder Crops.
 ISBN 978-94-010-8137-5.

Change of nutrient content by different slurry processing methods

Slurry aeration

Slurry aeration promotes aerobic decomposition processes. Depending on the kind and intensity of aeration there will be different rates of organic matter decomposition. Furthermore different gaseous nitrogen (N) losses can be observed. In a literature review of 63 trials with cattle and pig manure the average gaseous N losses from aeration were about 20% of total N and 30% of the ammonium N (NH_4^+-N) contents [23]. Gaseous N losses can be very high when slurry is aerated very intensively and with a strong increase of pH.

As shown by the example in Table 1, gaseous N losses can remain relatively low when slurry is aerated under controlled conditions. Losses are often well below 10% of the total N content. Initially a decrease of the NH_4^+-N content can be observed, whilst there will often be an increase of the organic N content due to microbiological N incorporation. The nitrate contents in slurry also normally remain very low even after very intensive aeration processes. The other nutrient contents are not influenced by aeration.

Anaerobic digestion

Compared with slurry aeration, anaerobic digestion normally leads to a more intensive degradation of the organic substances with lower associated gaseous N losses. In this way more of the organic bound nutrients are transformed into a plant available form. This holds especially for N. As shown in Table 2, anaerobic digestion leads to a decrease of organic N compounds and to an increase of the content of NH_4^+-N. The gaseous N losses are in the same range as under normal storage conditions.

Table 1. Change of the contents of organic matter and N in cattle slurry after storage and aeration [5]

	Contents in fresh slurry, kg m^{-3}	Change (in %) after	
		storage	aeration
Organic matter	49.2	−2.5	−20.7
Total N	1.86	−3.2	−6.9
NH_4^+-N	0.91	+7.9	−20.2
Organic N	0.95	−13.8	+6.3

Table 2. Composition of cattle slurry a) before and b) after anaerobic digestion

Reference		Dry matter	Organic matter	Total N	Organic N	NH_4^+-N
		in %				
[9]	a)	8.8	–	0.42	0.21	0.21
	b)	5.6	–	0.38	0.14	0.24
[3]	a)	–	4.7	0.202	0.095	0.107
	b)	–	3.2	0.196	0.083	0.113
own	a)	6.2	4.7	0.30	0.15	0.15
results	b)	4.7	3.4	0.28	0.10	0.18

Slurry separation

Depending on the type of slurry, its original dry matter (DM) content and the method of separation, 30–80% of total solids in slurry can be removed [2]. The higher the degree of separation, the greater is the change of the nutrient contents of the solid and the liquid fractions. After efficient separation of solids the liquid phase is essentially a NK-fertiliser with a high proportion of NH_4^+-N. The solid phase is a NPK-fertiliser with a high proportion of organic N. Table 3 shows an example of the change of nutrient contents after slurry separation. Similar results have been given by other authors [6, 14].

Slurry dilution

With thorough mixing the addition of water to slurry decreases the contents of all the slurry nutrients by the degree of dilution. Normally the form of the nutrients is not influenced.

Table 3. Change of the nutrient contents after slurry separation [12]. 1 m^3 slurry yielded 0.8–0.9 m^3 liquid fraction

		Cattle slurry			Pig slurry		
		original slurry	liquid fraction	solid fraction	original slurry	liquid fraction	solid fraction
Dry matter	%	6.85	3.33	21.30	7.73	4.69	29.0
Organic matter	%	4.59	2.36	14.20	4.48	3.21	19.91
Total N	%	0.27	0.24	0.35	0.54	0.50	0.78
P	%	0.05	0.03	0.14	0.23	0.06	0.46
K	%	0.25	0.24	0.23	0.30	0.32	0.28

Changes in fertilising effect after different slurry processing methods

Changes of slurry properties and of nutrient contents can have positive and negative effects upon the fertilising value. The number of fertiliser experiments with treated slurries is relatively small. There have been a number of experiments with aerated slurry, but there are only a few dealing with other processing methods.

Slurry aeration

In the literature we found 34 experiments, where the fertilising value of aerated slurry was compared with that of untreated slurry. Most of these experiments were conducted with cattle slurry, but some also with pig slurry [1, 4, 7, 8, 15, 21, 22, 25]. In most cases the yield differences between the treatments with aerated and unaerated slurries were not statistically significant. In 14 trials the yield with aerated slurry tended to be higher, whilst in 20 trials the yield was lower than with unaerated slurry. This shows, that the fertilising effect of slurry normally is not improved by slurry aeration.

The detailed results of one field trial are presented to represent many others (Fig. 1). In field trials on grassland Thalmann [23] applied slurry corresponding to 2 or 4 livestock units per ha (about 75 and 150 m^3 of cattle slurry $ha^{-1} yr^{-1}$) and compared the yield effects of normally stored slurry with surface aerated and deep aerated slurry (air injection below the slurry surface). Five cuts were harvested every year for three years. The first, third and fifth cuts had been fertilised with slurry. The second and fourth cuts of the treatments with the low amount of slurry received an additional amount of 60 kg mineral N ha^{-1}. Fig. 1 shows the yield results on a 3-year average. The following conclusions can be drawn:

- There was only a small difference between the yield effects of a normally stored and of an aerated slurry. In the treatments with the lower rate the yield effect of the aerated slurry was a little better, and with the high rate a little worse than the yield effect of unaerated slurry. The differences were not significant.
- The method of aeration had no influence upon the yield effects of the slurry. The differences between surface and deep aerated slurry were small and not significant.

An aerated slurry may have some advantages concerning its viscosity, but measurements about the influence of aeration upon this parameter are not available.

It has often been reported, that aerated slurry causes less scorching of grass than normal slurry. This statement could not be confirmed from most

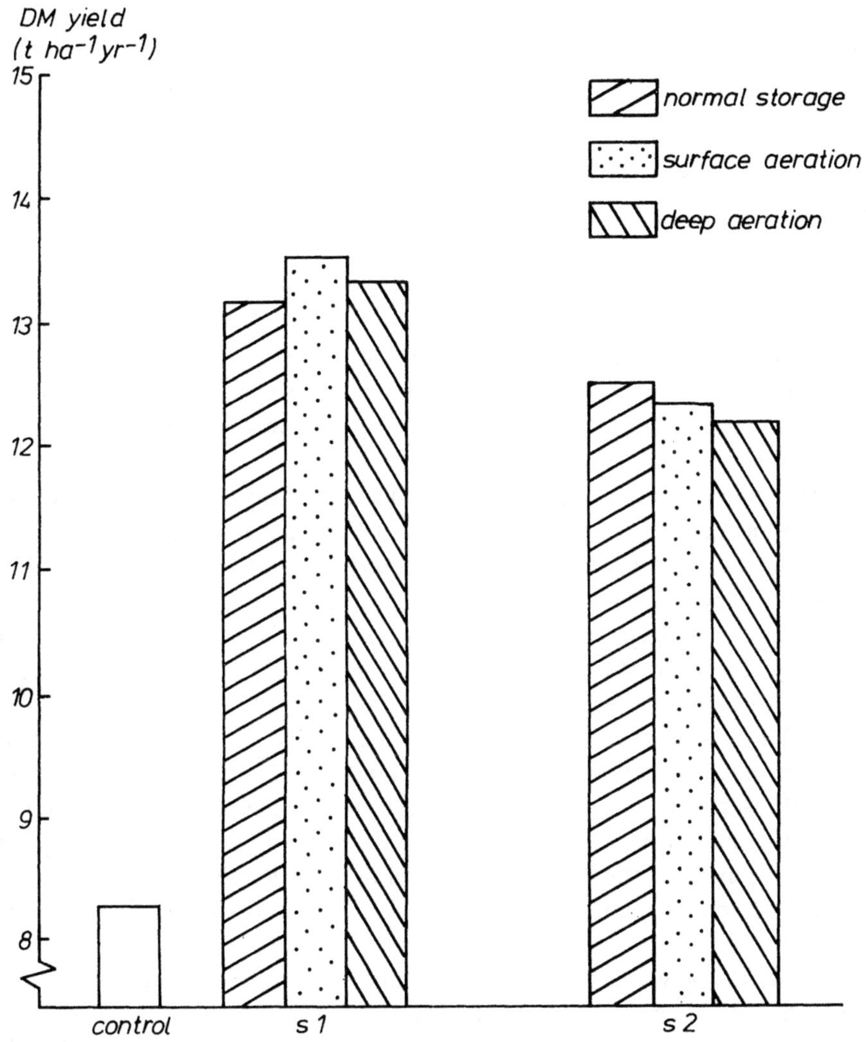

Figure 1. Dry matter yields of grass (t $ha^{-1} yr^{-1}$) after application of 2 rates of differently aerated cattle slurry [23]. The amounts of N applied were the same for each rate of application.
s 1: slurry and nutrient amounts ≙ 2 livestock units ha^{-1} (3 × 25 m^3 slurry ha^{-1}, 5% DM, 2.3 kg N m^{-3}); additionally 60 kg inorganic fertiliser N ha^{-1} for the second and fourth cuts.
s 2: slurry and nutrient amounts ≙ 4 livestock units ha^{-1} (3 × 50 m^3 slurry ha^{-1}, 5% DM, 2.3 kg N m^{-3}).

of the field trials with aerated and unaerated slurry. When correctly applied neither an unaerated nor an aerated slurry need cause scorching, which will result in yield decreases.

Yield decreases due to slurry aeration may be caused in first place by increases of gaseous NH_4^+-N loss and because an aerated slurry often has a higher organic N content than an unaerated slurry. After application of the same total-N amounts, the unaerated slurry often had a better yield effect than the aerated slurry.

Anaerobic digestion

Anaerobically digested slurries are normally obtained from a biogas plant. Besson *et al.* [4] carried out a pot experiment with digested and undigested cattle and pig slurry. The pots (5.8 kg of soil) received slurry corresponding to 1.5 and 3 g total N pot^{-1}. Fig. 2 shows the grass yields (*Lolium multiflorum*) for the different treatments as a 3-year average (slurry application before sowing).

In this experiment the digested cattle slurry had a greater yield effect, and the digested pig slurry a lower yield effect than the undigested slurries. As is known from other studies [11, 20, 26, 27] pig slurry gives a much better N response than cattle slurry.

The lower nitrogen response from cattle slurry is normally caused by its relatively low NH_4^+-N content, by its thickness, by its adhering to crop surfaces and – often combined with this – by relatively high ammonia (NH_3) losses. All these properties can be influenced positively by anaerobic digestion. For instance in the pot experiment the NH_4^+-N content of the undigested cattle slurry was 47%, and of the digested cattle slurry 54% of the total N amount. Therefore the better response from the digested cattle slurry may be caused partly by the higher amount of directly available N. In pig slurry the NH_4^+-N content was very high before anaerobic digestion (70% of the total N). We have no explanation for the slight negative effects.

In Oldenburg we carried out a 2 year field trial on grassland, where 20 m^3 ha^{-1} of anaerobically digested and of undigested cattle slurry were applied to each cut (3 cuts per year). Especially after the spring application of slurry we observed a faster N response on the plots which had received digested cattle slurry. In comparison to the plots receiving undigested cattle slurry the slurry adhered less to the grass surface so that the applied nutrients could be taken up relatively fast by the plants. In spite of the quicker nutrient effect, the yield differences between the plots receiving digested and undigested slurry were not significant (Fig. 3).

Similar results have been reported from other authors. In a pot experiment with grass (only one slurry application made at the beginning of growth) only the first cut had higher DM yields from treatments with digested rather than from treatments with undigested slurry [24]. The second

Figure 2. Dry matter yields in g pot^{-1} yr^{-1} (3 cuts) after fertilisation with digested and undigested cattle slurry (cs) and pig slurry (ps) (average yields of 2 soils of different pH) [4].

cut showed no yield differences between digested and undigested slurry and in the third cut the undigested slurry had a greater yield effect than the digested.

In a field trial on arable land [9] the response of barley showed no differences between digested and undigested slurry, for sugar beet the digested slurry had a slightly greater effect, but the yield differences were not significant.

In summary it can be stated, that anaerobic digestion improves some fertilising properties. Due to better flow characteristics and higher NH_4^+-N contents the slurry N is more rapidly available to the crop. On the other hand the long term effect of digested slurry is often smaller than that of

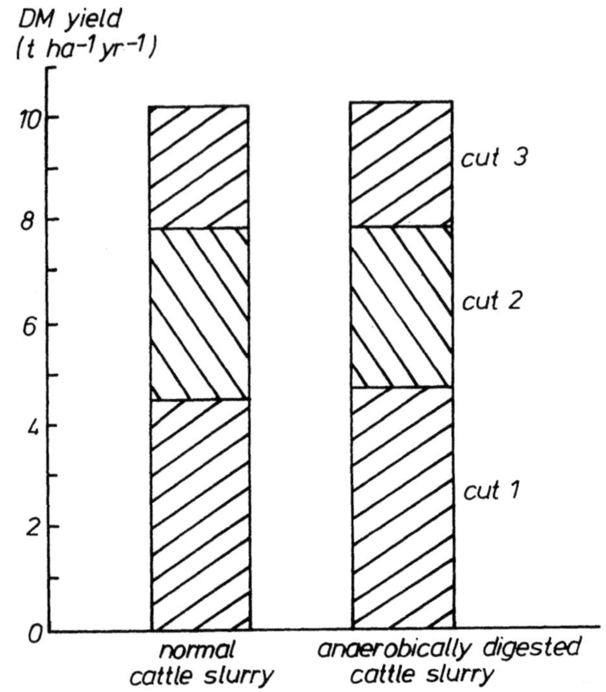

Figure 3. Dry matter yields of grassland after application of $20\,m^3$ cattle slurry ha^{-1} to each cut with and without anaerobic digestion (N content of digested and undigested slurry: $3\,kg\,m^{-3}$).

undigested slurry, due to lower contents of organic N. The fertilising effect of digested slurry therefore is often only slightly better in total than that of undigested slurry.

Slurry separation

Slurry separation leads to two different fertiliser components which can be very different in composition and fertilising effect: a liquid fraction which can be applied to agricultural fields not only with slurry tankers, but also by irrigation and a more or less solid fraction which can be distributed with a manure spreader.

Field trials have been carried out mainly with the liquid fraction of separated slurry. For example in a field trial on grassland [17] the liquid fraction of a separated cattle slurry gave a better yield response than an unseparated cattle slurry in all 4 years of the research. (Applications were of the same N amounts from the different processed slurries.) The yield effect of the liquid fraction was not as good as that of mineral fertilisers (Fig. 4).

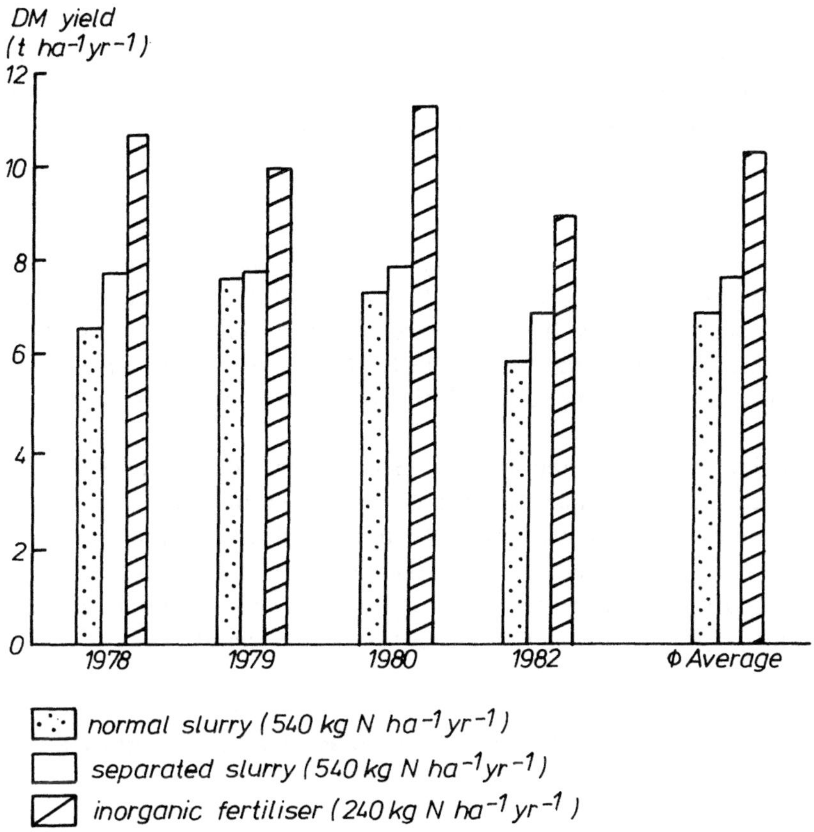

Figure 4. Effect of separation of cattle slurry upon the DM yield of grass (5 cuts per year) [17].

Similar results are known from other authors. In field trials on grassland in Great Britain [18] single cut yields of plots fertilised with the liquid fraction of separated cattle slurry were up to 30% higher than on plots fertilised with normal cattle slurry. In field trials on arable land (Oldenburg) we harvested 10% higher yields of winter barley on the plots receiving separated cattle slurry than on the plots with normal cattle slurry (both treatments received 150 kg total N ha^{-1} as slurry).

The greater yield effect of the liquid fraction of separated slurry can be explained partly by its higher NH_4^+-N content, and also by an improvement of flow characteristics. Positive effects of separation can be expected especially for cattle slurry. A quick flow of slurry from crop surfaces decreases the risks of negative effects to the sward, and also the risks of high N losses by volatilisation of NH_3.

We have no information on field trials with separated pig or poultry slurry.

The fertilising effect of the solid fraction has seldom been investigated. In field trials by Koriath [12] the solid fraction had a relatively low direct N effect, but a high residual effect in the following years. The nutrient effect of the solid fraction is similar to the nutrient effect of farmyard manure.

Slurry dilution

The positive yield effect of diluting slurry with water has been well known for a long time. The addition of 30% water to slurry considerably improves its flow characteristics. In a three year field trial on grassland Kretzschmar [13] achieved a yield increase of 0.9 t DM $ha^{-1} yr^{-1}$ with such a dilution. With further dilution he achieved still greater yield increases, up to 1.7 t DM $ha^{-1} yr^{-1}$ (Table 4).

Similar results are reported by Schechtner [20]. In his field trials he observed that slurry dilution has a positive effect on grass yield particularly for summer applications. Especially in hot and dry weather conditions the faster penetration of NH_4^+-N into the soil reduces gaseous N losses. When field distances are not too far, slurry dilution with water is normally economic, especially for slurry application to grassland.

Fodder quality, sward composition and fodder intake after grazing of slurried fields

Detailed examination of the effect of different slurry processing methods upon fodder quality, sward composition and fodder intake when grazing slurried fields, has only been reported for aeration. In the previously mentioned field trials [23] the following results have been obtained:

- The slightly lower N availability in the plots fertilised with aerated cattle slurry resulted in lower protein contents and slightly higher carbohydrate contents of the fodder compared with the plots fertilised with unaerated

Table 4. Yields of grass after application of cattle slurry with different dilutions; average of 3 years [13]. 3 slurry applications per year. First and second application: 20 m^3 cattle slurry ha^{-1}; third application: 10 m^3 cattle slurry ha^{-1} (DM content: 8–10%, N content: 4 kg m^{-3})

Dilution slurry:water	DM yield t $ha^{-1} yr^{-1}$	Additional yield t $ha^{-1} yr^{-1}$
1:0	7.6	–
1:0.3	8.5	+0.9
1:1	9.0	+1.4
1:3	9.3	+1.7

slurry. Other important fodder components were not influenced by slurry aeration.

- The sward composition was not influenced by fertilisation with slurries aerated by different methods. There was a difference only between plots fertilised with slurry and plots receiving inorganic fertilisers.
- In grazing trials it was observed, that on the plots fertilised with different aerated slurries there was no difference in the palatability of the grass. The palatability of the fodder was better on the plots fertilised with inorganic fertiliser.

Similar results were achieved in other experiments [10]. In summary, it can be stated that slurry aeration has only limited effects upon fodder quality, sward composition and fodder intake when grazing slurried fields.

Mott and Ernst [16] investigated fodder intake from grass plots fertilised with $15\,m^3\,ha^{-1}$ of undiluted slurry or $30\,m^3\,ha^{-1}$ of diluted slurry (1:1). They found no differences in fodder intake caused by dilution. We do not know of experiments with other slurry processing methods. In our opinion other processing methods will not have significant effects when slurry is properly applied, that means at low rates and on short swards.

Conclusions

Depending on the processing methods and the intensity of treatment, slurry properties will change to a greater or lesser extent. Changes in the nutrient contents, in their chemical bondings or in the flow characteristics of the slurry can influence the fertilising effect positively as well as negatively. Positive effects were caused in the experiments described in the first place by improving the flow characteristics of slurry so that it moves faster into the root zone of crops and increases nutrient efficiency. Any advantages of slurry processing methods in improving slurry spreading have not been dealt with in this report

The influence of different processing methods upon the fertilising effect of slurry can be summarized as follows:

- The influence of slurry aeration is uncertain. Some experiments showed positive effects, but most gave negative effects. Fodder quality was largely unchanged. The additional costs of slurry aeration can not normally be paid for by yield increases, improved fodder quality or improved fodder utilisation.
- Anaerobic digestion normally increases the content of NH_4^+-N in slurry. By this means the direct N effect of slurry can be increased, but the long

term effect will often be decreased. Therefore on grassland, the fertilising effect of a digested slurry over the whole year is often only slightly better than the fertilising effect of a normal stored slurry. The economy of a biogas plant can only be slightly improved by these positive yield effects.

- Slurry separation creates two fertiliser components. Provided that the same quantities of total N are applied, the liquid fraction normally has a higher fertilising effect than the unseparated slurry. The positive yield effects are often not high enough to pay for the processing costs. Slurry separation can generally be recommended, when the processing costs can largely be paid by selling the solids with or without composting. When the solids are used on the farm, their nutrient effect is similar to that of farmyard manure.
- Relatively high yield benefits can be obtained by slurry dilution, particularly when cattle slurry with a high DM content (about 10%) is applied in the summer or upon already well developed grass swards. Dilution of such slurries with water can improve their fertilising value considerably. The economics of slurry dilution is dependent on the distance to the fields. When the distance is not too far, a slurry dilution of 30% can be economic.

In the experiments considered in this report the fertilising effect of different processed slurries was calculated partly on the basis of similar slurry quantities before treatment, partly on the basis of similar amounts after treatment and partly on the basis of the same N supply by the slurries. A calculation on the basis of equal N supply does not take into account that slurry treatment can change the amount of slurry as well as the concentration of N fractions. This consideration applies for anaerobic digestion and still more for slurry separation. For a comparative test of separated and unseparated slurry it would be legitimate to ask whether the sum of the effects of the liquid and solid fractions is higher than the effect of the unseparated slurry. Such tests are not easy to carry out.

As a general rule it can be stated, that for slurry fertilisation of grassland proper slurry application is more important than slurry processing. Proper slurry application implies that:

- slurry amounts within the year are adequate to meet the fertiliser demands of the plants,
- slurry is applied in dressings $\leqslant 20\,m^3\,ha^{-1}$,
- application is restricted to the growing season and to short swards.

References

1. Abele U (1978) Ertragssteigerung durch Flüssigmistbehandlung. KTBL-Schrift 224
2. Ballssen E (1981) Separierung und Belüftung von Schweineflüssigdung. Forschungs-

bericht Agrartechnik des Arbeitskreises Forschung und Lehre der Max-Eyth-Gesellschaft (MEG) 61, Kiel

3. Besson J M (1981) Behandlung von Jauche und Gülle. Schriftenreihe Grünes Forum, Wagnersche Universitätsbuchhandlung Innsbruck, Folge 2, 87–120
4. Besson J M, Daniel R and Lischer P (1984) Comparaison de l'effet de l'azote dans les lisiers de bovins et de porcs sur le Ray-grass italien (*Lolium italicum*) après préparation selon trois méthodes: stockage, aération et méthanisation. Schweizerische Landwirtschaftliche Forschung 23 (3), 249–268
5. Besson J M, Lehmann V, Roulet M and Lischer P (1981) Güllebehandlung: Lagerung, Belüftung und Methangärung a) Wirkungen auf einige Gülleinhaltsstoffe. Bericht über 7. Arbeitstagung "Fragen der Güllerei", pp 29–52. Gumpenstein, Österreich: Bundesversuchshanstalt für alpenländische Landwirtschaft
6. Bölöni J and Mészáros G (1986) Latest chemical slurry handling methods. Working report of subnetwork 3 of the network on animal waste utilization. Gödöllö, Hungary: National Institute of Agricultural Engineering
7. Bonjour R (1977) Hofdüngeraufbereitung. Forschungsanstalt für Umweltschutz und Agrikulturchemie, Liebefeld
8. Buchgraber K (1983) Vergleich der Wirksamkeit konventioneller und alternativer Düngungssysteme auf dem Grünland hinsichtlich Ertrag, Futterqualität und Güte des Pflanzenbestandes. Diss. Universität Bodenkultur, Wien, Österreich
9. Dam Kofoed A and Søndergård Klausen P (1983) Gødningsvirkningen af afgasset ok ikke afgasset gylle. Tidsskrift for Planteavl 87, 119–128
10. Gjestang K E, Tjernshaugen O and Tveitnes S (1984) Grazing behaviour by heifers on pastures fertilised with different fertiliser categories. Applied Animal Behaviour Science 12, 33–41
11. Kiely P V (1980) Time and rate of application of animal manures. In Gasser J K R, ed. Effluents from livestock, pp 361–377. London, UK: Applied Science Publishers
12. Koriath H und Kollektiv (1975) Güllewirtschaft, Gülledüngung. Berlin, DDR: VEB Deutscher Landwirtschaftsverlag
13. Kretzschmar G (1984) Mit Gülledüngung den Grünlandertrag steigern. Landwirtschaftliches Wochenblatt Westfalen-Lippe, Nr. 8/84
14. Kroodsma W (1986) Separation as a method of manure handling and odours reduction in pig buildings. Working report of subnetwork 3 of the network on animal waste utilization. Gödöllö, Hungary: National Institute of Agricultural Engineering
15. Leskosek H and Sijanec V (1978) Düngewirkung der Gülleformen zu Grünland und Ackerland in Slowenien. Bericht über das 1. Versuchsjahr, Univerza v. Ljubljani, Biotechniska Fakulteta, Ljubliana, 1–16 (unpublished)
16. Mott N and Ernst P (1986) Results of field trials (unpublished)
17. Neuhaus H (1983) Düngen mit Rindergülle ohne Feststoffe. Landwirtschaftsblatt Weser-Ems Nr. 37, 12–14
18. Pain B F (1986) New concepts in the utilisation of slurry on grassland. RASE Cont., February
19. Pain B F, Thompson R B, Lande Cremer L C N de la and Holte L ten (1987) The use of additives in livestock slurries to improve their flow properties, conserve nitrogen and reduce odours (this volume)
20. Schechtner G (1981) Nährstoffwirkung und Sonderwirkung der Gülle auf dem Grünland. Bericht über 7. Arbeitstagung "Fragen der Güllerei", pp 135–196. Gumpenstein, Österreich: Bundesversuchsanstalt für alpenländische Landwirtschaft
21. Schechtner G (1982) Güllebelüftung und Güllezusätze. Der fortschrittliche Landwirt 60, 1–3
22. Schöllhorn J (1983) Versuch über die Wirkung von verschiedenen Behandlungsarten bei

Gülle auf Grünlandnarbe und Ertrag. Staatliche Versuchsanstalt für Grünlandwirtschaft und Futterbau Aulendorf (unpublished)
23. Thalmann H (1985) Wirkungen belüfteter und unbelüfteter Rindergülle unter Schnitt und Beweidung auf Dauergrünland. Bayerisches Staatsministerium für Ernährung, Landwirtschaft und Forsten, Dissertation
24. Theophile W (1981) Düngerwirkung von Faulschlamm. Diplomarbeit Kiel
25. Tjernshaugen O, Tveitnes S and Gjervan J O (1980) Forbehandling av husdyrk jodsel ok virkning av forbehandlet husdyrk jodsel til ulike vekster. Institut for bygningsteknik, Norges Landbruckshogskole, Stensiltrykk Nr. 166
26. Tunney H (1980) Effects of cattle slurry, pig slurry and fertiliser on yield and quality of grass silage. In Gasser J K R, ed. Effluents from livestock, pp 327–343. London, UK: Applied Science Publishers
27. Vetter H and Steffens G (1986) Wirtschaftseigene Düngung. DLG-Verlag

Utilisation of phosphorus and potassium from animal manures on grassland and forage crops

K A SMITH[a] and T A VAN DIJK[b]
[a] *ADAS Soil Science, MAFF, Woodthorne, Wolverhampton WV6 8TQ, UK*
[b] *Institute for Soil Fertility, P.O. Box 30003, 9750 RA Haren, The Netherlands*

Key words: animal manures, slurry, phosphorus, potassium, grassland, forage, injection

Abstract. A review is presented of experimental work to assess the phosphorus (P) and potassium (K) fertiliser value of animal manures, for grassland and forage crops.

Apparently conflicting information in the literature can often be, at least partly, accounted for by the specific conditions under which experiments are conducted. Agronomic and pot experiments have sometimes indicated a reduced short term efficiency of manures as sources of P and K but longer term studies have suggested that fertiliser and manure P and K can be regarded as equivalent.

Some current Dutch work on the effects of injection on the efficiency of slurry P and K is also described. Preliminary results suggest that grass uptake of P from injected slurry, is likely to be delayed until the second and third cut of a multi-cut system, whereas K is effective for all cuts.

Introduction

In contrast to the increase in fertiliser nitrogen (N), phosphorus (P) and potassium (K) use has remained relatively stable in many European countries [32]. On much of the intensively farmed agricultural land, P and K fertility is already at a satisfactory level, such that crop deficiencies of these major plant nutrients are unlikely to occur. Fertiliser policy is therefore generally aimed at maintaining soil fertility. Such practice, however, often fails to account adequately for the contribution from organic manures which may still be largely ignored on some farms.

A nutrient balance calculated for farms in Northern Ireland indicated an overall input surplus of $20\,kg\,P\,ha^{-1}\,yr^{-1}$ and $25\,kg\,K\,ha^{-1}\,yr^{-1}$ [23]. It was clear that very high excesses of P and K could develop on some farms, especially those with appreciable pig or poultry enterprises. The build-up of nutrients, particularly P and K has been observed in a number of studies [17, 23, 30] and has implications in terms of the health of grazing livestock and possible pollution of surface waters and ground waters.

Whether manures are used by the farmer on his own land or the surplus

H.G. v.d. Meer, et al. (eds.), Animal Manure on Grassland and Fodder Crops.
 ISBN 978-94-010-8137-5.

is transported to arable farms [37], it is important to know the fertiliser value in order to allow efficient utilisation and reduce the risks of nutrient deficiency, nutrient excess or water pollution.

Field experiments on grass

Crop responses to manure P and K are less common than those to N because of the high fertility of soils in many European countries. Relatively little work has therefore been undertaken and "availability" is often assessed on the basis of uptake by herbage. Published figures for manure P effectiveness compared with that of inorganic P fertilisers vary from 10% [2] to 100% [12, 24]. Some of the variation may be due to differences in the method of assessment or in the interpretation of results.

Slurry almost completely eliminated pasture response to P and K in work reported by Adams [1] and so the effectiveness of slurry P and K was assessed in terms of herbage analysis effects [2].

Despite the range in composition, slurry almost invariably increased herbage N and K levels (Table 1). Pasture magnesium (Mg) levels rose throughout the season but Mg status was consistently low in cut 1 and below the safe level (hypomagnesaemia) of 0.2% [4] for spring herbage dry matter (DM). Slurry, on balance, was found to slightly enhance the Mg status of herbage. On the basis of pasture uptake of K, slurry K was somewhat less than half as effective as fertiliser K (Table 2). The equivalent assessment of slurry P suggested 13–18% availability in the short term. Further evidence of the delayed or reduced efficiency of slurry P and K was obtained in subsequent years of these experiments [3] when both P and K surplus applied as slurry were less effective in raising herbage analysis than mineral fertilisers.

A number of authors have assumed the efficiency of manure P and K in relation to inorganic fertiliser P and K to be 100% [9]. Some supporting

Table 1. Main effects of fertiliser and slurry on grass mineral composition; number of cuts giving positive or negative effects; 3 cuts in each of 16 trials (1969–71) [2]

Positive or negative, significancy	% N		% P		% K		% Mg	
	Effect of:		Effect of:		Effect of:		Effect of:	
	Fert N	Slurry	Fert P	Slurry	Fert K	Slurry	Fert (K)	Slurry
pos., significant	40	37	21	22	43	36	0	18
pos., not sign.	6	9	27	18	5	9	2	15
neg., significant	2	1	0	6	0	2	17	12
neg., not sign.	0	1	0	2	0	1	29	2

Table 2. Effects of fertilisers and slurry on herbage analysis at cut 1, mean of 16 trials (1969–71) [2]

	Artificial fertiliser (no slurry)			Slurry (no fertilisers)
Nutrients applied ($kg\,ha^{-1}$)	120 N 0 K	120 N 50 K	120 N 100 K	115 NH_4^+-N 114 K
K removal ($kg\,ha^{-1}$)	95	109	120	108
apparent recovery of applied K (%)	–	28	25	11.4
Nutrients applied ($kg\,ha^{-1}$)	120 N 0 P	120 N 10 P	120 N 20 P	115 NH_4^+-N 86 P
P removal ($kg\,ha^{-1}$)	15.4	16.1	16.4	16.2
apparent recovery of applied P (%)	–	7	5	0.9

evidence is reported though sometimes in relation to crops other than grass. Ferwerda [12] observed that the slopes of potato yield response curves to fertiliser and farmyard manure (FYM) sources of P and K were the same, indicating equal efficiencies. Larsen [21] was unable to establish any difference in the effect of P in animal manures or fertilisers on the basis of a large number of Danish field experiments on barley and fodder beet crops. In these experiments, utilisation of P decreased with increased application rate, whether supplied as animal manure or fertiliser.

Field experiments in The Netherlands, to assess the P effect of FYM on grassland, indicated that physical smothering was responsible for reduced efficiency of manure P [31]. FYM at 40 $t\,ha^{-1}$ reduced the grass response to applied fertiliser P whereas, at only 20 $t\,ha^{-1}$, the manure P increased DM yield and appeared to be roughly half as effective as fertiliser P.

Slurry P and K was shown by McAllister [24] to be as effective as fertiliser P and K for barley in two series of experiments on soils of low or medium P and K status. In a long term (1970–75) slurry experiment on grassland [24] the responses to P were small, with little difference between fertiliser and slurry. Responses to applied K were more variable and the herbage K contents (Table 3) from all pig slurry treatments were lower than for fertiliser or cow slurry treatments. However, because of differential N supply and hence, yield between all of these treatments, it is not possible to draw firm conclusions on the relative efficiency of these sources of P and K.

In experiments in England reported by Unwin [36], herbage P levels were raised by slurry treatments to an extent influenced by soil type. Higher rates of slurry P were required to raise herbage P, on a poorly drained, silty clay loam soil compared to medium textured soils. There was no evidence of differential uptake from cattle or pig slurry. On the silty clay loam site, two

Table 3. Herbage uptake of P and K from different sources [24]

Nutrient source	Phosphorus		Potassium	
	Applied ($kg\,ha^{-1}\,yr^{-1}$)	Herbage analysis (% in DM)	Applied ($kg\,ha^{-1}\,yr^{-1}$)	Herbage analysis (% in DM)
Fertiliser	32	0.27	160	1.76
Pig slurry	29	0.27	53	1.26
	58	0.29	107	1.40
	117	0.30	214	1.59
Cattle slurry	18	0.27	140	1.68
	36	0.28	279	2.12
	73	0.28	557	2.53

years after slurry applications ceased, herbage P values returned to those on the nil slurry treatment (Table 4), even though 'available' soil P residues (sodium bicarbonate extraction) remained high.

The optimum concentration of P in forage DM for dairy cows and pregnant ewes is 0.4–0.5% [4]. High rates of slurry improved herbage P content but other minerals were sometimes adversely affected.

High rates of cattle slurry produced very large increases in available soil K and herbage K (Table 5) [35]. However, after two years with no further treatment, soil K had fallen appreciably and herbage K values were lower than on plots receiving regular fertiliser dressings.

Pot experiments

A large amount of information is available from pot experiments. The results need to be treated with caution since it might be expected that a poor correlation is likely between the response to P and K surface applied to an established grassland sward and response to P and K, possibly incorporated,

Table 4. Effect of cattle slurry P on grass herbage analysis and P offtakes [36]

Total slurry ($m^3\,ha^{-1}$)	0	292*	584*	810**	1620**
Total P ($kg\,ha^{-1}$)	189+	246*	492*	757**	1514**
% P in herbage DM in 1974	0.39	0.40	0.39	0.44	0.50
P offtake ($kg\,ha^{-1}$)	35	31	36	55	53
% P in herbage DM in 1976	0.28	0.26	0.24	0.26	0.29
P offtake ($kg\,ha^{-1}$)	16	13	16	15	17

+ inorganic fertiliser
* slurry applied 1972–1975
** slurry applied 1972–1974

Table 5. Effects of high rates of cattle slurry K on soil and herbage analysis [35]

Total K applied ($kg\,ha^{-1}$)		Herbage analysis		Soil analysis*	
		1974	1976	1974	1976**
		(% K in DM)		(available K, $mg\,l^{-1}$)	
NPK Fertiliser	81	1.6	1.5	57	80
1972–1976	475	2.1	2.0	64	85
	874	2.4	2.4	63	125
Slurry	4623	4.4	2.3	500	280
1972–1974	9246	5.1	2.0	847	450

* Ammonium nitrate extraction
** 5–15 cm depth soil sample

in a disturbed soil. Other differences e.g. temperature, may affect the response to manure nutrients in pot experiments when compared to field conditions.

The P contained in manures was found to be less effective than fertiliser P in growth chamber studies using a soil of high P status due to previous applications of organic manures [16]. In this case, monoammonium phosphate (MAP) was compared to P, surface applied, as cattle, pig and sheep faeces to an established sward. All treatments received the equivalent of 330 $kg\,N\,ha^{-1}$ as ammonium nitrate at the start of the experiment, *i.e.* the amount of N applied at the highest rate of MAP. No allowance was made for the N content of the faeces. Eight harvests were taken at 21-day intervals and, after each harvest, a further 120 $kg\,N\,ha^{-1}$ were applied. MAP outyielded all other sources of P only at the first and second harvest. At six harvests, source of P significantly affected herbage P content, with little difference between cattle, pig and sheep faeces, which were all lower than MAP.

Tunney and Pommel [34] incorporated dried pig slurry and monocalcium phosphate into both a sandy and a loamy soil and found almost equivalent results from both sources of P. All treatments received a basal nutrient solution (without P) at the start of the experiment and again after each harvest, with N and other nutrients applied at rates which were not limiting to plant growth. Slurry (dried) P compared well with fertiliser P for grass growth except at the very high rates of application when slurry P appeared to be more toxic than fertiliser P. These high rates tested 1 g P per pot of 900 g soil which would in fact, represent a rate in excess of 2000 $kg\,P\,ha^{-1}$ in the field situation. Similar results were obtained for grass P uptake (Table 6) except at the very high application rate.

Cumulative P uptake from the lower rates of addition indicated that slurry P was more efficient than fertiliser P on the sandy soil and about 90% as efficient as fertiliser P on the loamy soil.

Table 6. Relationship between P added and P uptake [34]

P added mg P per pot	Plant uptake, mg P per pot			
	Sandy soil		Loamy soil	
	Fert	Slurry	Fert	Slurry
Exp 1				
0	22	22	10	10
50	48	47	37	28
100	54	67	43	45
1000	134	43	110	87
Exp 2				
0	39	39	22	22
50	59	62	43	42
100	74	79	64	59
1000	137	52	115	89

Hanley and Murphy [19] compared surface-applied pig and cattle slurry with fertiliser nutrients for perennial ryegrass in pot experiments using a soil of low P and K status. In a P experiment all pots received an initial high rate of N and K and, in a K experiment, N and P, such that these nutrients were not limiting. From the DM yield – P response curve over three cuts (Fig. 1), the P content of cattle manure appeared to be only marginally less effective

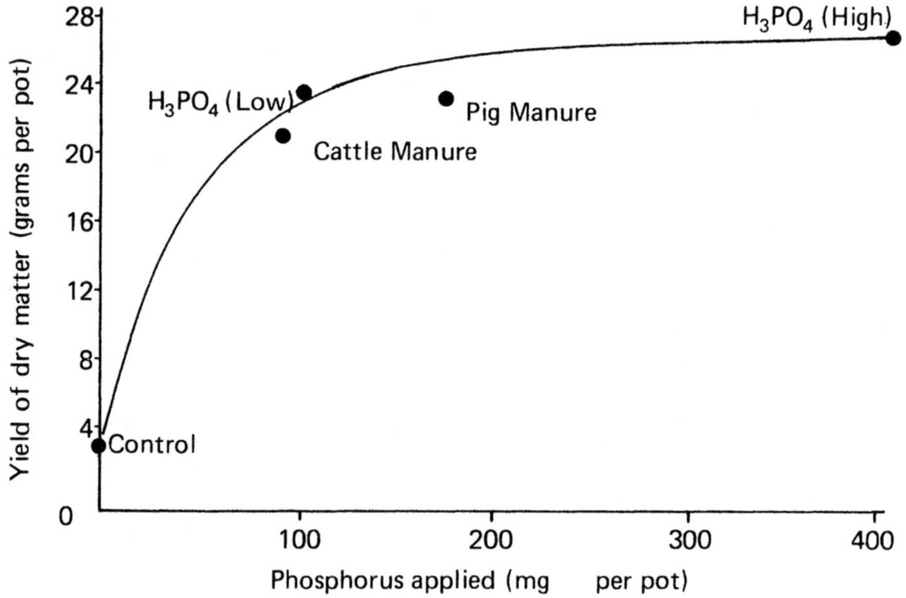

Figure 1. Comparative effects of phosphoric acid and animal manure-P on DM yield from three cuts of ryegrass [19].

than phosphoric acid (H_3PO_4). Pig manure appeared to be less effective than equivalent H_3PO_4 (estimate about 55%) but, because of the shape of the response curve at this point, a small error in the yield assessment would have had a large effect on the estimate of relative P efficiency. Data for all three cuts indicated that P in the manures became more effective in the later cuts, presumably as more organic P was mineralised and became available to plants. The K in cattle and pig manures and in potassium sulphate was equally available for DM production.

High P efficiencies in cattle, pig and poultry slurry relative to fertiliser P were confirmed in pot work by Amberger [7] and slurry K efficiency was almost equivalent to fertiliser K (Table 7).

It is generally recognised that plant species differ in their ability to utilise applied P. Jensen [20] used lettuce, beetroot and carrots as test crops in a pot assessment of the nutritional value of P in FYM. With beetroot, the nutrient value of P in FYM and superphosphate were equivalent but, with lettuce and carrots, the value of P in FYM appeared to be higher than in superphosphate.

Slurry use on forage crops

The relatively late sowing date often allows slurry to be spread on maize ground at high rates and at a time when there appears to be less chance of soil damage or water pollution. Investigating the effect of heavy dressing of slurry on forage maize, Pain and Phipps [26] noted a yield depression only at very high rates of slurry (500 t ha^{-1}). Slurry applications had little effect

Table 7. Partitioning and apparent recovery of total slurry P and K [7]

Type of manure	P fractions in % of total-P		Apparent recovery of P as % of total-P applied
	inorganic	organic	
Cattle slurry	78	22	34
Pig slurry	82	18	39
Poultry slurry	60	40	29
Fertiliser	100	–	42
	K fractions in % of total-K		Apparent recovery of K as % of total-K applied
	water-sol	$CaCl_2$-sol	
Cattle slurry	89	98	88
Pig slurry	83	97	84
Poultry slurry	81	94	89

on plant uptake of P but led to an increase in K content and a reduction in Mg levels. In practice, with slurry applied at sensible rates, a system can be devised to produce forage maize in which no inorganic fertiliser is required [27]. Five cows housed over winter will produce roughly 40 t slurry which will supply 100 kg available N, 35 kg P and 165 kg K [6]. This will be adequate to supply the requirements of 10 t maize DM ha^{-1}; sufficient forage for the winter housing of the 5 cows. Research in The Netherlands, however, has shown that the application of 22 kg P ha^{-1}, placed in the row at sowing, increased yields even when high rates of slurry were applied [8].

Gracey [15] observed a depression in red clover yield from cattle slurry applied to an initially pure red clover sward. Total DM yield however was not affected; presumably the slurry N had stimulated weeds and companion grass at the expense of clover. The effect on clover was most marked at first cut and the results showed that, by applying mineral P and K fertiliser in spring, with cattle slurry after the first cut, it was possible to maintain a high proportion of red clover in the sward and to produce clover DM yields similar to those obtained from inorganic fertilisers.

Recent work in The Netherlands (K. Dilz, unpublished) has examined the effects of large dressings of FYM on herbage mineral composition of a maize, maize, Italian ryegrass rotation. FYM was applied to maize crops at 0, 100 and 300 t ha^{-1} yr^{-1}, complementary to 250, 100 and 0 kg N ha^{-1} yr^{-1} as fertiliser, respectively and variable rates of P and K, and with only inorganic fertilisers applied to ryegrass. The experiment ran from 1972–1979

Table 8. Effect of FYM on yield and nutrient content of maize and Italian ryegrass (IRG)

	FYM rate t ha^{-1}	Fertiliser kg N ha^{-1} yr^{-1}	Yield t DM ha^{-1}	% P in DM	% K in DM
Maize					
1973	0 FYM	250	13.0	0.26	1.1
	100 FYM	100	14.4	0.30	1.5
	300 FYM	0	15.0	0.33	1.9
1979	0 FYM	250	10.2	0.22	1.3
	100 FYM	100	10.4	0.29	1.8
	300 FYM	0	10.6	0.33	1.7
IRG	*Residual effect*				
1973	0 FYM	300	10.6	0.40	3.8
	100 FYM	300	11.8	0.43	4.2
	300 FYM	300	12.8	0.46	4.4
1979	0 FYM	300	10.7	0.37	4.0
	100 FYM	300	12.8	0.45	4.8
	300 FYM	300	13.8	0.50	5.1

with FYM applied in two years in every three in the rotation. FYM consistently increased DM yield and P and K content of both maize and ryegrass (residual effect); sample results are shown in Table 8. Despite the increasing soil nutrient status, there appeared to be no increase in the P and K concentration in the maize crops over the years, whereas an increase in K concentration in Italian ryegrass was recorded. It is likely that there were substantial leaching losses, particularly of N and K on the light sandy soil.

Injection of slurry into grassland

During the last 10–15 years, new and improved techniques of injecting slurries and sludges into grassland have been developed [18]. The techniques have shown promise in terms of avoiding losses by surface run-off and by reducing odour nuisance [9, 22]. Experimental results have shown that injection can greatly increase the efficiency of slurry N [25]. However, this work has shown that a proportion of the slurry N only becomes effective later in the season [25, 28]. Some preliminary results (1985) of current experiments at the Institute for Soil Fertility in Haren (NL) to examine the availability of P and K from injected slurry, are discussed below (T.A. van Dijk, unpublished).

Materials and methods

Experiments were initiated in spring 1985 on 3 soils (two sandy soils and one clay over peat soil) of very low to low P status; P-AL (available P by extraction with ammonium lactate and acetic acid) < 30 mg P_2O_5 per 100 g dry soil. Phosphorus efficiency of injected cattle slurry was studied in combination with four rates of fertiliser P. On the sandy soils similar experiments on K efficiency were carried out. Experimental treatments consisted of either fertiliser P or K, at 0, $\frac{1}{2}$, 1 and $1\frac{1}{2}$ times the rate recommended on the basis of soil P or K status; these rates were applied with or without slurry injected at roughly 40 t ha^{-1}, in late April. All treatments received 460 kg N ha^{-1}.

The injection unit consisted of a pipe set behind a duckfoot tine and disc coulter. Depth is normally set at 15–20 cm and the distance between injector units, usually at 50 cm.

Dry matter yields and herbage analysis were determined on six cuts of grass during the season.

Results

Yield of DM was increased by P fertiliser on the lower P status soils (Table 9). Yields were enhanced by injection of slurry at all rates of fertiliser P. Phosphorus uptake was also increased by slurry, particularly at the lower rates of fertiliser P on the sandy soils. On the clay over peat soil, the higher available P content of the surface 5 cm of soil is the likely reason for the apparent lack of any contribution from the P in slurry placed at 15–20 cm. This result is in agreement with the observation in a survey of sandy grassland soils in The Netherlands that the P supply to the sward is primarily influenced by the P status of the soil layer immediately below the sod [11].

In the K experiments, DM yields and K uptake, particularly, responded to applied fertiliser K (Table 10). K uptake was also increased by injected slurry, especially at the lower rates of fertiliser K.

The seasonal distribution of response to P and K from injected slurry is also important. The pattern of P and K uptake by the sward according to the 6 cuts is indicated in Table 11; a + mark indicates a positive effect of injected slurry on the uptake of P or K. P uptake from injected slurry occurs mainly in cuts 2 and 3, whereas K is effective in all cuts. These observations are confirmed by the work of Tunney and Molloy [33] which demonstrated an increase in herbage N and K, but not P, content in cut 1 from injected slurry compared to surface-spread or nil slurry. By cut 2, these differences were less apparent, suggesting that differences in the effectiveness of slurry N, P and K had levelled out by that time.

Discussion

Experimental work to determine the fertiliser value of animal manures and slurries has been undertaken for 30 years. The variability of results obtained and sometimes apparently differing conclusions are reflected in the range of work reported here. Such variability, however, is hardly surprising in view of the variable nature of the materials tested and the conditions under which the experiments were conducted. Sometimes quite large differences may be explained on the basis of experimental conditions. The interaction of N supply with P and K uptake is likely to be an important factor, but this has not always been considered or reported by authors.

The low efficiencies of slurry P and K reported by Adams [2] were "short term" assessments on the basis of observations on first cut only. Moreover, P supply from the soil was high and the comparison was between a substantially higher rate of slurry P than fertiliser P.

Table 9. Effects of injected cattle slurry and fertiliser P on DM yield and P uptake by grass in three field experiments (1985)

Cattle slurry t ha^{-1}	Fertiliser P	DM yield, t ha^{-1} yr^{-1}			P uptake, kg ha^{-1} yr^{-1}		
		sand 1 P-AL = 10	sand 2 P-AL = 12	cl. peat P-AL = 24	sand 1 P-AL = 10	sand 2 P-AL = 12	cl. peat P-AL = 24
0	0	12.9	13.0	12.0	41	40	47
	½ × rec.*	13.6	14.0	12.2	56	64	55
	1 × rec.	14.5	14.2	12.6	70	76	62
	1½ × rec.	14.1	14.0	12.6	73	81	67
40	0	13.8	14.3	12.5	48	57	49
	½ × rec.	15.0	15.4	13.0	65	72	56
	1 × rec.	14.9	15.6	12.9	71	81	63
	1½ × rec.	15.2	15.6	12.6	78	82	64

* rec. = recommended rate; at low P-status 48 kg P ha^{-1} to the first cut, each following cut 13 kg P ha^{-1}

Table 10. Effects of injected cattle slurry and fertiliser K on DM yield and K uptake by grass in two field experiments (1985)

Cattle slurry $t\,ha^{-1}$	Fertiliser K	DM yield, $t\,ha^{-1}\,yr^{-1}$		K uptake, $kg\,ha^{-1}\,yr^{-1}$	
		sand 1	sand 2	sand 1	sand 2
0	0	11.8	12.1	238	212
	½ × rec.*	13.8	13.6	417	367
	1 × rec.	13.7	14.8	556	550
	1½ × rec.	13.4	14.5	605	602
40	0	12.9	13.3	375	304
	½ × rec.	14.1	15.4	538	493
	1 × rec.	14.0	14.4	618	559
	1½ × rec.	14.3	14.2	667	579

* rec. = recommended rate; at low K-status 149 kg K ha^{-1} to the first cut, each following cut 83 kg K ha^{-1}

It is interesting to note the differences in efficiency recorded in some pot studies. Slurry and fertiliser P were found to be equally effective, where dried slurry was incorporated into soils [34], in contrast to surface application of faeces to an established sward where lower efficiencies of manure P were observed [16]. FYM stimulated plant and root growth in pot studies reported by Jensen [20] and this led to increased P uptake and apparently higher efficiency of P when supplied as FYM, compared to fertiliser. In this study, timing and plant species were also shown to affect the response to applied P.

Some attempts have been made to include soil analysis into assessments of nutrient availability. McAllister [24] established relationships between P and K applied and removed and changes in extractable soil P and K. For P, about 15% of the excess P dressing could be recovered in extractable (ammonium acetate/acetic acid) P; the extent of P fixation could vary with soil type. This compares with advisory estimates in England and Wales [5] of only about 8–12% of excess P recovered in sodium bicarbonate extract-

Table 11. Seasonal distribution and relative responses to P and K from injected slurry

Cut	P-experiments		K-experiments		
	sand 1	sand 2	sand 1	sand 2	sand 3*
1	0	0	+	0	+++
2	++	+	++	++	++
3	+++	++	+++	++	+
4	+	+	++	+	++
5	+	0	+	+	+
6	+	+	+	+	+

* Experiment carried out in 1984 by J.J. Woldring (unpublished)

able P status of soils. McAllister [24] observed, at the only very high K application in a long term slurry experiment, that most of the excess K could be recovered as extractable K. The value of such assessments is questioned by the observations summarised in Table 5 [35]. Here a very large fall in soil K status was recorded within 2 years of the last slurry treatment, although this had been applied at sacrificially high rates. Differences in soil type may sometimes provide a partial explanation of such differing observations; the soil was a silty clay loam in the latter case but soil type was not recorded in the study by McAllister [24].

In studies involving high rates of slurry or manure application, small increases in extractable P have been noted at depth in the soil, although the majority is retained near the surface. Gerritse and Vriesema [14] showed that organic P in samples of cattle, pig and poultry slurries varied from about 1 to 15% of total P, with most of the rest being present as orthophosphate (inorganic) P. This agrees well with the data presented in Table 12 [29], although a generally higher proportion of organic P in cattle, pig and poultry slurries was reported by Amberger (Table 7) [7]. Only a small part of the organic P, however, appears to be mobile [13]; this is of microbial origin, of high molecular weight and is only slightly adsorbed by soils. This fraction represents only 0.2–2% of total slurry P and is of significance only when very high doses of slurry are applied to soils.

By contrast, soil K values can increase at depth following prolonged or high rates of slurry use. A two-fold increase in available K throughout the top 75 cm of a fine sandy loam soil was observed following high rates of pig and cattle slurry applied over a 15 year period [30].

Recent Dutch studies on soil analysis and manure P have shown only slightly reduced efficiency of manure P compared to fertiliser P, with increasing effectiveness in the long term (Table 12) [29]. These results, together with those of Hanley and Murphy [19] and the injection studies reported in this paper, emphasise the need to study the supply of nutrients from manures

Table 12. Inorganic P fraction of manure P and efficiency index of animal manure P compared to superphosphate (= 1.0) [29]

Source	% DM	Inorganic P % of total P	Effect on soil P status (P-AL) after:	
			0.75 years	1–1.5 years
Cattle slurry	12.5	81	0.8	0.9
Pig manure	24.5	95	0.8	0.9
Hen manure	22.7	86	0.8	0.9
Broiler manure	62.6	74	0.65	0.8

over an extended period and under different systems, to assess fully their value as sources of nutrients.

It is well known that large variations in soil nutrient status may occur within shallow layers of a soil, especially in long term grassland [24]. Depth of sampling therefore needs careful consideration where high rates of slurry or fertiliser have been applied or, in the context of recent cultivations or slurry injection (K.A. Smith, unpublished ADAS data).

Conclusions

Apparently conflicting information in the literature can often be at least partly accounted for by the specific conditions under which experiments are conducted.

When animal manures are used as a source of P for grass production or to raise the soil status, efficiency may be less than fertiliser P. This reduced short-term availability is unimportant in many grassland soils with existing satisfactory soil P status. Potassium in slurry is as available as K in inorganic fertiliser but may be liable to leaching when applied at high rates, especially on light soils.

In practice, reduced efficiencies of slurry P and K may result because of bad spreading practice, crop damage and nutrient loss by run-off and leaching (K). In the long run, however, there is no reason to regard the P and K efficiencies of animal manures as less than inorganic fertilisers, providing they are correctly applied.

Occasional soil and herbage analysis are worthwhile checks as part of a well designed slurry-fertiliser policy and will help to ensure optimum utilisation of all resources.

References

1. Adams S N (1973) The response of pastures in Northern Ireland to N, P and K fertilisers and to animal slurries. I. Effects on dry-matter yield. Journal of Agricultural Science (Cambridge) 81, 411–417
2. Adams S N (1973) The response of pastures in Northern Ireland to N, P and K fertilisers and to animal slurries. II. Effects on mineral composition. Journal of Agricultural Science (Cambridge) 81, 419–428
3. Adams S N (1974) The response of pastures in Northern Ireland to N, P and K fertilisers and to animal slurries. III. Effects on experiments continued for either two or three years. Journal of Agricultural Science (Cambridge) 82, 129–137
4. ADAS (1975) Important mineral elements in animal nutrition and their optimum concentration in forages. ADAS advisory paper No. 16. UK: MAFF

5. ADAS (1985) Phosphate and potash for rotations. Booklet 2496. UK: MAFF
6. ADAS (1986) Profitable use of farm manures. Booklet 2081. UK: MAFF
7. Amberger A (1982) Gülle – ein schlechtgenutzter Dünger. DLG – Mitteilungen 97 (2), 78–80
8. Arnold G H and Hag B A ten (1982) Rijenbemesting met fosfaat bij snijmais. Bedrijfsontwikkeling 13, 403–409
9. Dam Kofoed A (1981) Water pollution caused by run-off of manure and fertiliser. In Brogan J C, ed. Nitrogen losses and surface run-off from landspreading of manures, pp 70–75. The Hague: Martinus Nijhoff/ Dr Junk
10. Dijk T A van and Sturm H (1983) Fertiliser value of animal manures on the continent. Proceedings No. 220. London: The Fertiliser Society
11. Ehlert P A I (1985) Betekenis van de fosfaat- en kalitoestand van de onder de zode gelegen bodemlagen voor de fosfaat- en kalivoorziening van grasland. Report 11-85, 32 pp. Haren, The Netherlands: Institute for Soil Fertility
12. Ferwerda J D (1951) Over de werking van stalmest op bouwland, I. Verslagen Landbouwkundige Onderzoekingen No. 57.13, 59 pp. II. Verslagen Landbouwkundige Onderzoekingen No. 57.16, 75 pp.
13. Gerritse R G (1981) Mobility of phosphorus from pig slurry in soils. In Hucker T W G and Catroux G, eds. Phosphorus in sewage sludge and animal waste slurries, pp 347–369. Dordrecht, The Netherlands: Reidel
14. Gerritse R G and Vriesema R (1984) Phosphate distribution in animal waste slurries. Journal of Agricultural Science (Cambridge) 102, 159–161
15. Gracey H I (1981) Cattle slurry as a source of nutrients for red clover: herbage production and clover contribution. Grass and Forage Science 36, 291–295
16. Gracey H I (1984) Availability of phosphorus in organic manures compared with monoammonium phosphate. Agricultural Wastes 11, 133–141
17. Gracey H I and Stewart T A (1975) Effect of pig slurry on the accumulation of available soil phosphate and potash. Record of Agricultural Research, Ministry of Agriculture, N. Ireland 23, 65–66
18. Hall J E (1986) Soil injection research in the UK. In Dam Kofoed A, Williams J H and Hermite P l', eds. Efficient land use of sludge and manure, pp 78–89. London and New York: Elsevier Applied Science Publishers.
19. Hanley K P and Murphy M (1976) Comparative effects of animal manures and fertilisers on grass in pot experiments. Irish Journal of Agricultural Research 15 (1), 146–151
20. Jensen I (1982) Nutritional value of P in farmyard manure. Acta Agriculturae Scandinavica 32, 207–214
21. Larsen K E (1981) Phosphorus effect of animal manure and sewage sludge. In Hucker T W G and Catroux G, eds. Phosphorus in sewage sludge and animal waste slurries, pp 207–228. Dordrecht, The Netherlands: Reidel
22. Luten W, Geurink J H and Woldring J J (1983) Yield response and nitrate accumulation of herbage by injection of cattle slurry in grassland. In Corrall A J, ed. Efficient grassland farming. Proceedings of the 9th General Meeting of the European Grassland Federation, pp 185–191. Occasional Symposium No. 14 of the British Grassland Society. Hurley, UK: The British Grassland Society
23. McAllister J S V (1971) Nutrient balance on livestock farms. 1st Colloquium of the Potassium Institute Ltd, pp 113–121
24. McAllister J S V (1977) Efficient recycling of nutrients. In Voorburg J H, ed. Utilisation of manure by land spreading, pp 87–105. Modena, Italy: Commission of European Communities
25. Meer H G van der, Thompson R B, Snijders P J M and Geurink J H (1987) Utilization

of nitrogen from injected and surface-spread cattle slurry applied to grassland (this volume)

26. Pain B F and Phipps R H (1974) The effect of heavy dressings of slurry on forage maize production. Journal of the British Grassland Society 29, 263–267
27. Phipps R H and Pain B F (1978) The efficiency of energy use in forage maize production. In Bunting E S, Pain B F, Phipps R H , Wilkinson J M and Gunn R E, eds. Forage Maize, pp 323–336. UK: Agricultural Research Council
28. Prins W H and Boer D J den (1985) Aspects of nitrogen and phosphate fertilisation on grassland. Fertilisers and Agriculture No. 90 (October 1985), 27–36
29. Prummel J and Sissingh H A (1983) Fosfaatwerking van dierlijke mest. Bedrijfsontwikkeling 14, 963–966
30. Russell R D and Unwin R J (1982) Potassium in soils and herbage of intensive livestock farms. Internal ADAS report. London, UK: MAFF
31. Schechtner G, Tunney H, Arnold G H and Keuning J A (1980) Positive and negative effects of cattle manure on grassland with special reference to high rates of application. In Prins W H and Arnold G H, eds. The role of nitrogen in intensive grassland production. Proceedings of an international Symposium of the European Grassland Federation (Wageningen), pp 77–93. Wageningen: PUDOC
32. Tunney H (1984) Can agricultural practice reduce the amount of chemical fertiliser? In Winteringham F P W, ed. Environment and chemicals in agriculture, pp 285–305. London and New York: Elsevier
33. Tunney H and Molloy S P (1986) Comparison of grass production with soil injected and surface spread cattle slurry. In Dam Kofoed A, Williams J H and Hermite P l', eds. Efficient land use of sludge and manure, pp 90–98. London and New York: Elsevier Applied Science Publishers
34. Tunney H and Pommel B (1987) Phosphorus uptake by ryegrass from monocalcium phosphate and pig manure on two soils. Irish Journal of Agricultural Research (in press)
35. Unwin R J (1976) Slurry application to grassland. In ADAS Soil Science Reports of Experiments 1973–1976. MAFF S.W. Region, UK: MAFF
36. Unwin R J (1981) Phosphorus accumulation and mobility from large applications of slurry. In Hucker T W G and Catroux G, eds. Phosphorus in sewage sludge and animal waste slurries, pp 333–343. Dordrecht, The Netherlands: Reidel
37. Voorburg J H (1983) Utilisation of organic wastes as a fertiliser. Proceedings No. 219. London: The Fertiliser Society

Distribution of excreted nitrogen by grazing cattle and its effects on sward quality, herbage production and utilization

E A LANTINGA[a], J A KEUNING[b], J GROENWOLD[c] and P J A G DEENEN[a]

[a] *Department of Field Crops and Grassland Science, Agricultural University, Haarweg 333, 6709 RZ Wageningen, The Netherlands*

[b] *Research Station for Cattle, Sheep and Horse Husbandry (PR), Runderweg 6, 8219 PK Lelystad, The Netherlands (Seconded by the Netherlands Fertilizer Institute)*

[c] *Centre for Agrobiological Research (CABO), P.O. Box 14, 6700 AA Wageningen, The Netherlands*

Key words: grassland, grazing, nitrogen excretion, nitrogen response, botanical composition, herbage intake

Abstract. Grazing cattle exert positive and negative effects on pasture production. It is shown that at a fertilizer nitrogen (N) input of 200 kg N $ha^{-1} yr^{-1}$ and more, the benefit from N circulation through urine and dung is not of significance for the pasture as a whole. Of the negative effects, poaching has the greatest influence on the response of pasture production to N fertilization.

Two one-year experiments are described which showed that there were no significant differences in the response of grassland production to N fertilization between cutting and grazing usage above about 200 kg N $ha^{-1} yr^{-1}$.

Introduction

The supply of nitrogen (N) to a grazed sward is increased by the uneven return of N via dung and urine. When available N is below the level needed to support maximum herbage growth, this return can stimulate local herbage production. However, grazing animals also cause sward damage through treading, poaching, fouling, urine scorch and sod-pulling. The impact of these negative effects increases with increasing fertilization levels [4, 26, 27, 32]. This is associated with higher stocking rates and more vulnerable grass.

In this paper an attempt will be made to separate the positive and the negative effects of grazing cattle on sward quality, herbage production and utilization in pastures dominated by *Lolium perenne*. Furthermore, a comparison will be made between the response of herbage to fertilizer N under grazing or under cutting.

H.G. v.d. Meer, et al. (eds.), Animal Manure on Grassland and Fodder Crops.
 ISBN 978-94-010-8137-5.

Nitrogen contained in the faeces and urine of grazing cattle

The proportion of total N intake which is subsequently excreted or voided and its partition between urine and faeces depends on the type of cattle, the level of production, the intake of dry matter (DM) and the N concentration of the diet. The faeces of both beef and dairy cattle contain about 0.8 g N per 100 g DM consumed [30]. This agrees with the findings in a continuous grazing experiment of the Agricultural University in the Flevopolders. Throughout the grazing season it was found that the faeces of dairy cows, consuming 1 kg of concentrates and about 16 kg of herbage DM daily, contained on average 132 g N $cow^{-1} day^{-1}$. There appeared to be no influence of the level of N intake which varied between 450 and 775 g N $cow^{-1} day^{-1}$.

One kg of liveweight gain in growing cattle contains about 30 g N, whereas one kg of milk with a protein concentration of 3.4% contains about 5.5 g N. This implies that beef cattle normally retain a lower percentage of the N ingested than do dairy cattle (Table 1).

Dairy cows tend to urinate and to defaecate twelve times per day [9]. Assuming that dairy cows spend about four hours per day off the pasture, about ten urinations and ten defaecations will be made onto the pasture by one cow.

The area influenced by urine is often more than twice the area actually wetted owing to diffusion of N in the soil and the lateral spread of roots [31]. With dairy cows it has been found that the area on which growth was stimulated averaged 0.68 m^2 per urine patch and that a dung pat covered on average 0.05 m^2 [10].

From the above data it follows that at 750 cow-grazing days $ha^{-1} yr^{-1}$ the total area directly affected by urine is 5100 $m^2 ha^{-1} yr^{-1}$ and by faeces 375 $m^2 ha^{-1} yr^{-1}$. However, there may be overlap, especially with urine. Using the negative binomial function of Petersen *et al.* [19], it can be calculated that with a rather uniform distribution of the excretions about 3900 $m^2 ha^{-1} yr^{-1}$ is affected by at least one urine excretion in the above

Table 1. Calculated example of the partition of ingested N between animal products, faeces and urine by beef cattle (liveweight gain 800 g $steer^{-1} day^{-1}$) and dairy cattle (milk production 20 kg $cow^{-1} day^{-1}$). Nitrogen concentration of the diet is 0.04 g N per g DM

	Intake kg DM $head^{-1} day^{-1}$	N in: intake	animal products	faeces	urine	% in faeces and urine
		g N $head^{-1} day^{-1}$				
beef cattle	8	320	24	64	232	93
dairy cattle	16	640	110	128	402	83

example. In a situation of a more uneven distribution, this is somewhat lower due to more overlap. This is to be expected with continuous grazing where a typical feature of low stocking densities is 'camping areas'.

From the previous data it can also be calculated that the average rate of N applied to an area affected by a urination from a daily intake of 16 kg herbage DM (4% N) is about 500 kg N ha^{-1}. Similarly, it follows that one dung dropping is equivalent to an application in the order of 2000 kg N ha^{-1}.

Dung

Positive effects

Considering the extent to which N in faeces stimulates grass growth, MacLusky [14] observed that faeces affected the growth of herbage over an area about 6 times that actually covered. This agrees with the findings of MacDiarmid and Watkin [13]. They concluded that grasses growing up to 15 cm from the edge of a dung pat can derive N from the region under the pat owing to the lateral spread of grass roots. Assuming a mean area of a dung pat of 0.05 m^2, this is equivalent to an affected area of about 0.25 m^2. However, the direct response of herbage production to cattle dung is much smaller than that brought about by urine. This is because the release of inorganic N from dung is a relatively slow process. The visible effects of dung N may last for up to two years [24], but in total there is only a small beneficial effect on herbage production [33], certainly at moderate and high rates of fertilizer N (200 kg N ha^{-1} yr^{-1} and more). This is also clear from Table 2, which shows no influence of dung pats on the N concentration of adjacent herbage [28]. However, in the long term dung N contributes significantly to the typical increase in soil N content under permanent pastures [25]. This in turn results in a gradually increasing level of net mineralization of N. Unfortunately, data on this feature are lacking.

Table 2. Nitrogen concentration of herbage (g N per 100 g DM) growing inside and outside dung-affected areas at two N fertilization levels under continuous grazing

Month	N application, kg ha^{-1} yr^{-1}			
	214		427	
	dung-affected	not affected	dung-affected	not affected
June	3.3	3.3	4.0	4.1
August	2.9	2.9	3.3	3.3

Negative effects

The direct adverse effects of dung pats on herbage production are unimportant because of the small area covered and the possibility for exploitation of the space above the pats by adjacent plants [32]. The time required before dung has visually disappeared from a pasture depends on the weather and the activity of the soil fauna. On average, this is about three months and leaves a bare area. In pastures dominated by perennial ryegrass, recolonization takes place mainly by this species which has excellent tillering ability [2].

An indirect negative effect of dung is the rejection of herbage by grazing cattle in and adjacent to the contaminated area. This rejection is caused mainly by the smell of dung [16]. The size of the rejected area depends on the intensity of grazing, the rate of N fertilization [12] and possibly on the degree to which cows accustom themselves to eating dung-affected herbage [15]. On the other hand, it is remarkable that cows are quite prepared to graze close to fresh faeces, but that they avoid partly decomposed faeces [7]. Moreover, not every dung dropping causes rejection of surrounding herbage. Keuning [11] observed in September 1975 that after three rotational grazing periods in June, July and August, only 45 out of 98 dung pats gave rise to rejected patches, 38 in total with a mean area of $0.56\,m^2$. In particular, small dung pats (1–$2\,dm^2$) did not cause rejected patches. Visual observations indicated that on areas where dung pats were grouped, large rejected patches were found.

When there is a surplus of herbage, grazing cattle reject herbage around dung pats to a large extent. Eventually, when no silage cuts are taken or the fouled area is not topped, the pasture is covered with clumps of long rank herbage. This situation should be avoided because the ungrazed herbage becomes mature and will then be neglected due to differences in palatability rather than to the proximity of dung [18]. Moreover, tiller density in rejected areas will decrease gradually, giving rise to sward deterioration. It normally takes three to four months before herbage around dung pats is grazed down to the same level as surrounding pasture [29].

With carefully controlled continuous grazing, the rejected area can be minimized. The height of herbage around dung pats is kept at an acceptable level since the upper parts of the herbage in these areas are grazed regularly. In a current experiment by the Department of Field Crops and Grassland Science in the Flevopolders, where the mean sward height between dung affected areas is kept at about 5 cm, at maximum 15% of the area carrying herbage with a height not exceeding 15 cm is more or less rejected throughout the grazing season. Under this management herbage intake is sufficient for an average production of 22 kg fat-corrected milk $cow^{-1}\,day^{-1}$ between

May and October. The sward height is measured using a falling disc with a diameter of 50 cm and a weight of 340 g.

It is interesting to note that Greenhalgh and Reid [7] found that the daily intake of organic matter by grazing dairy cows was equally reduced by fouling at a low and at a high grazing intensity. At both grazing intensities the intake on previously grazed and topped paddocks was about 9% below that on previously cut paddocks. However, the daily intake of digestible organic matter and the fat-corrected milk production were not significantly reduced by fouling. This was because for some reason the herbage growing on the fouled pasture had a higher digestibility. The question remains what the influence of fouling would have been at equal digestibilities. In another experiment (Lantinga, unpublished) it was found that the relationship between daily herbage allowance and daily herbage intake per steer was the same for the first and the second grazing period in paddocks which were not topped after the first grazing. The results of these two experiments suggest that at corresponding herbage allowance levels cattle in fouled pastures can be forced to graze close to dung pats without reducing the individual intake significantly in comparison with clean pastures. However, this topic needs further research. Under a management system whereby cuts are taken for conservation, the alternation of mowing and grazing ensures better utilization of the herbage [14].

Urine

Positive effects

Normally, urine has a beneficial effect on herbage growth, especially in dry periods. The growth stimulation may be due either to urine potassium (K) or N [3]. In this paper only the N effects will be considered.

The N in cattle urine is present mainly as urea. After excretion there is a very rapid hydrolysis of urea into ammonium salts in the soil [8]. In the high pH environment generated by the hydrolysis of urea, ammonium is subject to ammonia volatilization. Hydrolysis is usually complete within one to three days of urine deposition [25]. The ammonium remaining in the soil is generally nitrified into nitrate. Both ammonium and nitrate can be absorbed and assimilated by the standing herbage and stimulate production of harvestable DM. The urine-N effects may last for about five months when applied in spring and for about ten months when applied in autumn [10, 18]. Except on very light soils, autumn depositions of urine may give rise to heterogeneity in pasture growth in the following spring (Keuning, unpublished). However, the increase in yield following autumn application has been found to be lower than that following spring application due to winter leaching of N [18].

The extent to which urine N can stimulate DM production is dependent on the level of fertilizer N input. To study this, the Department of Field Crops and Grassland Science has started research on the response of pasture production to N fertilization under continuous and rotational grazing. From the results collected in 1985 on a clay soil in the Flevopolders an estimation can be made of the positive effects of urine N at an annual fertilizer rate of 250 kg N ha^{-1}.

The N response was measured during one grazing season with continuous grazing at two N levels (250 and 550 kg N ha^{-1} yr^{-1}) and with weekly cutting over a range up to 700 kg N ha^{-1} yr^{-1}. It was hypothesized that the defoliation regime under continuous grazing could be simulated with weekly cutting. At 250 kg N ha^{-1} yr^{-1}, 695 cow-grazing days ha^{-1} yr^{-1} were attained. The potential effect of urine N on DM production was estimated by making use of dynamic simulation. The following assumptions were made: (i) one cow affects 6.8 m^2 of the pasture daily by urine; (ii) the urine N is effective for five months; (iii) during this period the DM production in urine-affected areas is enhanced by the measured difference in DM production under weekly cutting at 700 and 250 kg N ha^{-1} yr^{-1}. With 700 kg N ha^{-1} yr^{-1} the DM production appeared to be close to its maximum under weekly cutting (Lantinga, unpublished). The simulation yielded a potential increase in pasture production due to urine N at 250 kg fertilizer N ha^{-1} yr^{-1} of about 1200 kg DM ha^{-1}. Such an increase is noteworthy. However, the actual increase can only have been a fraction of this, since:

- it is now known that the response of DM production to a range of fertilizer N levels is greater with weekly cutting than with continuous grazing (Lantinga, unpublished);
- overlapping of urine patches occurs;
- the whole affected area will not give maximum production;
- maximum production will be realized only during a part of the five months period;
- not all excretions will stimulate DM production, with scorching as the extreme.

Taking these points into account, we estimate that in reality not more than about one-quarter of the calculated potential increase can have occurred. It will be clear that an increase in the order of 300 kg DM ha^{-1} yr^{-1} is not of significance and, moreover, that it is impossible to measure such an increase in whole-paddock trials.

Negative effects

In contrast with dung, urine-affected areas are not rejected at all by grazing cattle. Keuning (unpublished) observed in paddocks which were grazed for one day, that at the end of the day the urine-affected areas were grazed down to the same level as the non-affected areas. Norman and Green [18] observed that urine-affected areas were grazed in preference after some weeks of regrowth.

Scorching and death of grass within areas affected by urine is a serious problem in intensively managed pastures grazed by cattle. Research in The Netherlands by Keuning (reported in [26]) revealed that the occurrence of urine scorch is greatest if the soil is moist and the average daily temperature is between approximately 15 and 20 °C. Urine scorch is a more serious problem on sandy soils than on peat and clay soils and the extent is strongly correlated with the rate of fertilizer N (Table 3).

Groenwold and Keuning [9] observed that any cow can cause urine scorch. However, there appeared to be a large difference in extent of the damage between cows. This was related to differences in the chemical composition of the urine (Table 4). In another experiment it was found that the urinations during morning hours caused more damage than those later in the day (Table 5). This was also associated with differences in the chemical composition of the urine.

It was first shown by Richards and Wolton [23] that the scorching effect of urine is exerted on the root system of the grass and not on the foliage. Of the components which are formed during the hydrolysis of urine in the soil, ammonium-ions seem to be the most toxic to grass roots [8].

Urine scorch may lead to a deterioration in botanical composition. Richards and Wolton [23] observed that in patches where only a few grass tillers had been killed, the patches filled in with *Lolium perenne* before the end of the season. However, on the clay soil of their experiments, colonization of

Table 3. Effect of N fertilization on urine scorch (in % of the total area per year)

Soil type	kg N $ha^{-1} yr^{-1}$					
	0	150	200	300	400	600
Peat soil average 1977–1981	0.2	0.8		1.5		4.2
Sandy soil average 1980–1983			6.3		7.9	9.9

Table 4. Variations between cows in urine composition in relation to scorching. Urine samples were collected on 8 days during the grazing season of 1979. Permanent pasture on peat soil; 1000 kg fertilizer N $ha^{-1} yr^{-1}$

Cow	Conductivity ($mS\,cm^{-1}$)	Hippuric acid ($g\,l^{-1}$)	Urea ($g\,l^{-1}$)	Total N ($g\,l^{-1}$)	Patches with scorching (%)	Scale of scorching (%)			Average area damaged per scorched patch (dm^2)
						light	medium	severe	
47	16.4	6.8	16.0	8.9	46	27	27	46	10
89	12.3	5.3	10.9	6.1	24	33	54	13	18
156	15.1	6.6	14.2	7.4	69	55	36	9	7
167	16.7	8.6	17.6	9.7	86	12	44	44	18

Table 5. Effect of time of the day on urine composition in relation to urine scorching. Three periods of 36 to 48 hours were studied during July and August 1981. Permanent pasture on peat soil; 1000 kg fertilizer N $ha^{-1} yr^{-1}$

Time of the day (h)	Conductivity ($mS\,cm^{-1}$)	Hippuric acid ($g\,l^{-1}$)	Urea ($g\,l^{-1}$)	Total N ($g\,l^{-1}$)	Patches with scorching (%)	Scale of scorching (%)			Average area damaged per scorched patch (dm^2)
						light	medium	severe	
06–12.00	21.3	5.7	16.0	11.1	79	21	26	53	20
12–18.00	20.6	6.8	13.5	9.8	59	22	43	35	17
18–24.00	16.4	6.0	11.7	8.4	54	33	33	33	17
24–06.00	19.4	6.0	14.9	10.4	68	29	47	24	15

Table 6. Botanical composition after four years redevelopment in completely killed urine patches

Species	Percentage of total DM	Percentage ground cover
Lolium perenne	44	55
Poa trivialis	16	13
Poa annua	13	9
Alopecurus geniculatus	4	5
Elymus repens	–	1
Other grass species	1	3
Taraxacum officinale	21	13
Other herbs	1	1

completely killed patches was mainly by *Poa annua, Plantago major, Rumex obtusifolius* and *Sonchus oleraceus.* Keuning and Groenwold [10] followed the change in botanical composition of scorched patches on a sandy soil over a period of four years. At the beginning of the experiment the pasture comprised more than 90% *Lolium perenne* on a dry weight basis. The regrowth on scorched patches consisted mainly of *Poa annua* and partly of *Stellaria media* and *Taraxacum officinale.* Fifteen months after development of the patches *Lolium perenne* increased to 25%, but *Poa annua* was still the most important species. In the following years the botanical composition improved gradually, but did not reach the previous sward quality (Table 6).

Urine scorch will diminish if water intake by the cattle is increased. To this end, Keuning and Groenwold (unpublished) examined the effect of adding molasses to the drinking water. In 1983, when a 2%-solution of molasses was supplied *ad lib.* to grazing yearlings, the results were very promising. The average daily intake of molasses-water was 57.8 l head^{-1}, whereas the controls consumed only 21.6 l water head^{-1} day^{-1}. This resulted in a decrease of the scorched area by 57%. The N concentration of the urine from the yearlings drinking molasses-water averaged 3.68 g l^{-1} and that from the control group 8.65 g l^{-1}. In 1984 and 1985 the experiment was repeated with milking cows. Surprisingly no differences were established in the water consumption and consequently neither in the extent of urine scorch. In the autumn of both years the experiment was continued with yearlings and the same effect was obtained as in 1983. It is not known why yearlings and milking cows showed such different reactions.

Response of pasture production to nitrogen fertilization

Fertilizer N plays an important role in intensive grassland management. The economic use of this input depends upon the response to the applied N, either in terms of herbage DM production or animal products. Experimental data in the literature on the yield response to N under grazing are scanty and often conflicting. This is due partly to the fact that measurement of herbage DM production under grazing can only be done using indirect techniques.

At present, the Dutch standard fertilizer recommendation for grassland on clay and sandy soils is 400 kg N $ha^{-1} yr^{-1}$ [20]. This recommendation is based principally on trials where the response of DM production to N fertilization was evaluated under a mean 4-weekly cutting interval. However, grazing animals exert positive effects on DM production due to the recycling of nutrients, as well as negative effects. In order to maintain individual animal performance and to counteract sward deterioration when sward damage occurs, stocking density has to be decreased which tends to reduce annual pasture yields. This effect is well illustrated by the results of field experiments carried out in the 1970s and 1980s at North Wyke (Devon, UK), a site with a high risk of poaching. Both Richards [22] and Wilkins (personal communication) observed on grazed swards no further response to N above 300 kg N $ha^{-1} yr^{-1}$. However, on swards cut at 4-weekly intervals at the same site, the DM yield reached a maximum at about 600 kg N $ha^{-1} yr^{-1}$. In both experiments there was considerable herbage soiling and poaching damage on the grazed swards at high rates of N. This was most obvious during wet conditions in the spring.

Other factors which may influence the response of herbage yield to N fertilization are the supply of phosphorus (P) and K and the botanical composition. At increasing concentrations of herbage N, the required concentrations in herbage of both P and K also increase [21]. In England and Wales, a K concentration of 20 g $(kg\,DM)^{-1}$ is regarded as sufficient for good grass growth [21]. However, trials by Arnold [1] revealed that at N concentrations around 30 g $(kg\,DM)^{-1}$, a K concentration of 20 g $(kg\,DM)^{-1}$ depressed annual herbage yield by more than 5%. According to Prins *et al.* [21], at an average N concentration of 30 g $(kg\,DM)^{-1}$ a K concentration of over 30 g $(kg\,DM)^{-1}$ is required. For this reason, it is possible that in some of the research carried out in England and Wales, the response of grassland production to fertilizer N supply has been reduced by a suboptimal K supply.

In the research by the Department of Field Crops and Grassland Science in the Flevopolders, mentioned before, the swards comprised more than 90% *Lolium perenne* on a dry weight basis, the P and K supply was good and there had been no poaching damage and hardly any urine scorch since the start in 1984. Only damage due to sod-pulling has appeared to be of significance, at least visually and especially during late summer at high N fertilizer. However, Tallowin *et al.* [27] found that, even at high N levels, sod-pulling does not lead to a serious loss of herbage DM. In 1984 and 1985 the N response was evaluated in one-year trials under continuous grazing or cutting. In one of the cutting treatments, the herbage was harvested when a yield of about 2000 kg DM ha^{-1} above a cutting height of 5 cm was reached. This cutting regime is comparable with that applied by Prins [20]. In both years the response of pasture production to N fertilization was evaluated at two N levels (1984: 214 and 427 kg N $ha^{-1} yr^{-1}$; 1985: 250 and 550 kg N $ha^{-1} yr^{-1}$). The number of animals on each area was adjusted throughout the grazing season when necessary to maintain a mean sward

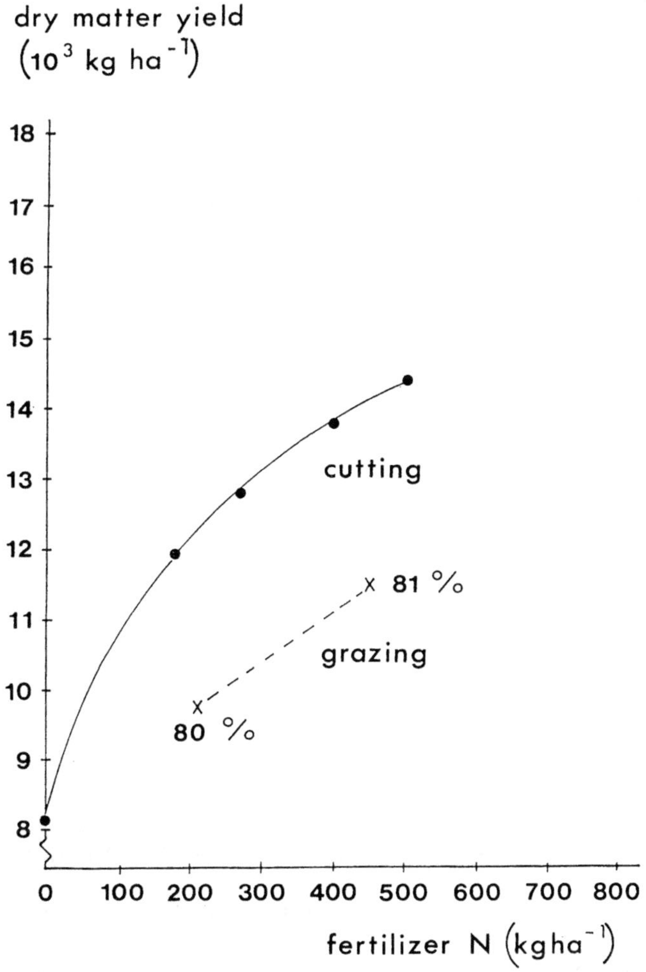

Figure 1. Effect of N fertilization on herbage DM yield under continuous grazing (weekly moved cages) and cutting (mean 4-weekly cutting interval). Yields under grazing are also expressed as a percentage of those under cutting at corresponding N levels. Swifterbant, 1984.

height of about 6 cm. In 1985 the yields under grazing and cutting were compared on the basis of herbage DM production. Under continuous grazing the production was measured by using exclusion cages which were moved weekly [6]. In 1985 the N response under grazing and cutting was compared on basis of the Dutch net energy system [5]. In this system the energy value of the herbage and the animal requirements are expressed in Dutch feed units for lactation (VEM). One VEM contains 6.9 kJ net energy for lactation (NEL). Under cutting the VEM yield was calculated from

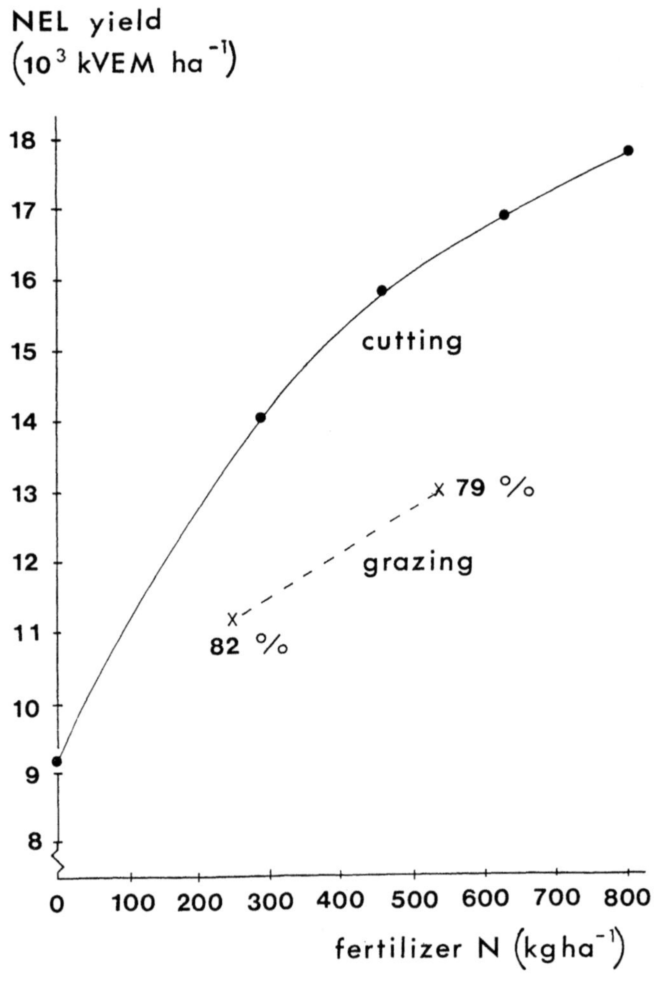

Figure 2. Response of NEL yield to N fertilization under continuous grazing (animal performance) and cutting (mean 4-weekly cutting interval). NEL: Net Energy for Lactation expressed in kVEM; 1 kVEM = 6.9 MJ. Yields under grazing are also expressed of those under cutting at corresponding N levels. Swifterbant, 1985.

harvested DM and chemical analyses. The VEM yield under grazing was calculated from animal production data using established standards for production and maintenance [17]. In both years the yields under grazing at both the low and the high levels of fertilizer N were about 80% of those obtained under cutting (Figs. 1 and 2). This suggests that:

(i) at the low fertilizer N rates there were no significant positive effects of the N returned by the grazing stock, as concluded before;

(ii) at the high fertilizer-N rates the negative effects exerted by the grazing cattle were not of significance.

The difference of about 20% in yield reflects the lower utilization efficiency of produced DM under grazing with respect to cutting [12].

In optimizing N fertilization, it is important to know the limit above which the yield increase does not pay for the extra N input. Prins [20] assumed that the 'optimum' rate of N application is at a marginal profitability of 7.5 kg DM per kg N applied. In long-term cutting trials on sand and clay soils the 'optimum' application appeared to be 420 kg N $ha^{-1} yr^{-1}$. Corré and Dijkman (personal communication) observed in two consecutive years on a clay soil, that under a management system involving alternate grazing and silage cutting the marginal yield response was 7.5 kg DM per kg N applied at a fertilizer rate of about 400 kg N $ha^{-1} yr^{-1}$. In this experiment there were also no problems with sward deterioration. The measured N response curves between 250 and 750 kg fertilizer N $ha^{-1} yr^{-1}$ were very similar to those given by Prins [20] for clay soils under cutting only.

From the above experiments it may be concluded that for at least one growing season the response of grassland production to N fertilization will not differ significantly between cutting and grazing above about 200 kg N $ha^{-1} yr^{-1}$.

References

1. Arnold G H (1978) Potassium regime, dry matter yield, and magnesium content of grass. Proceedings of the 7th General Meeting of the European Grassland Federation (Ghent), pp 9.42–9.53
2. Castle M E and MacDaid E (1972) The decomposition of cattle dung and its effect on pasture. Journal of the British Grassland Society 27, 133–137
3. During C and McNaught K J (1961) Effects of cow urine on growth of pasture and uptake of nutrients. New Zealand Journal of agricultural Research 4, 591–605
4. Edmond D B (1966) The influence of animal treading on pasture growth. Proceedings of the 10th International Grassland Congress (Helsinki), 453–458
5. Es A J H van (1978) Feed evaluation for ruminants. I. The system in use from May 1977 onwards in the Netherlands. Livestock Production Science 5, 331–345
6. Frame J (1981) Herbage mass. In Hodgson J, Baker R D, Davies A, Laidlaw A S and

Leaver J D, eds. Sward measurement handbook, pp 39–69. Hurley, UK: The British Grassland Society

7. Greenhalgh J F D and Reid G W (1969) The effects of grazing intensity on herbage consumption and animal production. III. Dairy cows grazed at two intensities on clean and contaminated pasture. Journal of Agricultural Science (Cambridge) 72, 223–228
8. Groenwold J and Heringa J W (1981) Urinebrandplekken in grasland. 2. Samenstelling van koeie-urine en omzettingen in de grond. Bedrijfsontwikkeling 12, 459–462
9. Groenwold J and Keuning J A (in prep.) Urinebrandplekken in grasland. III. Chemische processen
10. Keuning J A and Groenwold J (in prep.) Urinebrandplekken in grasland. I. Botanische aspecten
11. Keuning J A and Vellinga Th V (1986) Zodekwaliteit en botanische samenstelling bij intensieve graslandexploitatie. Gebundelde Verslagen No. 27, pp 1–14. Wageningen, The Netherlands: Nederlandse Vereniging voor Weide- en Voederbouw
12. Lantinga E A (1985) Productivity of grasslands under continuous and rotational grazing. Doctoral thesis, Agricultural University, Wageningen
13. MacDiarmid B N and Watkin B R (1972) The cattle dung patch. II. Effect of a dung patch on the chemical status of the soil, and ammonia losses from the patch. Journal of the British Grassland Society 27, 43–48
14. MacLusky D S (1960) Some estimates of the areas of pasture fouled by the excreta of dairy cows. Journal of the British Grassland Society 15, 181–188
15. Marsh R and Campling R C (1970) Fouling of pastures by dung. Herbage Abstracts 40, 123–130
16. Marten G C and Donker J D (1966) Animal excrement as a factor influencing acceptability of grazed forage. Proceedings of the 10th International Grassland Congress (Helsinki), 359–361
17. Meijs J A C (1981) Herbage intake by grazing dairy cows. Agricultural Research Report 909. Wageningen, The Netherlands: Centre for Agricultural Publishing and Documentation
18. Norman M J T and Green J O (1958) The local influence of cattle dung and urine upon the yield and botanical composition of permanent pasture. Journal of the British Grassland Society 13, 39–45
19. Petersen R G, Lucas H L and Woodhouse W W (1956) The distribution of excreta by freely grazing cattle and its effect on pasture fertility. I. Excretal distribution. Agronomy Journal 48, 440–449
20. Prins W H (1983) Limits to nitrogen fertilizer on grassland. Doctoral thesis, Agricultural University, Wageningen
21. Prins W H, Boer D J den and Burg P F J van (1986) Requirements for phosphorus, potassium and other nutrients for grassland in relation to nitrogen usage. In Cooper J P and Raymond W F, eds. Grassland Manuring. Occasional Symposium No. 20 (London), pp 28–45. Hurley, UK: The British Grassland Society
22. Richards I R (1977) Influence of soil and sward characteristics on the response to nitrogen. In Gilsenan B, ed. Animal production from temperate grassland. Proceedings of an international Meeting (Dublin), pp 45–49. Dublin, Ireland: Irish Grassland and Animal Production Association
23. Richards I R and Wolton K M (1975) A note on urine scorch caused by grazing animals. Journal of the British Grassland Society 30, 187–188
24. Richards I R and Wolton K M (1976) The spatial distribution of excreta under intensive cattle grazing. Journal of the British Grassland Society 31, 89–92
25. Ryden J C (1984) The flow of nitrogen in grassland. Proceedings No. 229. London, UK: The Fertiliser Society
26. Schechtner G, Tunney H, Arnold G H and Keuning J A (1980) Positive and negative

effects of cattle manure on grassland with special reference to high rates of application. In Prins W H and Arnold G H, eds. The role of nitrogen in intensive grassland production. Proceedings of an international Meeting of the European Grassland Federation (Wageningen), pp 77–93. Wageningen, The Netherlands: PUDOC

27. Tallowin J R B, Kirkham F W and Brookman S K E (1986) Sward damage by sod-pulling – the effect of nitrogen. In Frame J, ed. Grazing. Occasional Symposium No. 19 (Great Malvern), pp 44–48. Hurley, UK: The British Grassland Society
28. Valk H and Ramshorst H van (1986) Bepaling van de grasopname van melkkoeien bij standweiden op twee verschillende stikstofniveaus. Doctoraal verslag. Wageningen, The Netherlands: Agricultural University
29. Weeda W C (1967) The effect of cattle dung patches on pasture growth, botanical composition and pasture utilisation. New Zealand Journal of agricultural Research 10, 150–159
30. Whitehead D C (1970) The role of nitrogen in grassland productivity. Commonwealth Bureau of Pastures and Field Crops, Bulletin 48. Hurley, UK: Commonwealth Agricultural Bureaux
31. Whitehead D C (1986) Sources and transformations of organic nitrogen in intensively managed grassland soils. In Meer H G van der, Ryden J C and Ennik G C, eds. Nitrogen fluxes in intensive grassland systems, pp 47–58. Dordrecht, The Netherlands: Martinus Nijhoff Publishers
32. Wilkins R J and Garwood E A (1986) Effects of treading, poaching and fouling on grassland production and utilization. In Frame J, ed. Grazing. Occasional Symposium No. 19 (Great Malvern), pp 19–31. Hurley, UK: The British Grassland Society
33. Wolton K M (1979) Dung and urine as agents of sward change: a review. In Charles A H and Haggar R J, eds. Changes in sward composition and productivity. Occasional Symposium No. 10 (York), pp 131–135. Hurley, UK: The British Grassland Society

Negative effects of animal manure on grassland due to surface spreading and injection

W H PRINS[a] and P J M SNIJDERS[b]

[a] *Netherlands Fertilizer Institute, seconded to the Institute for Soil Fertility, P.O. Box 30003, 9750 RA Haren, The Netherlands*

[b] *Research Station for Cattle, Sheep and Horse Husbandry, Runderweg 6, 8219 PK Lelystad, The Netherlands*

Key words: slurry, farmyard manure, variable nutrient content, smothering, scorching, injection damage

Abstract. The effects of animal manure generally are a combination of positive and negative effects. Negative effects are apparent when they are larger than the positive effects or when the positive effects are masked. The paper deals with a selection of negative effects, mainly of cattle manure:

1) Variability in nutrient content, with an example of surface spreading from storage tanks containing unmixed and well-mixed slurry;
2) Uneven distribution of nutrients from surface spreading and injection. With surface spreading the unevenness may result in reduction in herbage yield and quality. With injection it has been observed that the concentrations of nitrogen, nitrate and phosphorus in the herbage decreased with increasing distance from the injection slit.
3) Sward damage caused by surface spreading. This concerns smothering, scorching and possibly other, as yet unknown effects, often resulting in a decrease in herbage yield. This negative effect is most apparent in the first cut after spreading and strongest at the highest rate of manure application. There were no essential differences in yield decrease between cattle slurry and farmyard manure. Dilution of slurry and addition of water immediately after surface spreading reduced or even prevented the negative effects.
4) Sward damage caused by injection. The injection equipment damages the sward, resulting in a yield decrease in the first cut(s). After passage under dry conditions a strip of grass along the injection slit may die and there may be an ingress of unwanted species.

The paper gives possible ways to reduce the negative effects.

Introduction

The effects of animal manures generally are a combination of positive and negative effects. Positive effects such as the supply of nutrients or of organic matter are well documented, but specific research into negative effects has started only recently [2, 26]. Schechtner [25] listed negative effects and ranked them according to frequency of occurrence and degree of nuisance. His list includes negative effects with respect to soil, water, plant, animal and man.

This paper deals with recent data on a selection of negative effects and indicates possible ways of overcoming these effects. The negative effects considered, particularly in relation to farmyard manure (FYM) and slurry from dairy cattle, are:

- Variability in nutrient content
- Uneven distribution of nutrients upon surface spreading and injection
- Sward damage caused by surface spreading
- Sward damage caused by injection.

Variability in nutrient content

Whereas mineral fertilizer application can be tailored to suit specific crops and timed exactly to suit plant requirement, manure supplies a variable range of nutrients (Table 1). The variability is related to kind of animal, kind of feed, moisture content, dilution and losses during storage.

Even within one dairy farm there may be large variations due to the degree of mixing of manure in the storage tank prior to land spreading. Fig. 1 shows the nutrient content of unmixed and well-mixed cattle slurry. The large variation among slurry tankers when slurry had not been mixed is clear. It is well-known that the upper layer in the storage tank has the highest dry matter (DM) content. Beneath this layer the slurry becomes more liquid until more solid material is reached at the bottom of the tank. This is reflected in the DM content in Fig. 1. The concentration of nutrients varies according to DM content. This is especially true for phosphorus (P), calcium (Ca) and magnesium (Mg), which are almost completely excreted in the faeces and are therefore, more closely related to the DM content than nitrogen (N) and potassium (K).

The variation among slurry tankers greatly affects the distribution of nutrients, as is shown in Table 2: while the average application appeared to be satisfactory, the area fertilized by tankers 4 and 5 received far less and the area fertilized by tankers 1, 2 and 3 received far more than the average

Table 1. Animal slurry composition on Irish farms [29]

	DM, %	N, kg t^{-1}	P, kg t^{-1}	K, kg t^{-1}
Cattle slurry (33 farms)				
mean	8	2.8	0.6	4.2
range	1–14	0.8–5.6	0.1–1.2	0.8–6.4
Pig slurry (25 farms)				
mean	4	3.0	0.9	1.5
range	1–13	0.4–7.0	0.1–3.4	0.2–3.3

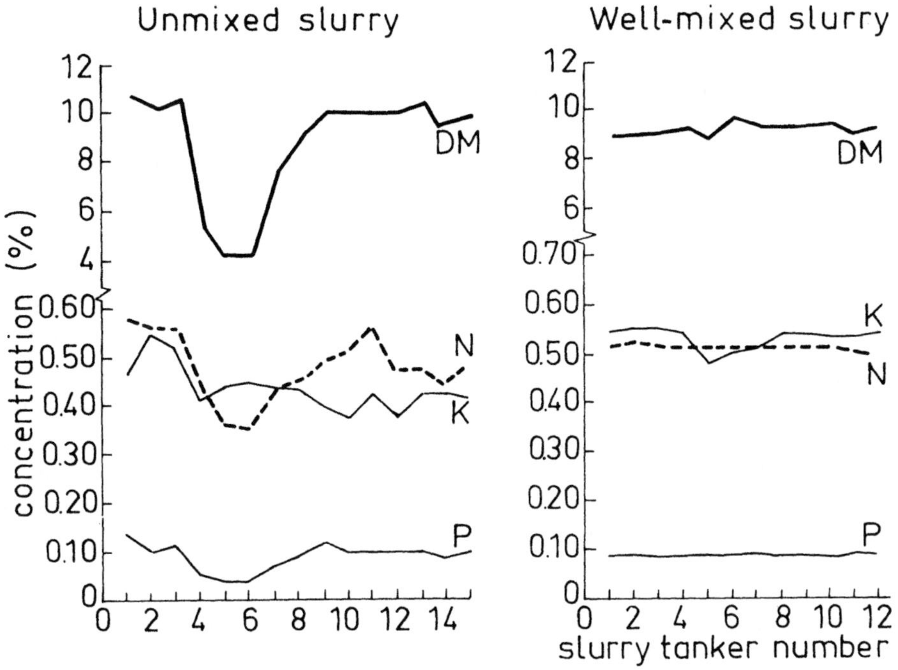

Figure 1. Differences in variation of DM, N, P and K in unmixed slurry from one farm and well-mixed slurry from a second farm, as shown by analysis of samples from a series of slurry tankers loaded consecutively from the storage tanks of both farms. The average values for DM, N, P and K for the unmixed slurry were 8.9, 0.49, 0.09 and 0.43% and for the well-mixed slurry 9.4, 0.52, 0.07 and 0.52%, respectively; adapted from [15].

amount. The effect of good mixing is reflected in a small variation of the nutrient concentration between the slurry tankers (Fig. 1). Thus, one sample suffices as an indication of the quantity of nutrients applied [15].

The problem for the farmer is that he does not know the exact nutrient concentration of the manure. The available quick tests are promising but not yet suitable for use by farmers [16, 29].

Table 2. Variation in nutrient application (kg ha^{-1}) upon surface spreading of unmixed slurry at a rate of 15 t ha^{-1} with five slurry tankers, as shown in Fig. 1 [15]

	Tanker number					Average
	1	2	3	4	5	
DM	1,620	1,560	1,620	825	630	1,250
N	87	85	85	68	56	76
P	20	15	17	7	4	13
K	71	82	78	63	66	72
Ca	36	30	34	15	11	25
Mg	23	16	20	9	7	15

Uneven distribution of nutrients from surface spreading and injection

In addition to the problem of uneven distribution of nutrients in pasture from dung and urine patches [21, 31], problems exist with the distribution of nutrients from surface spreading and injection of manures.

Distribution from surface spreading

Differences in nutrient distribution from spreading owing to incomplete mixing of slurry in the storage tank have already been mentioned. However, in farm practice uneven spreading of slurry from one tanker also often occurs. An example is given in Table 3, where in testing 17 types of slurry spreaders the aim was to apply 25 $m^3 ha^{-1}$. However, the extreme rates of slurry applied varied from 17 to 55 $m^3 ha^{-1}$. Only 5 (out of the 17) spreaders had a deviation from the mean of less than 30%. That these deviations are far too large is clear when the minimum and maximum quantities are converted into quantities of N (see last column of Table 3).

Table 3. Test results of 6 types of slurry spreaders out of a total of 17 types, each type aiming at an average of 25 $m^3 ha^{-1}$. Longitudinal and lateral deviations are combined [30]

Type of slurry spreader	Mean quantity of slurry applied, $m^3 ha^{-1}$	Observed deviation from mean %, + and −	Minimum and maximum application	
			$m^3 ha^{-1}$	N, $kg ha^{-1}$
A	17	28	12–22	60–110
B	25	37	16–34	80–170
C	27	27	20–34	100–170
D	27	46	15–39	75–195
E	32	28	23–41	115–205
F	55	37	35–75	165–375

Table 4. Variation in chemical composition of herbage at different distances from the injection slit at two sampling dates. Spring injection of cattle slurry at a rate of 80 $t ha^{-1}$, no N fertilizer applied [27]

Sampling date and distance, cm	DM, % in fresh material	N, % in DM	NO_3^--N % in DM
27 May 1980			
0–5	16.3	3.73	0.14
5–15	16.5	3.47	0.08
15–25	16.7	3.08	0.06
9 June 1981			
0–5	13.5	4.18	0.51
5–15	13.6	3.79	0.32
15–25	15.0	3.67	0.23

As with spreading of mineral fertilizers [10], uneven spreading of manure can also result in reduction in crop yield and quality. Adequate results may be achieved, however, by using slurry spreaders with properly set distribution equipment, taking into account corrections for wind effects [5]. Whilst equipment for improving the lateral distribution is coming onto the market, there is still no adequate solution offered to the farmer for applying the exact amount or for proper longitudinal distribution. At present the best option is a tractor with sufficient power together with a slurry spreader equipped with a screw pump [18].

Distribution from injection

In The Netherlands, there has been substantial research into the injection of slurry [6, 22]. At present the distance between injection tines is 50 cm. This does not give a good distribution of nutrients. An interesting example relating to the N availability is given in Table 4: the closer to the injection slit, the higher the concentrations of total N and nitrate N in the herbage [27].

In other experiments the effects of injection of cattle slurry on the availability of P were assessed. In the herbage of the 1st cut, in May, there was a significant decrease in P concentration of the herbage with increasing distance from the injection slit (Fig. 2). This decrease was also significant over the whole season but not for each of the 2nd, 3rd and 4th cuts.

Experiments with injection of ammonia into grassland have shown that for an optimum N supply the distance between the tines should be less than 30 cm [8]. For practical reasons, however, for slurry injection a distance of at least 50 cm has been adopted. Shorter distances carry the risk of "rolling up" the sward.

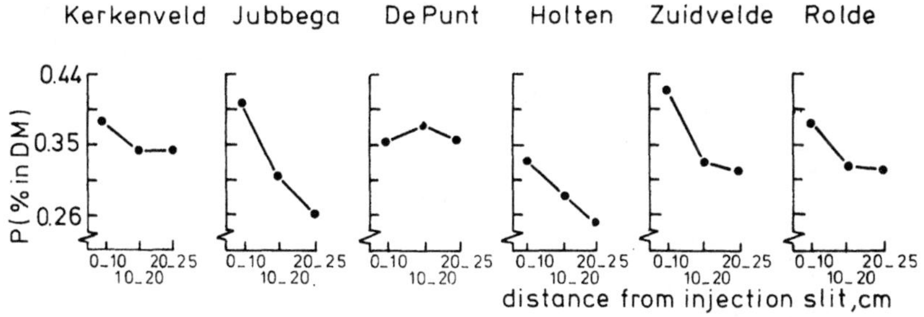

Figure 2. Effect of distance from the injection slit of cattle slurry, injected at a rate of 40 t ha^{-1} in March/April, on P concentration of herbage cut in May at distances of 0–10, 10–20 and 20–25 cm from the injection slit in P deficient sandy soils at six locations (M. van Noordwijk, unpublished).

Sward damage caused by surface spreading

Negative effects have been demonstrated in many trials with surface application of FYM or slurry [2, 24, 26]. It should be stressed that in these trials nutrients other than the one studied were supplied in adequate quantities. Recent experiments in The Netherlands, designed to study the N or P effect of cattle manure, have shown the negative effect of the manures on DM yield to be greatest in the first cut after spreading and at the highest rates of application (Tables 5–7). No essential differences existed between the results for the first cuts of short-term trials with FYM (Table 5) and cattle slurry (Table 6) and long-term trials with both FYM and slurry (Table 7), despite the differences in material and in distribution methods with FYM spread by hand and slurry spread by machine.

In the one-year experiments, there was over the whole season still a negative effect with 40 t FYM (Table 5) but not with slurry, even at application rates of 60–70 t ha^{-1} (Table 6).

In the long-term experiments (Table 7), the annual yields did not show negative effects from 40 t FYM or slurry, but there was still a negative effect on DM yield after surface spreading of 80 t slurry, resulting in a difference in DM yield of 0.53 t $ha^{-1} yr^{-1}$. This was determined at a high rate of fertilizer N application of 540 kg ha^{-1} both in 1982 and 1983, so as to mask the effects of slurry N. The question is whether this high rate of fertilizer might have promoted the negative effect. This would not seem likely since over the years 1982–1983 the maximum herbage DM yield was obtained

Table 5. Herbage DM yield (t ha^{-1}) of first cuts and for the whole season without and with surface spreading of farmyard manure (FYM) by hand in January/February. One-year experiments on clay soil at Hedel and on sandy soil at Vries. Fertilizer range of N: 300–360 kg $ha^{-1} yr^{-1}$ and of P: 105–131 kg $ha^{-1} yr^{-1}$

FYM t ha^{-1}	Year, location and month of application						Average percentage change
	1974 Hedel Jan.	1975 Hedel Jan.	1975 Vries Jan.	1976 Hedel Feb.	1977 Hedel Jan.	Average	
	First cut						
0	4.13	3.84	5.09	3.97	3.15	4.04	0.0
20	3.97	3.58	5.07	3.87	3.25	3.95	−2.2
40	3.83	3.43	4.74	3.07	3.14	3.65	−9.6
	Whole season						
0	11.69	12.90	13.19	8.32	12.35	11.69	0.0
20	11.96	12.79	13.34	8.96	12.42	11.89	+1.7
40	11.97	12.52	12.95	7.04	12.54	11.40	−2.5

Table 6. Herbage DM yield ($t\,ha^{-1}$) of first cuts and for the whole season without and with surface spreading of cattle slurry in one-year experiments on sandy soil. Fertilizer N, P and K shown are applications in $kg\,ha^{-1}\,yr^{-1}$

	Year, location and month of application					
	1978 Ydermade February	1979 Den Ham* May	1980 Ruurlo* March	1984 Selmien* March	Average	Average percentage change
Fert. N:	480	540	660	480		
Fert. P:	66	76	96	98		
Fert. K:	465	244	345	481		
Slurry $t\,ha^{-1}$	First cut					
0	2.31	2.99	3.87	2.32	2.87	0.0
20		3.49	3.55		2.96	+3.1
30–40	1.94	3.17	3.17	2.76	2.76	−3.8
60–70	1.47			2.30	2.44	−15.0
	Whole season					
0	11.79	10.57	16.21	13.50	13.02	0.0
20		11.07	16.84		13.58	+4.3
30–40	11.27	10.91	16.33	14.93	13.36	+2.6
60–70	11.91			13.86	13.26	+1.8

* Slurry was spread with a machine that was driven over the plots. Distance between tyre tracks was 1.05 m, width of the cutting bar was 1.54 m, so sward damage from slurry may have been aggravated by the tractor tyres.

with fertilizer applied at a rate of 540 $kg\,ha^{-1}\,yr^{-1}$: 12.71 $t\,ha^{-1}$ without slurry and 12.18 t with 80 t slurry. With 360 $kg\,ha^{-1}\,yr^{-1}$ of fertilizer N the DM yields were 12.14 $t\,ha^{-1}$ without slurry and 11.88 t with 80 t slurry (still a negative effect). With 180 $kg\,ha^{-1}\,yr^{-1}$ of fertilizer N the DM yields were 9.83 $t\,ha^{-1}$ without slurry and 10.44 t with 80 t slurry. In the last case the positive N effect of the slurry exceeded the negative effect.

The variation in results of Tables 5–7 is most likely due to weather conditions after spreading. The negative effects of FYM were almost absent in the wet spring of 1977 (Table 5) and in the wet spring of 1983, thus offsetting the long-term negative effects at Den Ham and Ruurlo (Table 7).

The results of Tables 5–7 clearly demonstrate a close relationship between quantity of manure and yield reduction; with about 70 $t\,ha^{-1}$ of slurry the yield of the first cut was reduced by about 15% on average.

Another recent example of negative effects can be taken from the UK where surface applications of cattle slurry to an Italian ryegrass sward were made in October, January or March (K.A. Smith, private communication). The sward showed no visible effect of October slurry application and yields were not significantly affected. Growth on all plots receiving slurry in

Table 7. Herbage DM yield of first cuts and for the whole season without and with surface spreading of cattle manure. Month of application in different years: always in January at Hedel; May, March, April, April, March at Den Ham, and March, April, April, April, March at Ruurlo, respectively. Fertilizer N, P and K shown are average applications in kg ha^{-1} yr^{-1}

Cattle manure t ha^{-1} yr^{-1}	Type of manure, period, location and soil type				Percentage change	
		FYM 1974–1976 Hedel clay	Slurry* 1979–1983 Den Ham sand	Slurry* 1980–1984 Ruurlo sand	FYM	Slurry
	Fert. N:	360	552	579		
	Fert. P:	105	78	44		
	Fert. K:	398	267	371		
		First cut				
0		4.27	3.65	2.96	0.0	0.0
20		4.13	3.76	2.92	−3.3	+0.8
40		4.04	3.34	2.56	−5.4	−11.0
80			3.05**			−16.0**
		Whole season				
0		11.50	12.69	13.42	0.0	0.0
20		11.77	13.07	13.64	+2.3	+2.3
40		11.65	12.79	13.76	+1.3	+1.7
80			12.16**			−4.2**

* see note in Table 6.
** 1982 and 1983 only (after 10 t slurry ha^{-1} yr^{-1} in 1979–1981). Change relative to no slurry in 1982 and 1983.

January and March was visibly retarded and the yield of the April cut was significantly reduced. In the May cut the negative effect of the March application was about double that of the October application (Table 8). It is notable that in this experiment 1:1 dilution of slurry with water did not increase the yield, in contrast to the results of experiments in Austria [25].

Table 8. Herbage DM yield (t ha^{-1}) of April and May cuts without slurry plus 240–300 kg ha^{-1} of fertilizer N and with surface spreading of cattle slurry in October 1984, January and March 1985. Average of 25 and 50 m^3 ha^{-1} of undiluted slurry (8% DM) and 50 and 100 m^3 ha^{-1} of diluted slurry (4% DM) plus 240 kg ha^{-1} of fertilizer N to mask the effects of slurry N (K. A. Smith, private communication)

Date of slurry application	Cut on 17 April 1985		Cut on 14 May 1985	
	DM yield	Percentage change	DM yield	Percentage change
–	1.88	0.0	5.68	0.0
15 October 1984	1.74	−7.4	5.48	−3.5
15 January 1985	1.50	−20.2	5.41	−4.8
14 March 1985	1.25	−33.5	5.26	−7.4

– no slurry.

In the experiments of Tables 5–7, manure applications *without adding fertilizer N or P* did not show negative effects, because of the stronger positive nutrient effect of the manure. One should always be aware that the effect of manure is a combination of positive and negative effects.

Sward damage by surface spreading is caused by smothering, scorching and possibly other, unknown, factors, or a combination of these.

Smothering

An example of smothering effects is given in Fig. 3, which demonstrates the results of a trial with increasing rates of P fertilizer [2]. In comparison with surface-spread slurry, the highest DM yield was achieved when no cattle slurry was applied (Fig. 3a). However, the situation was reversed when the slurry was washed off immediately after spreading (Fig. 3b). From these results it was concluded that the negative effect observed when the slurry was not washed off was due to smothering of the sward by the slurry.

The smothering effect varies from a reduction of plant photosynthesis to complete kill of grass species, leaving bare patches. Depending on the damage, the sward recovers more or less rapidly. Thus, smothering may cause a delay in growth resulting in a yield decrease of the first cut after

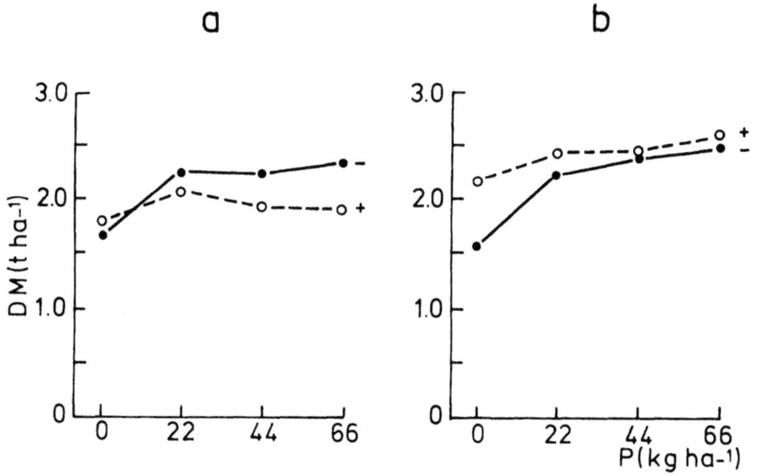

Figure 3. DM yield of the first cut of phosphate-deficient grassland at increasing rates of superphosphate application without (−) and with (+) slurry at a rate 38 m^3 ha^{-1}.
a = without and b = with 4.4 mm water to wash the slurry off the grass. Application of superphosphate on 3 January and application of slurry (0.09% P) on 3 February 1978. Fertilizer N: 160 kg ha^{-1} in three split applications, fertilizer K: 199 kg ha^{-1} in four split applications [2].

surface spreading. However, over a season this yield reduction may disappear completely (Tables 5–7).

The smothering effect may be prevented by spreading dried FYM instead of fresh FYM, because dried and crumbled FYM does not stay on the tillers but falls on to the ground [2]. The physical nature of different manures is thus relevant. It has been observed that at similar total N contents pig slurry yielded more grass than cattle slurry. This is because pig slurry, being more granular in nature, tends to fall to the soil between the tillers and thus results in less severe smothering than the more fibrous cattle slurry [28]. However, others claim that the difference between pig slurry and cattle slurry is mainly due to a higher N recovery from pig slurry because of less ammonia volatilization [23] (J.H. Geurink, private communication).

That factors other than smothering may also have negative effects was shown in a trial where smothering was avoided by applying the cattle slurry in autumn [2, 3]. After application, sufficient rain fell to wash the slurry off the grass. There were no visible signs of scorch. Throughout the winter, however, the slurry plots showed a brown haze while the plots without slurry were fairly green. The outcome was that in the following spring, yields from the slurry plots were substantially lower than those from the non-slurry plots, especially on poorly drained grassland (Table 9). The reason for these negative effects is not known, but root damage due to ammonia [20] or toxins [1] are possibilities.

Scorching

It is well-known that the sward may be scorched by liquid manure (*i.e.* the liquid fraction of stored cattle manure) when spread in warm sunny weather

Table 9. Herbage DM yield ($t\,ha^{-1}$) of early and late cuts in 1982 without and with surface spreading of cattle slurry in early November 1981 on well-drained and poorly drained grassland. Overall mineral fertilizer N: 200 $kg\,ha^{-1}$, P: 23 $kg\,ha^{-1}$, and K: 199 $kg\,ha^{-1}$ [3]

Cattle slurry $t\,ha^{-1}$	Clay, well-drained		Clay on peat, poorly drained	
	DM yield	Percentage change	DM yield	Percentage change
	Early cut, 29 April		4 May	
0	2.56	0.0	1.63	0.0
20	2.35	−8.2	1.19	−27.0
40	2.10	−18.0	0.88	−46.0
	Late cut, 18 May		27 May	
0	5.81	0.0	5.31	0.0
20	5.74	−1.2	4.76	−10.4
40	5.48	−5.7	4.53	−16.6

or on frozen soil [17], and by urine from grazing cattle under warm and moist conditions in summer [26]. However, little is known of the negative effect of scorching on herbage yield. A recent example is shown by field experiments in The Netherlands (Table 10) from which it appears that the substantial negative effect from liquid cattle manure may be as large as that from an equal amount of cattle slurry. Visible scorching was least in the case of the slurry application and it may be assumed that the negative effects of the slurry were due to a combination of smothering and scorching. Damage was least at the lower levels of application of liquid manure and slurry. With pig slurry the damage was small. In these experiments liquid manure and slurry were applied to a short stubble (5 cm). In an experiment with applications on taller grass (about 10 cm), damage was more severe. Damage was particularly visible in tyre tracks where the crushed plant cells could easily be entered by the damaging constituents [4]. As the distance between tractor tyres was 1.12 m while the width of the cutting bar was 1.40 m, it seems likely that the negative effect was aggravated by the damage in the tyre tracks, especially with the liquid manure treatments.

Yields were also reduced by scorching in a recently described pot trial: where the slurry stuck to the grass and could not be washed off there was substantial visible damage due to scorching [19].

The main soluble constituents of urine are urea, K, Na, Cl and hippuric acid while those of slurry are ammonium, K, Na, Cl, benzoic acid from the urine fraction and P, Ca and Mg from the faeces fraction of the slurry. The higher the concentration of these constituents the larger the chances of plasmolysis, which is observed as scorching [17]. It is still not known which

Table 10. Herbage DM yield without and with surface spreading of animal manure and slurry of different origins. Average results of the first cut after application, in four experiments conducted on sandy soil in the dry summer of 1986 at Bosma Zathe. All plots received mineral fertilizer N: 120 kg, P: 16 kg, and K: 83 kg ha^{-1} [4]

Treatments	Manure			Herbage	
	DM %	N %	Quantity t ha^{-1}	DM yield t ha^{-1}	Percentage change
control			0	2.84	0.0
liquid cattle manure	3.9	0.46	10	2.46	−13.4
liquid cattle manure	3.9	0.46	20	2.32	−17.4
cattle slurry	9.9	0.47	10	2.50	−12.1
cattle slurry	9.9	0.47	20	2.35	−17.3
liquids from fractionated pig slurry	3.7	0.38	20	2.74	−3.5
pig slurry	5.5	0.42	20	2.69	−5.3

constituents cause the damage. Damage to the plants may occur via two pathways: the leaves and the roots.

Pathway via leaves

In field experiments with topdressing of urea solution it was found that N applications as low as 30–45 kg ha^{-1} could cause serious scorch [7]. Leaves of laboratory-grown perennial ryegrass exposed to extreme concentrations of 10% ammonia almost all darkened within minutes and desiccated after a few hours. At concentrations of 0.1% ammonia only a few leaf tips were damaged (H. A. Lee and A. N. Davison, private communication). Visible damage due to slurry application to grass was confined to those parts of the leaves which were in direct contact with the slurry [12].

Pathway via roots

In greenhouse experiments, perennial ryegrass plants were placed in pots containing slurry in different dilutions with water and in pots with solutions of N, as ammonium or nitrate. Scorching manifested itself in darkening of the leaves and a decrease in water content (Fig. 4). In 1:1 and 1:3 ratios of slurry to water, leaves completely desiccated and growth was seriously retarded. Pots with N fertilizer solutions showed similar but less serious symptoms when ammonium fertilizer was used, whereas symptoms were almost absent with nitrate fertilizer. The water content of the leaves proved to be a good measure of the damage; a decrease of the water content of 5% in comparison with the control gave a decrease in DM yield of about 40% [12]. The osmotic potential of the slurry solution did not influence the symptoms. It was concluded that scorching via the roots may be caused by

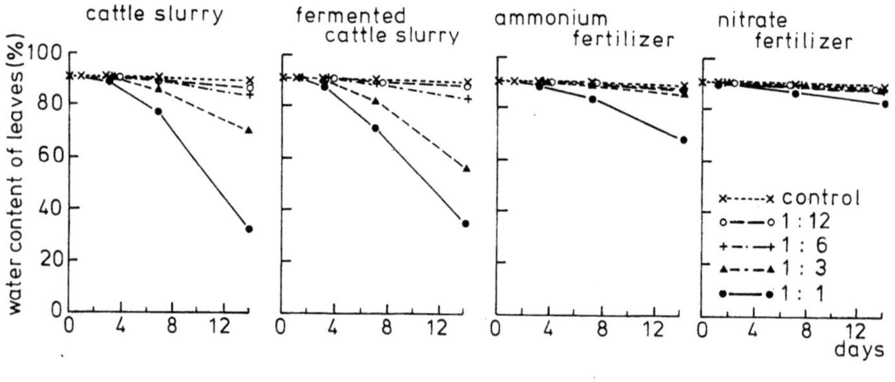

Figure 4. Changes in water content of leaves as a result of the addition to the root environment of different manures and fertilizers undiluted or diluted with water. N concentrations in the 1:12 dilution of the four materials were (as NH_4^+-N) 240, 280, 290 and (as NO_3^--N) 340 mg kg^{-1}, respectively (W. Douma, E. Hamwijk and R. Wortelboer, unpublished).

a number of toxic substances, ammonium being an important one [12]. In laboratory experiments with cattle urine, ammonium was also regarded as the most damaging agent to grass roots [13].

Scorching due to urine may be decreased by diluting the urine through increased water intake by the cattle [21] while scorching due to urea may be decreased by immediate addition of water as was the case when sufficient rain fell in trials with foliar spraying of urea (K. Dilz, private communication). Similarly, scorching by slurry may be decreased due to using diluted slurry [19] or by immediate irrigation after surface spreading of slurry [11].

Scorching of the leaves may be prevented by applying the slurry to the soil surface in between the grass plants [19]. Naturally, smothering may be prevented in the same way.

Sward damage caused by injection

The injection equipment itself may have a negative effect on grassland. Experiments with injection of ammonia have shown that the passage of injection tines damages the sward significantly. Damage was least on sandy soil and at an injection depth of about 5 cm [8].

In The Netherlands, for slurry injection at 15–20 cm depth, 15–20 cm wide wings are attached to the tines and substantial damage may therefore be expected. In three long-term experiments with injection of slurry into sandy soil the sward-cutting effect alone (*i.e.* injection without slurry application) could be studied at different rates of N fertilizer application. At low fertilizer N rates (0 and approximately 200 kg ha^{-1} yr^{-1}) there was initially a negative effect due to cutting which generally turned into a small positive effect in the course of the season. At the higher rates (approximately 400 and 600 kg N ha^{-1} yr^{-1}) the negative effect often continued throughout the season. It is assumed that the change from a negative effect at the lower rates of N to a positive effect was due to stimulation of the mineralization of soil N, as was indicated by the extra N uptake by the plants. This extra N uptake was on average 15 kg ha^{-1} yr^{-1} [22]. It can be concluded that at the higher rates of N fertilizer application this extra mineralization of N was masked.

When injecting slurry the effect due to cutting is more or less neutralized by the positive nutrient and "specific" effects of the slurry. Thus, negative effects can best be observed when ample amounts of mineral fertilizer are supplied, because only then do the nutrients in the slurry not contribute to the yield increase. It is assumed that there are no, or only minor, harmful effects of these fertilizers. Table 11 shows the effect in three long-term trials, of injection from 0 (sward-cutting effect alone) up to 80 t ha^{-1} of cattle

Table 11. Herbage DM yield (t ha^{-1}) of first cuts and for the whole season without injection and without slurry (0), with passage of injection equipment without slurry (0^+) and with spring injection of increasing amounts of slurry in three long-term experiments on sandy soils. Overall mineral fertilizer N: 690, 552 and 579 kg, P: 166, 78 and 44 kg, and K: 353, 267 and 371 kg ha^{-1} yr^{-1} at Hemrik, Den Ham and Ruurlo, respectively [27]

Cattle slurry t ha^{-1}	Location and period				Average percentage change
	Hemrik* (1978–1981)	Den Ham* (1979–1983)	Ruurlo* (1980–1984)	Average	
	First cut				
0	3.29	3.65	2.96	3.30	0.0
0^+	3.22	3.32	2.44	2.99	−9.4
20–30	3.35	3.46	2.59	3.18	−3.6
40–60	3.39	3.21	2.66	3.09	−6.4
80–90	3.33	3.27	2.67	3.09	−6.4
	Whole season				
0	15.45	12.68	13.42	13.85	0.0
0^+	15.52	12.82	12.82	13.72	−0.9
20	15.51	12.92	13.64	14.02	+1.2
40	15.79	12.87	13.79	14.15	+2.2
80	15.58	13.32	13.40	14.10	+1.8

* Distance between tyre tracks was 1.05 m, width of the cutting bar was 1.40 m at Hemrik and 1.54 m at Den Ham and Ruurlo, respectively, so sward damage from slurry may have been aggravated by the tractor tyres.

slurry, relative to a control treatment with no slurry and no injection. All treatments received fertilizer N at a rate of about 600 kg ha^{-1} yr^{-1}, sufficient to give near-maximum DM yields. The sward-cutting effect alone resulted in an almost 10% decrease in yield of the first cut. When slurry was injected in spring, the negative effect on the first cut was only about 5%. On an annual basis there was, on average, still a slight negative effect without slurry and a slight positive effect with slurry. Results with lower N fertilizer rates are not shown, but at N fertilizer rates of about 400 kg ha^{-1} yr^{-1} the DM yields of the treatments with 0, 20–30, 40–60 and 80–90 t injected slurry changed by −7.9, −4.6, −0.3 and 4.3% for the first cut and −0.7, −0.1, 5.1 and 5.3% for the whole season, respectively, in comparison with the control treatment. At still lower rates of N fertilizer the percentage changes became increasingly positive at all rates of slurry application.

In the UK, experiments with injection equipment have shown that the disturbance of grassland in spring significantly reduced yield by about 15%. Injection in autumn, however, did not affect production as the crop had the chance to recover before the onset of spring growth [14].

In the experiments reported in Table 11, variation between years was large; considerable damage from the tines was observed in dry years [27].

The negative effect appears to be largest in the first cut after injection. Table 11 gave results of injection in spring, but similar effects have been obtained with injection in summer. Under dry conditions in summer a small strip of grass along the injection slit may die and there may be an invasion of less productive species like *Stellaria media* or *Elymus repens* (syn. *Agropyron repens*) [32]. This relationship with the weather fully agrees with the outcome of the trials with ammonia injection. The slits became wide cracks, especially in clay soil, along which desiccation occurred [8].

Damage increased strongly when the frequency of injection during the same season was increased from one to three times [9]. This may also depend on the soil conditions at injection. To decrease the risk of damage, it is necessary to inject only in soils which are in a moist condition [14, 32].

Final remarks

This paper has not dealt with negative effects of animal manure on flora (supply of weeds), fauna (carryover of diseases like botulism, eradication of worms), soil (structure deterioration due to heavy traffic), herbage quality (excess nitrate or K concentrations), herbage intake (odour and taste), although these are also important.

The effects discussed have led to a number of conclusions:

1. The variability in nutrient content makes it difficult for the farmer to know the quantities he supplies. Reliable quick tests need to be developed to improve precision.
2. The farmer has to be aware of the risks of uneven distribution with manure application by surface spreading and injection as this will influence crop yield and quality.
3. Sward damage from surface spreading, due to smothering and scorching, generally has a negative effect on herbage production of the first cut after application of the manure. This damage, which increases with increasing quantities of manure applied, may be partly prevented by washing the manure particles off the grass and by injecting the manure into the soil.
4. Sward damage from injection equipment also occurs mostly in the first cut(s) after application. Damage is least in soils which are in a moist condition and therefore spring is generally the best time for injection.

Acknowledgements

Thanks are due to K. A Smith for supplying UK data and to G. H. Arnold and J. H. Geurink for critically reading the manuscript.

References

1. Abele U (1978) Ertragssteigerung durch Flüssigmistbehandlung. KTBL-Schrift Nr. 224, pp 134
2. Arnold G H (1981) Beobachtungen von negativen Nebenwirkungen von Gülle auf Dauergrünland. Bericht über die 7. Arbeitstagung "Fragen der Güllerei" (Gumpenstein, 1981), pp 627–632
3. Arnold G H, Hakvoort B J and Prins W H (1982) Invloed van fosfaatbemesting in het voorjaar op kleigrasland met een hoge fosfaattoestand. Jaarverslag ROC Bosma Zathe 1982, 53–58
4. Boer P J de (1987) De gevolgen van organische mest voor de grasgroei in de eerste snede na aanwending. Jaarverslag ROC Bosma Zathe 1986 (in press)
5. Bosma A H and Buitink W J (1985) Bovengrondse verdeling van mengmest. Landbouwmechanisatie 36 (2), 166–169
6. Buitink W J (1985) De ontwikkeling van een praktijkmachine voor het injecteren van mengmest. IMAG Publikatie 214, pp 30
7. Burg P F J van (1958) De overbemesting van granen door middel van ureumbespuiting. Stikstof no. 18, 182–188
8. Burg P F J van and Brakel G D van (1964) Ammoniakinjectie als methode van stikstofbemesting. 1. De bemestingswaarde van ammoniak. Stikstof no. 44, 260–273
9. Dam Kofoed A (1981) Water pollution caused by run-off of manure and fertiliser. In Brogan J C, ed. Nitrogen losses and surface run-off from landspreading of manures. Proceedings of an EEC Workshop (Wexford, 1980), pp 70–75. The Hague: Martinus Nijhoff
10. Dilz K, Brakel G D van and Richards I R (1985) Effects of uneven fertiliser spreading on crop yield and quality. Proceedings no. 240, pp 51. London: The Fertiliser Society
11. Geneijgen J van (1980) Bij inregenen van mengmest geen schade aan het gras. Jaarverslag ROC Zegveld, 24
12. Gijsman A and Hamwijk E (1986) Een vergelijkend onderzoek naar de bemestende waarde en gebruiksmogelijkheden van drijfmest en gegiste mest op een rundveehouderij, pp 48. Utrecht: Rijks Universiteit Utrecht
13. Groenwold J and Heringa J W (1981) Urinebrandplekken in grasland, 2. Samenstelling van koeie-urine en omzettingen in de grond. Bedrijfsontwikkeling 12, 459–462
14. Hall J E (1986) Soil injection research in the UK. In Dam Kofoed A, Williams J H and Hermite P l', eds. Efficient land use of sludge and manure. Proceedings CEC Seminar (Askov, 1985), pp 78–89. London: Elsevier
15. Keuning J A (1986) Mengen is nodig voor goede verdeling voedingsstoffen. Boerderij 71 (12 mrt), 64–65
16. Kjellerup V (1986) Agros Nitrogen Meter for estimation of ammonium nitrogen in slurry and liquid manure. In Dam Kofoed A, Williams J H and Hermite P l', eds. Efficient land use of sludge and manure. Proceedings CEC Seminar (Askov, 1985), pp 216–223. London: Elsevier
17. Kolenbrander G J and Lande Cremer L C N de la (1967) Stalmest en gier, waarde en mogelijkheden, pp 188. Wageningen: Veenman
18. Krause R (1986) Flüssigmist met dem Tankwagen besser dosieren und verteilen. Landtechnik 41, 114–116
19. Küntzel U, Krause R and Jonuscheit C (1987) Scorching of *Lolium perenne* caused by cattle slurry (this volume)
20. Kutschera L and Sobotik M (1981) Gülleflora – Unterschiede durch Klima und Boden. Bericht über die 7. Arbeitstagung "Fragen der Güllerei" (Gumpenstein, 1981), pp 79–119

21. Lantinga E A, Keuning J A, Groenwold J and Deenen P J A G (1987) Distribution of excreted nitrogen by grazing cattle and its effects on sward quality, herbage production and utilization (this volume)
22. Meer H G van der, Thompson R B, Snijders P J M and Geurink J H (1987) Utilization of nitrogen from injected and surface-spread cattle slurry applied to grassland (this volume)
23. Molloy S P and Tunney H (1983) A laboratory study of ammonia volatilization from cattle and pig slurry. Irish Journal of Agricultural Research 22, 37–45
24. Pain B F (1977) Slurry utilisation on the dairy farm. In Voorburg J H, ed. Utilisation of manure by landspreading. Proceedings of an EEC Seminar (Modena, 1976), EUR 5672e, pp 51–64
25. Schechtner G (1986) Economical use of slurry on grasslands without causing damages on soil, water and plants and with small odour emissions. Proceedings of the Consultation of the FAO European Network on Waste Utilization (Burgos, 1986), pp 79–98
26. Schechtner G, Tunney H, Arnold G H and Keuning J A (1980) Positive and negative effects of cattle manure with special reference to high rates of application. In Prins W H and Arnold G H, eds. The role of nitrogen in intensive grassland production. Proceedings international symposium European Grassland Federation, pp 77–93. Wageningen: Pudoc
27. Snijders P J M, Woldring J J, Geurink J H and Meer H G van der (1987) Stikstofwerking van geïnjecteerde runderdrijfmest op grasland. Proefstation voor de Rundveehouderij, Schapenhouderij en Paardenhouderij, Rapport no. 103 (in press)
28. Tunney H (1981) Nitrogen uptake by crop – grassland. In Brogan J C, ed. Nitrogen losses and surface run-off from landspreading of manures. Proceedings of an EEC Workshop (Wexford, 1980), pp 161–166 and 190. The Hague: Martinus Nijhoff
29. Tunney H (1986) Slurry meter: rapid field test for estimating dry matter and fertiliser value of slurry. In Dam Kofoed A, Williams J H and Hermite P l', eds. Efficient land use of sludge and manure. Proceedings CEC Seminar (Askov, 1985), pp 206–215. London: Elsevier
30. Vetter H, Steffens G and Kowalewsky H H (1981) Bessere Gülleverteilung notwendig. Landwirtschafsblatt Weser Ems, nr. 19
31. Weeda W C (1977) Effect of cattle dung patches on soil tests and botanical and chemical composition of herbage. New Zealand Journal of Agricultural Research 20, 471–478
32. Woldring J J (1977) Injecteren van drijfmest in blijvend grasland. Proefstation voor de Rundveehouderij, Intern Rapport no. 76, pp 31

Cattle slurry and farmyard manure as fertilizers for forage maize

J SCHRÖDER[a] and K DILZ[b]

[a] *Research Station for Arable Farming and Field Production of Vegetables, P.O. Box 430, 8200 AK Lelystad, The Netherlands*

[b] *Netherlands Fertilizer Institute, seconded to the Institute for Soil Fertility, P.O. Box 30003, 9750 RA Haren, The Netherlands*

Key words: manure, slurry, nitrogen, forage maize, specific effect, residual effect

Abstract. The effects of large amounts of animal manure on forage maize have been investigated in a number of long-term field trials. Further, their impact on soils and their residual effect has been studied.

Yield levels were mainly determined by temperature and water supply, and were favoured by large amounts of cattle slurry up to 200–300 t $ha^{-1} yr^{-1}$ or 200 t $ha^{-1} yr^{-1}$ of farmyard manure. Even with annual manure rates of 100–200 t ha^{-1}, the crops responded positively to 75–100 kg of inorganic nitrogen (N) fertilizer ha^{-1}. Combined with manure, yields were higher than with mineral fertilizers only. This points to a specific effect of animal manure which especially occurred on sandy soils. This specific effect was attributed partly to improved soil conditions and partly to a better supply of minerals, especially potassium (K). The use of animal manure enabled a more efficient utilization of the N that had been absorbed by the crop.

Animal manure hardly affected the feeding value of the maize and led to only a moderate increase of the mineral contents. With large amounts of animal manure the input of minerals by far exceeded the uptake of the crops. Of the total input of N, roughly 20–30% accumulated in soil organic matter. Inorganic N content of soil did not increase over the years as considerable amounts of N are lost through various processes; in one of the trials it was demonstrated that about 25% of the total N input applied as cattle slurry, was lost by leaching. This was independent of its rate of application. As a result N recovery from manure was poor; suggestions for improvement are made.

From the response of a first year grass crop it was concluded that the residual effect of manure applications could amount to 4–6 kg N per 10 t $ha^{-1} yr^{-1}$ that had been applied. This was only 1% of the total N input from the preceding 9–11 years. As with maize, manure had a specific effect on grass yields. The composition of the grass, more than that of maize, responded to the various fertilizer treatments and should be taken into account.

Introduction

Specialization and intensification are the main characteristics of the agricultural industry in The Netherlands. Specialization has taken place both on a farm and on a regional level. On sandy soils in the eastern and southern

H.G. v.d. Meer, et al. (eds.), Animal Manure on Grassland and Fodder Crops.
 ISBN 978-94-010-8137-5.

parts of the country, the agricultural industry is almost completely focussed on livestock farming. In these regions mainly permanent grass and forage maize are grown. The area of maize rose from a few thousands of hectares in 1970 up to almost 200,000 ha in 1986. An important agricultural activity in these regions, next to dairy farming, is the fattening of pigs, poultry and calves, predominantly fed with externally produced compound feeds. As a result excessive amounts of animal manure are produced [31].

The use of large amounts of manure on permanent grassland may lead to an imbalance of nutrients and may cause damage to the sward. As a result especially forage maize is often amply supplied with animal manure. This may have consequences for the yield and quality of the crop and will affect the amount of mineral fertilizers needed for optimum growth. Further, there will be a residual effect on the following crops. The use of excessive amounts of animal manure raises concern about the accumulation and emission of nutrients to the environment through leaching or volatilization. Most of these aspects have been studied in a number of long-term field trials. Special attention was paid to nitrogen (N), both because of its dominating effect on plant growth and its mobile character. In this paper the results of these trials are presented.

Materials and methods

The effects of annually applied cattle slurry and farmyard manure (hereafter, where possible, both referred to as manure) on the yield and quality of forage maize, have been studied in three experiments on sandy soils at Heino, Maarheeze and Gortel. These three soils initially had an organic matter content in the top soil (0–20 cm) of 4.2, 2.9 and 4.2% and a potential rooting depth of 70–90, 50–70 and 30–40 cm, respectively. In Gortel water was supplied, if necessary, by sprinkler irrigation. Another experiment was carried out on a marine clay at Lelystad, which had a clay content of 25% in the layer of 0–60 cm and a potential rooting depth of more than 100 cm.

Table 1 shows the manure treatments in the experiments. Maize was cropped continuously, with the exception of the experiment at Heino where three consecutive maize crops were alternated by one year grass which did not receive manure treatments (1975, 1979). In 1983 grass was grown in the three experiments on sandy soil, to assess the residual effect of the preceding manure treatments.

Slurry and manure were surface-spread during autumn and winter (Lelystad), autumn, winter and spring (Heino and Maarheeze) or spring only (Gortel). Applications at Heino, Maarheeze and Lelystad were made with

Table 1. Manure treatments on forage maize in the experiments at Heino, Maarheeze, Lelystad and Gortel

Site	Type of organic fertilizer	Years of application	Application rate (t ha^{-1} yr^{-1})						
			0	50	100	150	200	250	300
Heino	cattle slurry	1972–1974		+	+	+	+	+	+
		1976–1978	+	+	+	+	+	+	+
		1980–1982	+	+	+	+	+	+	+
Maarheeze	cattle slurry	1974–1982		+	+	+	+	+	+
Lelystad	cattle slurry	1976–1980	+			+			+
Gortel	farmyard manure	1972–1982	+	+	+	+	+		

trailed vacuum spreader tanks, in Gortel manure was applied with a conventional spreader. Where the total amount of manure was split over more applications, it was proportionally divided over two dates (Lelystad) or distributed according to fixed schedules (Heino and Maarheeze; Table 2). The cattle slurry was incorporated into the soil within 3 days after application, the farmyard manure was ploughed under within 3 weeks.

In the experiment at Heino inorganic fertilizer treatments (0, 75, 150, 225 kg N ha^{-1} yr^{-1}) were superimposed on some of the manure treatments (0, 100, 200, 300 t ha^{-1} yr^{-1}) from 1976 onwards. From the first year, rates of fertilizer N were also included in the experiment at Lelystad (0, 75, 150, 225 kg N ha^{-1} yr^{-1}) and in the experiment at Gortel (0, 50, 100, 150, 200, 250 kg N ha^{-1} yr^{-1}). The fertilizer N, applied as calcium ammonium nitrate, was broadcast immediately after sowing. Some manure treatments received supplementary mineral fertilizers (Table 3).

Normally maize was sown during the last week of April at a density of 100,000 plants per ha. The variety LG 11 was used except for the 1972–1974 period at Heino and the 1972–1982 period at Gortel when Capella was sown.

The experiments at Heino, Maarheeze and Lelystad were set up as randomized blocks (3 replicates) with fertilizer N treatments as subplots. The plot size was at least 60 m^2, 30 m^2 of it being harvested. The experiment at Gortel also had a split-plot design and was set up in 2 replicates. The plot size was 56 m^2, 20.5 m^2 being harvested.

Table 2. Time schedule for slurry application in the experiments at Heino and Maarheeze

Application period	Application rate (t ha^{-1} yr^{-1})					
	50	100	150	200	250	300
November–December	0	0	0	50	100	150
January–February	0	50	100	100	100	100
March–April	50	50	50	50	50	50

Table 3. Supplementary mineral fertilizers (kg ha^{-1} yr^{-1}) in the experiments at Heino, Maarheeze, Lelystad and Gortel

Element	Site and manure treatment (t ha^{-1} yr^{-1})					
	Heino		Maarheeze	Lelystad	Gortel	
	0	50	50	0	0	50
N	0	100*	90**	0	0	0
P	50	15***	10**	45	65	60****
K	200	0	0	0	155	135****

*until 1978; **until 1979; ***until 1977; ****these amounts have been added to all other manure treatments too.

Composition of the manure, yield and quality of the forage and soil fertility were determined annually. In the Maarheeze experiment the quality of upper groundwater has also been monitored from 1977. The average compositions of the manures are presented in Table 4. Table 5 shows the average annual input of N from these organic fertilizers. The inputs are not always exactly proportional to the rate as a consequence of the schedule given in Table 2. More detailed information is given elsewhere [29].

Results

Yields of dry matter

None of the treatments had an adverse effect on plant emergence or increased the incidence of plant diseases or of pests. With increasing manure rates, crops showed a better appearance from early June. This was reflected by the dry matter (DM) yields at harvest with an exception for the highest rates (Table 6).

There were great differences in annual yields which were mainly caused by soil type and climatic factors such as temperature and rainfall. Thus in the extremely dry and hot summer of 1976, maximum yields were high on the

Table 4. Average composition of the manure during the experiments

Site	Dry matter (%)	Organic matter (%)	Content in fresh material (in %)						(in ppm)	
			N	P	K	Ca	Mg	Na	Cu	B
Heino	9.4	6.9	0.50	0.08	0.54	0.15	0.07	0.08	3.6	3.8
Maarheeze	11.4	8.5	0.51	0.10	0.46	0.15	0.07	0.09	5.0	3.7
Lelystad	8.6	6.5	0.40	0.08	0.46	0.14	0.08	0.06	3.7	3.6
Gortel	20.0	16.2	0.55	0.13	0.32	0.29	0.08	0.04	–	–

Table 5. Average annual input of nitrogen (kg N ha^{-1}) with manure

Site	Application rate (t ha^{-1} yr^{-1})					
	50	100	150	200	250	300
Heino	260	525	796	1048	1299	1551
Maarheeze	252	505	758	1013	1267	1522
Lelystad	–	–	589	–	–	1178
Gortel	276	553	829	1105	–	–

clay soil at Lelystad and on the irrigated light sandy soil at Gortel. Without irrigation they were lower than average, particularly where the organic matter content of the soil and the rooting depth were lower. The crop at Maarheeze suffered more from drought than that at Heino.

Maximum yields in all the trials were generally obtained at manure rates of 150 t ha^{-1} yr^{-1}. Above this rate yield increases were not significant. Notwithstanding the excessive amounts of nutrients being applied annually, the rate at which maximum yields were obtained, did not change over the course of the experiments. With 0, 50 or 100 t manure ha^{-1} yr^{-1}, yields gradually decreased in the Heino experiment, at least relatively. On the untreated plots this could be explained by the gradual exhaustion of soil N reserves. To some extent this was valid as well for the manure rate of 50 t ha^{-1} yr^{-1} where no supplementary N was given after 1978. The same trend could be detected in the Maarheeze experiment. In the Gortel trial no such decrease was observed, presumably because of the low soil fertility at the onset of the trial.

The positive effect of manure on maize yields could not only be explained

Table 6. Relative DM yields of forage maize as affected by manure

Site	Period	Application rate (t ha^{-1} yr^{-1})							(100 = ... t DM ha^{-1} yr^{-1})
		0*	50*	100	150	200	250	300	
Heino	1972–1974	–	93a	97ab	97ab	100b	97ab	94a**	(13.4)
	1976–1978	80a	86ab	92bc	98c	100c	99c	98c	(14.5)
	1980–1982	65a	74ab	81bd	90cd	94cd	92cd	100c	(15.5)
Maarheeze	1974–1976	–	90a	92ab	95bc	97cd	100d	97cd	(13.7)
	1977–1979	–	81a	86a	94b	99b	100b	98b	(12.9)
	1980–1982	–	76a	89b	98c	99c	100c	99c	(14.8)
Lelystad	1976	96a	–	–	100a	–	–	96a	(16.8)
	1977–1980	79a	–	–	97b	–	–	100b	(12.4)
Gortel	1972–1975	40a	72b	83bc	93cd	100d	–	–	(13.3)
	1976–1979	40a	80b	94c	97c	100c	–	–	(11.3)
	1980–1982	29a	71b	93c	98c	100c	–	–	(14.7)

* See Table 3 for supplementary fertilizers

** Different letters denote significant ($P < 0.05$) differences within experiments and periods according to Tukey-test.

by its N input as can be demonstrated from those trials where fertilizer N was superimposed on manure treatments. In Fig. 1 data from the experiments at Heino, Lelystad and Gortel are presented in the three quadrants diagram. The upper left quadrant (II) shows the relationship between fertilizer N and DM production at different levels of manure, the lower right quadrant (IV) gives the relationship between fertilizer N and the N uptake of the crop (N yield). The slope of the lines in this quadrant indicates the apparent recovery of the fertilizer N. The upper right quadrant (I) gives the relationship between N uptake and DM yield and shows to what extent additional uptake of N by the crop leads to extra DM production.

It appears from Fig. 1 that yield levels were higher for crops that had been grown with inorganic fertilizer combined with manure than for those that had been grown with inorganic fertilizer only. This specific effect of manure [9] was approximately 5, 20 and 25% in the Lelystad, Heino and Gortel experiments, respectively, as can be derived from quadrant II. Quadrant I, however, discriminates between N and non-N explanatory factors for the specific effect. From that quadrant it can be seen that in the experiment at Lelystad, N alone, either by its rate or by the way it was offered to the crop, could explain the observed yield differences. In the Heino and Gortel experiments however, additional N uptake did lead to a diminishing response in DM yields if only inorganic fertilizer had been used. In combination with manure, additional N uptake led to an increase in DM yields which points to a specific effect of manure, apart from N. In Gortel particularly, this was expressed in the N content of the crop. The increase due to fertilizer N was smaller with, than without manure (Table 7). Restricted to these non-N factors, the specific effect was roughly estimated to be 0, 5 and 10% in the Leystad, Heino and Gortel experiments, respectively. The possible causes of this specific effect will be discussed later in more detail.

As far as fertilizer N is concerned, it was concluded that without manure, the average economic optimum N rate in the Heino experiment was 75 kg N $ha^{-1} yr^{-1}$. In the last two years of the experiment, as soil N reserves

Table 7. Average N content (%) of forage maize as affected by fertilizer N and manure

Site	Fertilizer N ($kg\,ha^{-1}\,yr^{-1}$)	Manure treatment ($t\,ha^{-1}\,yr^{-1}$)		
		0	100	150
Heino (1976–1982)	0	1.14	1.40	–
	225	1.47	1.57	–
Lelystad (1976–1982)	0	1.22	–	1.34
	225	1.38	–	1.44
Gortel (1972–1982)	0	0.91	1.21	–
	250	1.36	1.48	–

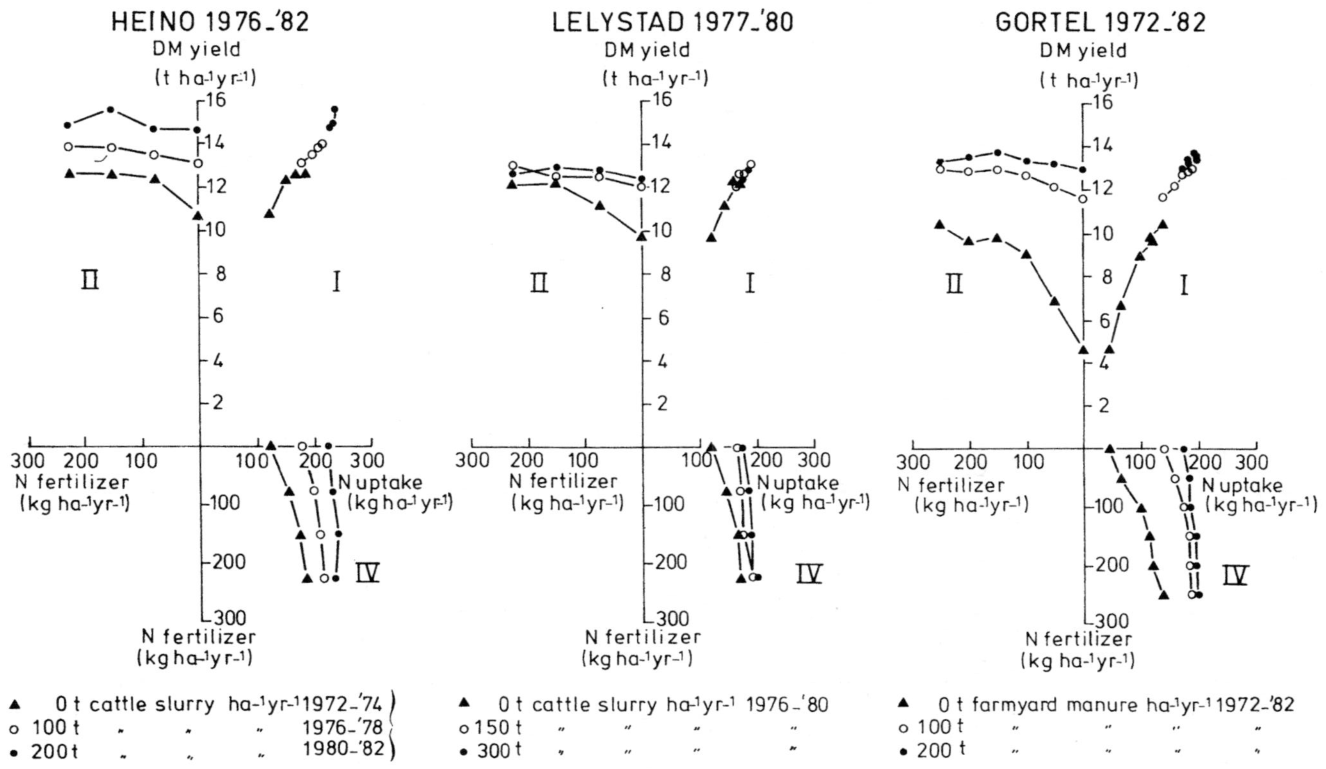

Figure 1. The effect of fertilizer N on N uptake and DM yield of forage maize at three rates of manure.

Table 8. The amount of inorganic N in the soil (kg N ha^{-1}) in spring as affected by manure; Heino 1976–1982

Soil layer (cm)		Application rate (t ha^{-1} yr^{-1})				
		0	50	100	200	300
0–20	mean	36	61	118	140	199
	(range)	(20–53)	(22–126)	(36–218)	(86–236)	(116–287)
20–60	mean	40	52	91	149	239
	(range)	(20–66)	(32–89)	(37–166)	(54–254)	(143–373)
0–60	mean	76	113	209	289	438
	(range)	(54–105)	(58–193)	(118–384)	(161–431)	(333–660)

became exhausted, the optimum N rate was 150 kg N ha^{-1} yr^{-1}, however. At manure rates of 100 and 200 t ha^{-1} yr^{-1} too, supplementary N fertilizer at a rate of 75 kg N ha^{-1} yr^{-1} was worthwhile, the necessary response of 5 kg DM per kg N being achieved. The increments of 75 kg N ha^{-1} yr^{-1}, however, were too large to assess the need for fertilizer N more precisely. From Table 8 it can be seen that this positive response may be explained by the fact that in some years the total amount of inorganic soil N was too small to meet the demands of the crop despite manure applications. This was especially the case for the upper soil layer of 20 cm. Soil N reserves in spring were not decisive for the response to fertilizer N as net mineralization during the growing period and the general yield level were determining the need for fertilizer N too [29].

In the experiment at Lelystad the highest yields without manure were obtained at N rates of 150 kg ha^{-1} yr^{-1}. Even with large amounts of manure, maize yields responded positively to fertilizer N at rates of 75 kg N ha^{-1} yr^{-1}. With these rates the required response of 5 kg DM per kg N was only just obtained.

At Gortel yields were maximum at N rates of approximately 150 kg N ha^{-1} yr^{-1} without manure. With 100 and 200 t manure ha^{-1} yr^{-1}, mineral N supplementations of 100 and 50 kg N ha^{-1} yr^{-1}, respectively, seemed to be justified.

Quality

The effects of increasing rates of manure on the quality of forage maize are given in Table 9. The Heino experiment may serve as an example, trends being similar in the other trials. Dry matter (DM) content tended to be somewhat lower at higher rates of manure. The same holds for the feeding value (NEL). Trends were weak, however. The content of N, digestible crude protein (DCP), nitrate-N (NO_3^--N), phosphorus (P), and potassium

Table 9. Dry matter content (%), net energy content (NEL), digestible crude protein (DCP) content and mineral composition (% in DM) of forage maize as affected by manure; Heino 1976–1982

	Application rate ($t\,ha^{-1}\,yr^{-1}$)			
	0	100	200	300
DM content	31.2	30.7	30.5	30.3
NEL*	971	955	954	948
N	1.14	1.40	1.54	1.55
DCP	4.0	5.5	6.4	6.4
NO_3^--N	< 0.01	0.12	0.14	0.16
P	0.17	0.24	0.28	0.32
K	1.39	1.82	1.90	2.21
Mg	0.11	0.13	0.12	0.12

* Net Energy for Lactation in VEM per kg DM; 1 VEM = 6.9 kJ

(K) increased with increasing rates of manure, whereas the magnesium (Mg) concentration hardly changed. Health problems caused by too high nitrate contents or poor K/Mg balances would not be probable [3]. Leaves and stalks responded more markedly than ears to increasing rates of manure as can be seen from Table 10. In the stover there was a distinct increase in the nitrate and K content, whereas these were hardly affected in the ear. If ear development is impeded e.g. by drought and its proportion is far below 50%, the composition of the stover should be taken into account.

Table 10. Mineral composition (% in DM) of stalks and leaves, ears and whole plant of forage maize as affected by fertilizer N and manure; Gortel 1972–1982

Mineral	Crop parts	Treatment				
		Manure	0	200	200	($t\,ha^{-1}\,yr^{-1}$)
		Fertilizer N	250	0	250	($kg\,ha^{-1}\,yr^{-1}$)
N	stalks and leaves		1.17	1.14	1.37	
	ears		1.55	1.55	1.65	
	whole plant		1.36	1.34	1.52	
NO_3^--N	stalks and leaves		0.06	0.07	0.17	
	ears		< 0.01	< 0.01	< 0.01	
	whole plant		0.03	0.04	0.09	
P	stalks and leaves		0.11	0.26	0.28	
	ears		0.26	0.34	0.34	
	whole plant		0.18	0.30	0.31	
K	stalks and leaves		1.71	2.60	2.72	
	ears		0.58	0.60	0.60	
	whole plant		1.16	1.66	1.73	
Mg	stalks and leaves		0.22	0.17	0.17	
	ears		0.12	0.12	0.13	
	whole plant		0.17	0.15	0.15	

Table 11. Average surplus of N (kg N $ha^{-1} yr^{-1}$; input with manure minus uptake by maize)

Site	Period	Application rate (t $ha^{-1} yr^{-1}$)						
		0	50	100	150	200	250	300
Heino	1976–1982	−124	73	313	540	752	979	1194
Maarheeze	1974–1982	–	105	340	577	814	1069	1321
Lelystad	1976–1980	−134	–	–	415	–	–	994
Gortel	1972–1982	−43	173	413	663	932	–	–

The fate of mineral surplusses

At higher manure rates, the offtake of nutrients by the crop was greatly exceeded by the nutrient input in spite of the higher yields and contents of the harvested plant material. The annual N surplus is given in Table 11. Large differences between the input and offtake of minerals, raise the question of the fate of these surplusses.

As a consequence of annual treatments of 300 t slurry ha^{-1}, the organic matter content of the top soil layer (0–20 cm) increased by 0.8% for both the Heino and Maarheeze trials; in the 20–60 cm layer the increase was 0.5%. For lower slurry rates the increases were proportional. In the Gortel trial the annual application of 200 t manure ha^{-1}, increased the organic matter content of the top soil from about 4% up to about 7% and the total N content from 0.11 to 0.21%. The total N content of the soil at Heino and Maarheeze was also raised as shown in Table 12. After seven years the total soil N reserves in the top soils of these two experiments had been augmented by approximately 600 kg N ha^{-1} for the lower slurry rates and 1200 kg N ha^{-1} for the higher rates. The increases in total soil N reserves that took place in the 20–60 cm layer were similar. These retentions only account for a part of the excess of more than 7000 kg N ha^{-1} at the highest slurry rate. Because half of the slurry N is water soluble and part of the organic N is easily decomposable into inorganic N [15], this indicates that substantial losses by ammonia (NH_3) volatilization, denitrification or leaching must have occurred.

Table 12. Total N in soil (%, 0–20 cm layer) as affected by annual slurry applications

Site	Year	Application rate (t $ha^{-1} yr^{-1}$)						
		0	50	100	150	200	250	300
Heino	1975	0.14	0.14	0.15	0.15	0.13	0.12	0.13
	1982	0.14	0.16	0.17	0.19	0.18	0.18	0.18
Maarheeze	1975	–	0.11	0.12	0.11	0.11	0.11	0.11
	1982	–	0.13	0.13	0.14	0.16	0.15	0.15

Table 13. The amounts of inorganic soil N (kg N ha^{-1}, 0–60 cm layer) in autumn as affected by slurry applications; Maarheeze

Application rate t ha^{-1} yr^{-1}	Years							
	1975	1976	1977	1978	1979	1980	1981	1982
50	149	164	102	97	110	77	64	118
100	122	200	86	97	123	99	106	170
200	166	359	118	189	202	164	143	342
300	230	562	182	240	262	289	211	317
precipitation between sampling dates (mm)	424	773	671	660	941	700	665	

The amounts of inorganic N in the soil as assessed after harvest in autumn, reflected the rate of slurry but, apart from annual variations, there was no increase in this accumulation of inorganic N over the years. This is demonstrated in Table 13 with data from the Maarheeze trial which is representative for the other trials. Periodic soil sampling [29] revealed that N losses occurred partly during the growing season. During winter leaching played an important role as demonstrated with data from the Maarheeze trial (Table 14). Losses by leaching increased at higher manure applications. The relative losses, however, were almost constant. Phosphorus surplusses accumulated in the soil, slowly moving downward. At the end of the experiments P enrichment could be detected at depths of 40 cm or more [29].

As expected, K was even more mobile. For this element a certain surplus was necessary to keep soil fertility at the required level (8 mg K_2O per 100 g air dry soil for the trials on sandy soil and 16 mg K_2O per 100 g for the trial on clay soil [2]). At Lelystad a surplus was not really needed because marine clays have a high natural K supply. In Heino, however, it was difficult to maintain the initial status at the 50 and 100 t manure rate despite the annual surplus of 65 and 250 kg K ha^{-1}, respectively (Fig. 2). This could probably

Table 14. Average leaching of N (kg N ha^{-1} yr^{-1}) as affected by slurry applications; Maarheeze 1977–1982

	Application rate (t ha^{-1} yr^{-1})						
	(0)	50**	100	150	200	250	300
N leached*** in kg N ha^{-1} yr^{-1}	(51)*	145	150	230	316	406	430
N leached as % of manure N + fertilizer N	–	36	19	23	25	27	24

* calculated with regression analysis
** see Table 3 for supplementary fertilizers
*** average precipitation surplus 309 mm per year

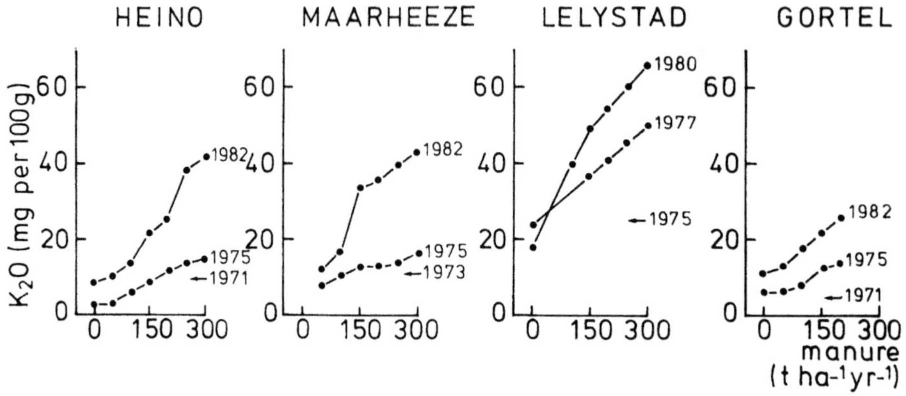

Figure 2. K level (mg K_2O per 100 g) in upper soil layer (0–20 cm) as affected by manure (← = initial value at the beginning of the experiments).

be explained by leaching losses and by the growing of grass in 1975 and 1979 when no supplementary K fertilizer was offered. In the experiment at Gortel the poor K status became visible in the lower K content of the maize which received only mineral N, P and K (Table 15). In the Heino trial, such a decrease did not occur.

For the Heino and Maarheeze trials, soil pH values were gradually lowered during the experiments. Manure applications could only partly alleviate this decrease. As a consequence, it was felt necessary to lime the Heino site in the spring of 1981. Such a decrease did not occur at Lelystad where soils have a high calcium carbonate content and in the Gortel trial which was limed annually from 1973 (Fig. 3).

Residual effect on grass

The residual effect of manure applications has been assessed with grass. Fig. 4 shows the results for the Heino, Maarheeze and Gortel trials. Even the

Table 15. Average K content (%) of forage maize as affected by fertilizer N and manure

Site	Fertilizer N ($kg\,ha^{-1}\,yr^{-1}$)	Manure treatment ($t\,ha^{-1}\,yr^{-1}$)		
		0	150	200
Heino (1976–1982)	0	1.39	–	1.90
	225	1.43	–	–
Lelystad (1977–1980)	0	1.36	1.68	–
	225	1.45	1.67	–
Gortel (1972–1982)	0	1.63	–	1.66
	250	1.16	–	1.73

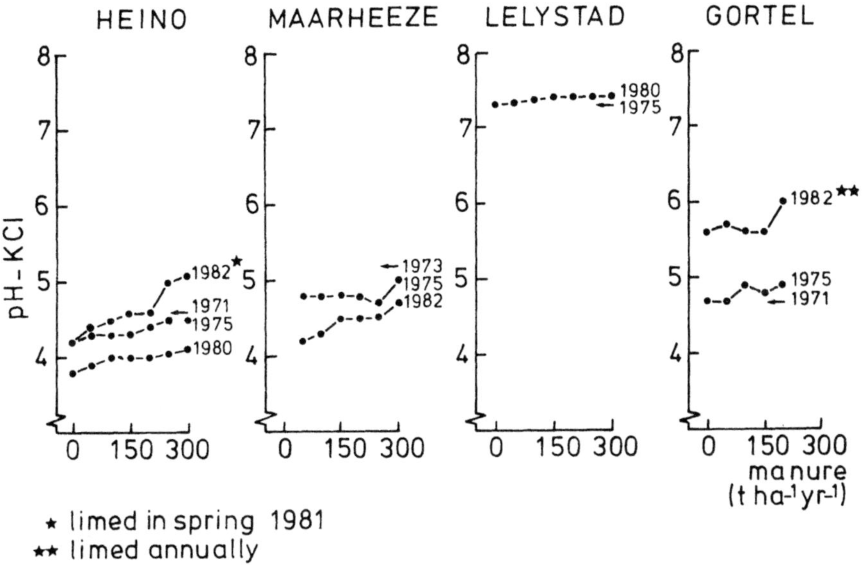

Figure 3. pH-KCl values in upper soil layer (0–20 cm) as affected by manure (← = initial value at the beginning of the experiments).

highest rates of fertilizer N were insufficient for maximum grass yields (quadrant II), due to excessive rainfall in the spring of 1983.

As far as residual effects are concerned, grass in the Heino trial recovered almost 6 kg N $ha^{-1} yr^{-1}$ per increment of 10 t slurry (range 50–250 t $ha^{-1} yr^{-1}$) after 9 applications in the preceding 11 years. At Maarheeze grass hardly absorbed 4 kg N $ha^{-1} yr^{-1}$ per increment of 10 t slurry after this slurry (range 50–200 t $ha^{-1} yr^{-1}$) had been applied for 9 consecutive years. In the Gortel trial grass absorbed 4 kg N $ha^{-1} yr^{-1}$ per increment of 10 t farmyard manure (range 0–200 t $ha^{-1} yr^{-1}$) after 11 years of annual application. Grass in the Heino, Maarheeze and Gortel trials took up 1.3, 0.8 and 0.7%, respectively, of the total input of manure N during the experimental period.

Grass yields were higher where manure rates had been higher in preceding years. According to the results this was mainly caused by an increased N uptake (quadrant IV). At high fertilizer N levels on grass (quadrant I), the absorbed N is somewhat more efficiently used for DM production the higher the preceding manure rate. As with maize, grass benefitted from a specific effect of the manure. This effect was estimated to range from 6% (Maarheeze) to 10% (Heino and Gortel).

Residual effects were also expressed in the composition of the harvested grass. This was especially demonstrated in the early cuts. Results from the Heino trial serve as an example (Table 16). At N fertilizer rates of 320 and

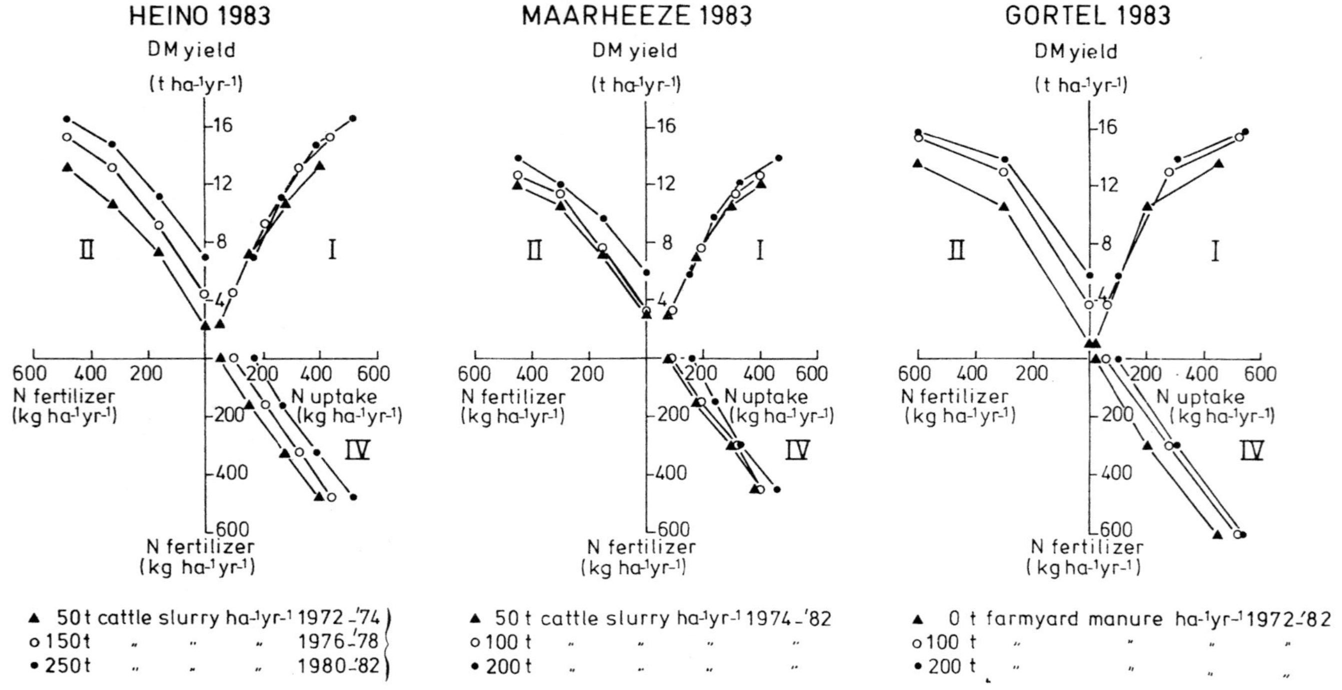

Figure 4. Residual effect of manure (applied on maize) on N uptake and DM yield of grass.

Table 16. Mineral composition of grass (% in DM), second cut, after 9 years of manure application on preceding maize crops; Heino 1983

Rate of manure on maize ($t\,ha^{-1}\,yr^{-1}$)	Rate of fertilizer N on grass ($kg\ N\ ha^{-1}\,yr^{-1}$)	Mineral			
		K	Mg	N	NO_3^--N
50	0	3.02	0.16	2.29	< 0.01
	160	3.58	0.22	3.09	0.01
	320	3.69	0.25	4.00	0.16
	480	3.73	0.24	4.46	0.36
150	0	3.42	0.21	2.45	< 0.01
	160	3.96	0.24	3.25	0.02
	320	4.43	0.28	4.18	0.23
	480	4.37	0.27	4.59	0.42
250	0	3.97	0.21	2.61	< 0.01
	160	4.71	0.26	3.63	0.05
	320	4.93	0.29	4.51	0.30
	480	5.16	0.29	5.14	0.54

480 kg N $ha^{-1}\,yr^{-1}$, problems of grass tetany and nitrate poisoning [3] could be expected regardless of the manure given in preceding years. Only at the highest rate of slurry (250 t $ha^{-1}\,yr^{-1}$) there would already be some risk of grass tetany at a fertilizer rate of 160 kg N $ha^{-1}\,yr^{-1}$.

Discussion

The N demand of forage maize under the conditions of North-West Europe will vary between 150 and 250 kg N $ha^{-1}\,yr^{-1}$ depending on yield level [4, 7, 8, 13, 19]. These demands can be met by animal manure provided that volatilization losses are minimized [16, 21, 22, 25]. Opinions about the effects of excessive amounts of animal waste on maize yields, are conflicting. Some authors report adverse effects at higher rates [7], especially when these treatments are repeated over several years [17]. Others found a positive relationship between manure rates and maize yields even when rates of more than 1500 kg N $ha^{-1}\,yr^{-1}$ were applied during 10 consecutive years or more [20].

The trials presented in this paper showed that even on a long term, excessive amounts of animal waste favoured maize yields. Remarkably, even at rates of 100–200 t $ha^{-1}\,yr^{-1}$, moderate amounts of additional fertilizer N were still beneficial for DM yields. This can probably be attributed to uncertainties linked with the use of manure, its effectivity being strongly affected by weather conditions during and after land spreading. Losses by volatilization of ammonia and denitrification and leaching of nitrate, and also the rates of mineralization and nitrification, will largely determine the ultimate effect of the use of manure. These uncertainties also played a role in the present trials. In one of the trials with cattle slurry for example, 14–25% of the manure N that had been applied in the preceding months, was detected in spring as inorganic soil N in the 0–60 cm layer (Table 8). Assuming that 50% of the total N content in cattle slurry is inorganic [12], it can be concluded that considerable losses had occurred. However, the possibility that some inorganic N had been immobilized temporarily [27] can not be excluded.

During the first weeks after sowing, maize derives its nutrients from a rather shallow soil layer which may contain insufficient N, due to leaching. During this stage additional mineral N dressings (broadcast or row placement), can assure unhampered initial growth [13, 19]. This may be a further explanation for the positive response of maize to fertilizer N, even if amply manured.

In the present trials, maize grown with a combination of organic and mineral fertilizer, outyielded maize with only mineral N, P and K. This specific effect of organic fertilizers [9] was more pronounced on sandy than on clay soil (Fig. 1). It should be asked to what extent other nutrients were limiting.

In the trials on sandy soils the supply of K might have been inadequate (Fig. 2). In the Heino trial this possibility was supported by a low K content of the grass in 1983. When preceded by maize with an annual rate of 50 t slurry ha^{-1}, the grass of the last 3 cuts had an average K content of 2.18%. It decreased from 2.68 to 1.78% with increasing rates of fertilizer N on grass. In general these contents are considered as sub-optimal for grass production [3]. It is not clear whether the K contents as observed in maize (Table 15), were also too low for optimum growth. Critical values from the literature mostly refer to specific plant parts at younger crop stages [e.g. 10, 26]. A K content of 1.5–1.6% is considered as typical for whole plant forage maize at maturity [1, 13]. From this point of view, K may have been a growth limiting factor especially in the Gortel trial.

The increasing effect on soil pH (Fig. 3) and the slow release of inorganic N linked with repeated applications of manure, may partly explain the specific effect. Finally, manure had a substantial effect on the organic matter content of the sandy soils, thus improving the water holding capacity. On the marine clay soil at Lelystad with a high natural K supply, a high pH value, a permanent groundwater table (150 cm) and a high capillary conductivity, specific effects were small. This smaller effect of manure on heavier soils has been experienced elsewhere [23].

Maize exhibited a low recovery of the N that had been put in either as inorganic N or as manure N (Table 17). The recoveries were lower as N rates increased. This was partly due to high natural N mineralization rates: maize on unfertilized plots at Heino, Lelystad and Gortel had average uptakes of 124, 134 and 43 kg N $ha^{-1} yr^{-1}$, respectively. Recovery of manure N was worse than that of fertilizer N because only part of the N in manure is inorganic. The relatively high recovery at the 50 t $ha^{-1} yr^{-1}$ rate at Heino and Maarheeze, may therefore not only result from its late spring application (Table 2), but probably also from the additional fertilizer N that had been given during the first years (Table 3). The poor recovery of manure N can also be explained by losses due to the methods and timing of applications. At Gortel where the manure was not immediately incorporated, volatilization probably played a role; in the other trials on sandy soils, leaching too evidently played its part.

It was remarkable that an almost constant fraction of the manure N (approximately 25%) was lost by leaching despite differences in N surplus and differences in application dates. Possibly, denitrification losses increased more than proportionally at higher rates of manure [14]. The observed leaching at the 50 t $ha^{-1} yr^{-1}$ rate may partly have resulted from additional fertilizer N (Table 14).

A part of the N surplus was preserved in the organic matter. For the highest manure rate the retention was about 30% of the total N input at Gortel and Heino and about 20% at Maarheeze. For the lowest manure rates it was calculated that almost 70% was retained in the organic matter. An N balance however [29] does not agree with such large retentions in

Table 17. Apparent recovery (%) of manure N (without fertilizer N) and fertilizer N (without manure N) by forage maize

Site	Period	Application rate of manure (t $ha^{-1} yr^{-1}$)				
		50*	100	150	200	300
Heino	1976–1982	17	12	11	10	7
Maarheeze	1974–1982	(15–31)	(11–19)	(9–15)	(9–13)	(6–9)**
Lelystad	1977–1980	–	–	7	–	5
Gortel	1972–1982	22	18	15	12	–

Site	Period	Application rate of fertilizer N (kg N $ha^{-1} yr^{-1}$)						
		50	75	100	150	200	225	250
Heino	1976–1982		44		35		28	
Lelystad	1977–1980		35		30		22	
Gortel	1972–1982	44		55	48	40		39

* see Table 3 for supplementary fertilizers

** estimated natural N mineralization of 110 or 70 kg N $ha^{-1} yr^{-1}$, respectively organic matter, unless N transport between plots has taken place. Given the background of a very large natural organic N pool, the accuracy of measured changes can be questioned at the lower manure rates.

Organic and mineral fertilizers only had a minor effect on forage maize quality. The effect on leaves and stalks was quite clear, but that on ears was negligible. Delayed maturation [7, 19] scarcely occurred, although in the present trials only two varieties were used. In agreement with research elsewhere [7, 11, 20], the risks of nitrate poisoning or hypomagnesaemic tetany were both insignificant [3].

Manure can affect crop yields some years after application [17, 24, 33]. In the present trials grass benefitted from the manure given to maize in preceding years. As experienced elsewhere [20], the mineral composition of grass responded more than that of maize to differences in soil fertility. To avoid nitrate poisoning and tetany [3], residual effects should be taken into account. The residual effect on grass was not only attributable to an increased uptake of N, but resulted as well from a specific effect of the manure (Fig. 4).

To avoid accumulation of plant nutrients in the soil [5, 6, 11, 20, 24] and increased emissions to the environment, the nutritional demands of a crop, including some inevitable losses by fixation and leaching, should not be exceeded by the inputs of fertilizer. The use of manure therefore requires a careful and accurate evaluation of its nutritional ability. The recovery of manure N by the maize crop can also be improved by the immediate incorporation of manure into the soil, thus minimizing volatilization losses. Furthermore, recovery can be improved by avoiding leaching risks; land spreading should therefore be postponed to the spring or, if the latter is impossible, leaching can be minimized by growing catch crops [28, 30, 32]. Where maize is cropped continuously, however, its late harvest restricts successful catch crops. Nitrification inhibitors may then be an alternative measure [18]; more research is needed though. Nitrogen efficiency is also determined by other growth factors; additional mineral fertilizers should be regarded as tools to establish these optimal growth conditions as far as nutrients are concerned. They can compensate for uncertainties and assure an unimpeded initital growth.

Reported results originate from joint research at the following institutes:

- Research Station for Arable Farming and Field Production of Vegetables, Lelystad (B A ten Hag, H M G van der Werf and J Schröder).
- Research Station for Cattle, Sheep and Horse Husbandry, Lelystad (G Krist, J Dapper and H Everts).
- Institute for Soil Fertility, Haren (L C N de la Lande Cremer).
- Netherlands Fertilizer Institute, Den Haag (K Dilz, A Brak and J Postmus).

– Institute for Land and Water Management Research, Wageningen (J H A M Steenvoorden and H P Oosterom).

References

1. Anonymus (1973) Mineralstoffgehalte in Futtermitteln. Frankfurt: DLG Verlag
2. Anonymus (1984) Adviesbasis voor de bemesting van landbouwgronden. Wageningen: Consulentschap voor Bodemaangelegenheden in de Landbouw
3. Anonymus (1984) Handboek voor de rundveehouderij. Lelystad: Proefstation voor de Rundveehouderij, Schapenhouderij en Paardenhouderij
4. Anonymus (1985) Gülledüngung zu Mais – umweltgerecht. Bonn: Deutsches Maiskomitee e.V.
5. Antoun H, Visser S A, Cescas M P and Joyal P (1985) Effect of liquid hog manure application rates on silage corn yield and nutrient uptake. Canadian Journal of Plant Science 65 (1), 63–70
6. Asmus F, Görlitz H and Klocke M (1982) Wirkung langjähriger hoher Güllegaben auf Pflanzen und Boden in Futterbau-Fruchtfolgen. Archiv für Acker- und Pflanzenbau und Bodenkunde (Berlin) 26 (11), 725–732
7. Borstel U von and Nitsch A (1984) Stickstoffsteigerungsversuch zu Silomais, N-min Methode in Verbindung mit Gülledüngung. Mais 84/1, 22–23
8. Buchner A and Sturm H (1985) Gezielter düngen - intensiv, wirtschaftlich, umweltbezogen, pp 247–254. Frankfurt: DLG Verlag
9. Cooke G W (1977) The roles of organic manures and organic matter in managing soils for higher crop yields. Proceedings international Seminar on Soil, environment and fertility management in intensive agriculture, pp 53–64. Tokyo: The Society of the Science of Soil and Manure
10. Cornforth I S and Steele K W (1981) Interpretation of maize leaf analyses in New Zealand. New Zealand Journal of Experimental Agriculture 9, 91–96
11. Culley J L B, Phillips P A, Hore F L and Patni N K (1981) Soil chemical properties and removal of nutrients by corn resulting from different rates and timing of dairy manure applications. Canadian Journal of Soil Science 61 (1), 35–46
12. Dijk T A van and Sturm H (1983) Fertiliser value of animal manures on the continent. Proceedings No. 220, pp 45. London: The Fertiliser Society
13. Dilz K, Hag B A ten, Lammers H W and Lande Cremer L C N de la (1984) Fertilization of forage maize in the Netherlands. Netherlands Nitrogen Technical Bulletin nr. 14, pp 25. Den Haag: Nederlands Meststoffen Instituut
14. Dowdell R J (1981) Denitrification in soils treated with animal slurry. In Brogan J C, ed. Nitrogen losses and surface run-off from landspreading of manures, pp 397–408. Den Haag: Martinus Nijhoff/Dr Junk
15. Faassen H G van and Dijk H van (1987) Manure as a source of nitrogen and phosphorus in soils (this volume)
16. Freney J R, Simpson J R and Denmead O T (1983) Volatilization of ammonia. In Freney J R and Simpson J R, eds. Gaseous loss of nitrogen from plant-soil systems, pp 1–32. Den Haag: Martinus Nijhoff/Dr Junk
17. Görlitz H, Herrman V and Jauert R (1985) Ertrag und Nährstoffnutzung nach ein- und mehrjährigen hohen Güllegaben zu Silomais sowie ihre Nachwirkung auf sandigen Böden. Archiv für Acker- und Pflanzenbau und Bodenkunde (Berlin) 29 (1), 55–60
18. Gutser R und Amberger A (1985) Nitratauswaschung nach Gülledüngung mit Didinzusatz. Landwirtschaftliche Forschung Kongressband 1984, 137–145

19. Hepting L (1985) Mais macht keine Unterschiede. DLG-Mitteilungen 7, 389–391
20. Ito Y and Miyazawa K (1984) Effect of long-term heavy application of fresh farmyard manure on yield and nutrient status of forage crops. Japan Agricultural Research Quarterly 17 (4), 242–247
21. Klausner S D and Guest R W (1981) Influence of NH_3 conservation from dairy manure on the yield of corn. Agronomy Journal 73, 720–723
22. Kolenbrander G J (1981) Effect of injection of animal waste on ammonia losses by volatilisation on arable land and grassland. In Brogan J C, ed. Nitrogen losses and surface run-off from landspreading of manures, pp 425–430. Den Haag: Martinus Nijhoff/Dr Junk
23. Kürten P W and Saalbach E (1961). Über den Einfluss von Stallmist auf die Wirkung der Stickstoffdüngung. Zeitschrift für Acker- und Pflanzenbau 114, 23–31
24. Lund Z F and Doss B D (1980) Residual effects of dairy cattle manure on plant growth and soil properties. Agronomy Journal 72, 123–130
25. Meincke J (1985) Kvæggylle til majs i vækstperioden. Tidsskrift for Planteavl 89, 25–29
26. Melsted S W, Motto H L and Peck T R (1969) Critical plant nutrient composition values useful in interpreting plant analysis data. Agronomy Journal 61, 17–20
27. Neeteson J J, Greenwood D J and Habets E J M H (1986) Dependence of soil mineral N on N-fertilizer application. Plant and Soil 91 (3), 417–420
28. Nielsen N E and Jensen H E (1985) Soil mineral nitrogen as affected by undersown catch crops. Proceedings symposium on assessment of nitrogen fertilizer requirement, pp 101–110. Haren: Instituut voor Bodemvruchtbaarheid
29. Schröder J (1985) De invloed van grote giften runderdrijfmest op de groei, opbrengst en kwaliteit van snijmais en op de bodemvruchtbaarheid en waterverontreiniging. Verslagen nrs. 30–32, 353 pp. Lelystad: Proefstation voor de Akkerbouw en de Groenteteelt in de Vollegrond
30. Steffens G and Vetter H (1984) Stickstoffverlagerung nach Gülledüngung mit und ohne Zwischenfruchtanbau. Landwirtschaftliche Forschung Sonderheft 40, 354–362
31. Wadman W, Sluijsmans C M J and Lande Cremer L C N de la (1987) Value of animal manures: changes in perception (this volume)
32. Wantulla A, Vollmer F J and Kühbauch W (1986) Ganzjährige Beobachtung des Stickstoffantrages unter begülltem Mais. Vortragsreihe der 39. Hochschultagung der Landwirtschaftliche Fakultät, pp 57–72. Bonn: Universität Bonn
33. Werner W, Scherer H W and Drescher D (1985) Untersuchungen über den Einfluss langjähriger Gülledüngung auf N-Fraktionen und N Nachlieferung des Bodens. Zeitschrift für Acker- und Pflanzenbau 155, 137–144

Utilization of nitrogen from slurry applied to fodder crops. Research report from the Nordic countries, Denmark, Finland and Sweden

G SIMAN[a], A DAM KOFOED[b], K E LARSEN[b], E KEMPPAINEN[c] and S STEINECK[a]

[a] *Dept. of Soil Sciences, Swedish University of Agricultural Sciences, Box 7014, S-750 07 Uppsala, Sweden*

[b] *Askov Experimental Station, DK-6600 Vejen, Denmark*

[c] *Dept. of Agricultural Chemistry and Physics, Agricultural Research Centre, SF-31 600 Jokioinen, Finland*

Key words: slurry, nitrogen, slurry injection, surface spreading, response of fodder crops

Abstract. Results from a recently started long-term Nordic slurry experiment, consisting of seven field trials, located at four places are reported. The aim of the project is to study how slurry from intensive pig and dairy cattle holdings can be utilized in crop production in an economic and environmentally acceptable manner. The applications correspond to a slurry production of 0, 1, 2 and 4 dairy cow units $ha^{-1} yr^{-1}$, *i.e.* about 0, 50, 100 and 200 kg NH_4^+-N $ha^{-1} yr^{-1}$ in slurry. In the farming system adapted to pig husbandry, spring barley is used in monoculture. In milk production, two different rotations are used: *alternative 1* in Finland and Sweden, 1) barley, 2) ley I, 3) ley II; and *alternative 2* in Denmark, 1) barley, 2) grass, 3) fodder beet. Slurry is applied prior to sowing of barley and fodder beet and at the onset of growth and after first cuts of the swards of grass and ley. Comparisons are made between injected and surface-spread slurry and also of the effect of NH_4^+-N in slurry with calcium ammonium nitrate nitrogen (CAN-N).

The largest yields in all crops were obtained at rates of 200 kg NH_4^+-N $ha^{-1} yr^{-1}$ in either pig slurry or cattle slurry, apart from one trial with barley. The response to NH_4^+-N in slurry compared with CAN-N varied from 78 to 106% in barley, from 79 to 81% in grass, from 14 to 34% in leys and from 85 to 103% in fodder beet (roots). Injection at the second rate of applied slurry gave 14% yield increases of barley grain, 12% of grass and 20% of fodder beet roots in comparison with surface-spread slurry. In leys, injection gave yield increases at two places and reductions at one place, the latter resulting from mechanical damage caused by the tines of the injector.

Introduction

From the late 1940s, increasingly rapid economic and social developments have led to radical changes in agriculture towards enterprises specializing either on crops for sale or on animal production. With specialization the natural recirculation of nutrients for plant production was broken. Plant

H.G. v.d. Meer, et al. (eds.), Animal Manure on Grassland and Fodder Crops.
 ISBN 978-94-010-8137-5.

production without livestock became strongly dependent on the use of fertilizers, whereas farms with large numbers of animals per unit of area encountered problems in finding ways of using their manure.

With the specialization in animal production there has been also a change in the methods of handling the manure. The larger pig and dairy cow holdings especially have shifted towards slurry systems. At present 50, 19 and 35% of manure is handled as slurry in Denmark, Finland and Sweden, respectively.

The total nutrient content in the faeces and urine produced annually, and the average amount of plant nutrients expressed in $kg\,ha^{-1}$ are given in Table 1 [4, 6, 16, 17]. The table shows that the average load on agricultural land is moderate even in Denmark, the Nordic country with the densest population of animals. The problem is that animal production has become concentrated on fewer farms which has resulted in a high number of animals per unit area.

As in many other countries, the total number of animals in relation to the area available for spreading is restricted. For some years now, a threshold value of two animal units per ha has been used as the maximum recommended animal density when giving planning permission for animal producing enterprises [13, 14]. In these guidelines one animal unit corresponds to 1 dairy cow, 2 young cattle, 10 pig places or 67 layers. The guideline is based principally on the amount of nitrogen (N) in the manure. Two animal units per ha correspond to a production of 150–200 kg total N or about 100 kg ammonium-N (NH_4^+-N) $ha^{-1}\,yr^{-1}$. It should be possible to use such an amount of N, together with the other nutrients in manure, in an acceptable manner in plant production. However, the manure is annually applied only to a certain part of the total spreading area on the farm. This implies that the rates per ha on individual fields will be very high. Other problems arise when, as is often the case, the manure is spread at unsuitable times of the year and when the spreading is uneven which may lead to losses of plant nutrients. Losses of N during handling, storage, spreading and in the field not only cause environmental problems but also lead to uncertainty about

Table 1. Annual production of N, P and K in faeces and urine from cattle, pigs and poultry in the three Nordic countries

Country	In 1000 tons			In kg per ha agricultural land		
	N	P	K	N	P	K
Denmark	310	70	230	106	24	79
Finland	102	20	93	42	8	40
Sweden	95	18	101	32	6	34

the nutrient effect of the manure [1, 2, 10, 20, 21]. This strongly contributes to the prevailing lack of interest in farmyard manure as a source of nutrients.

In 1983 a working group was appointed by the Nordic Association of Agricultural Scientists (NJF) to study how large amounts of animal manures can satisfactorily be used in crop production. Since the greatest problem is found on farms with intensive animal production using slurry handling techniques, the main effort has concerned them. The project is based on seven field experiments and started in 1984. It aims to gain insight into the agronomic aspects of application of large amounts of slurry in crop production.

The objectives of this research are to study:

1. Possibilities of high plant production using only slurry as source of nutrients.
2. The highest annual amount of slurry without leading to negative effects on yield and quality or on the environment.
3. The effect of N, phosphorus (P) and potassium (K) in slurry compared to inorganic fertilizer.
4. Injection compared to surface spreading.
5. Spreading of slurry in spring compared to autumn.
6. The long-term effect on content of soil nutrients and nutrient balance.
7. Suitable cropping plans on farms with intensive animal production.

Materials and methods

Seven long-term field trials were started jointly by the Askov Experimental Station, Denmark, The Agricultural Research Centre, Jokioinen, Finland and the Swedish University of Agricultural Sciences, Uppsala, Sweden, on four different sites – Askov, Jokioinen, Uppsala and Everöd. The locations of the sites are shown in Fig. 1. The project includes two types of farm management at each experimental site (at Everöd only milk production). One type concerns intensive milk production and the other intensive pig production.

Crop rotations

Pig holdings. The crop consists of spring barley in monoculture. The aim is to produce feed for use on the farm and to provide an area for spreading the slurry produced.

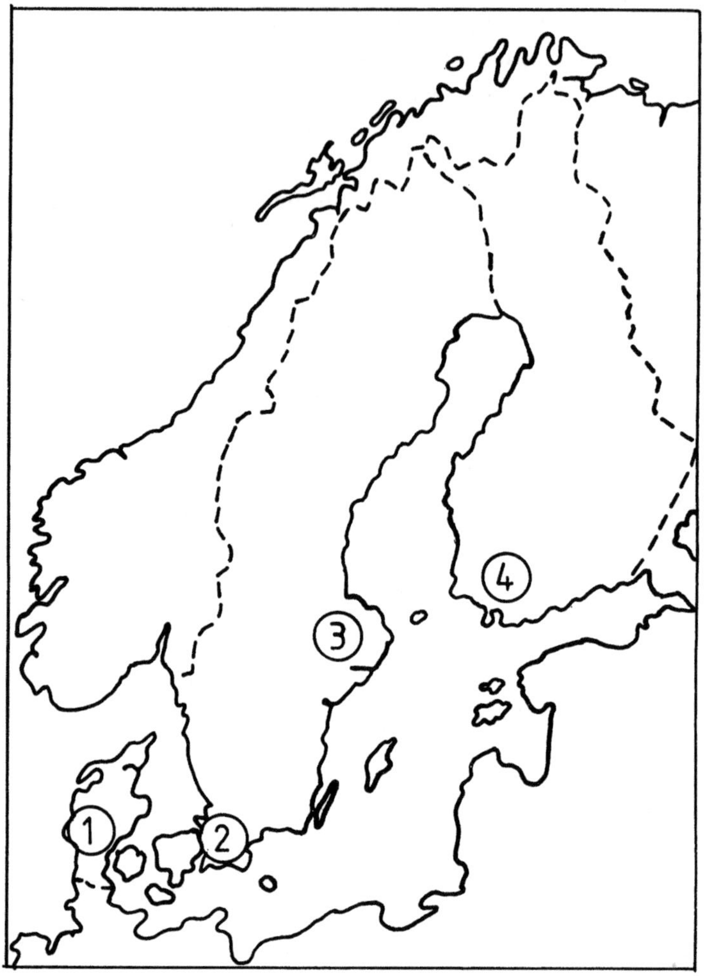

Figure 1. Location of the experimental sites. 1. Askov, 2. Everöd, 3. Uppsala, 4. Jokioinen.

Dairy holdings. Two different crop rotations are studied because of different climate and local practices in the areas where the trials are located. In Finland and Sweden the crop rotation consists of 1) spring barley, 2) ley I, and 3) ley II, and in Denmark of 1) spring barley, 2) grass, and 3) fodder beet. Each crop included in the rotation is grown annually.

Treatments

The treatments in both crop rotations consisted of an untreated control, autumn applied slurry, spring/summer applied slurry, injected slurry, surface-spread slurry, one or three rates of slurry and four (at Askov only three) rates of fertilizer. The rates of slurry in the spring/summer application treatments are comparable to an annual production from 1, 2 and 4 livestock units per ha. The slurry doses are based on the NH_4^+-N content. On the assumption that the NH_4^+-N in slurry produced annually by one livestock unit is 50 kg, the slurry rate will be 50, 100 and 200 kg NH_4^+-N $ha^{-1} yr^{-1}$, respectively. The autumn applied slurry treatment consisted only of the second slurry rate, comparable to 100 kg NH_4^+-N $ha^{-1} yr^{-1}$. The fertilizer rates are adjusted to provide the same amounts of N, P and K as NH_4^+-N, P and K in the corresponding slurry rates. Calcium ammonium nitrate (CAN), superphosphate and muriate of potash are used as fertilizers. In spring/summer applied slurry treatments, the slurry added to the ley and grass is split into two (Finland and Sweden) or three (Denmark) applications, one in the spring before the onset of growth and one after the first cut (and second cut in Denmark). When the crops are spring barley and fodder beet the slurry is applied to the seedbed before sowing. At Askov the slurry is injected into the plots at 15 cm depth and at Jokioinen, Uppsala and Everöd it is spread in bands on the soil surface using a spreader-mounted injector system consisting of injector tines spaced at 50 cm (at Everöd and Uppsala 40 cm) from each other [18]. The surface-spread slurry applied before sowing of barley and fodder beet is immediately followed by harrowing. In addition, both injected and surface-spread treatments are included on all the sites at a slurry rate of 100 kg NH_4^+-N $ha^{-1} yr^{-1}$. In autumn applied slurry treatments, the slurry is injected at Askov and spread in bands onto the soil surface at Jokioinen, Uppsala and Everöd. This is followed by ploughing into the soil on the same day when the crop in the next year is planned to be barley.

Prior to each application the stored slurry is agitated. Two slurry samples are taken from each load applied. One of them is immediately analysed for NH_4^+-N at the farm using the Agros Nitrogen Meter [5]. The NH_4^+-N values are used to adjust the flow-meter on the spreading equipment to spread a defined amount of slurry. Further analyses of dry matter (DM), ash and plant nutrient contents are carried out on another sample taken from slurry which had actually been applied. No statistically significant difference was found between Agros Nitrogen Meter and the official laboratory method in measuring NH_4^+-N and easily oxidized organic N in slurry. The accuracy of the flow-meter is stated to be $\pm 0.5\%$ (Fischer & Porter, Mod. 10D 1420, NW 80).

The application of slurry began during the spring of 1985.

Some basic data on the soils at each location are given in Table 2. The duration of the trials is intended to be long-term. Some results from the first experimental years, 1985–86, are given below.

Table 2. Analysis of soil, 0–20 cm depth, from experimental sites. P, K and Mg values in mg per 100 g air-dried soil

Site	Texture	Org. C (%)	pH (H_2O)	P-AL*	K-AL*	Mg-AL*
Askov	Sandy clay loam	1.5	6.2	5.8	7.1	3.9
Jokioinen	Silty clay	3.0	6.1	13.4	27.0	53.0
Uppsala	Silty clay loam	3.0	6.3	6.3	8.3	12.7
Everöd	Loamy sand	1.2	5.8	12.3	7.4	3.1

* according to Egnér *et al.* [3]

Results and discussion

Analysis of slurries

The composition of pig and cattle slurries used in the experiments is given in Table 3. Variations in DM and nutrient contents of slurries applied at different times were large. On average, the NH_4^+-N content of pig and cattle slurries were 0.30% and 0.18%, respectively. The target rates supplying 50, 100 and 200 kg NH_4^+-N ha^{-1} yr^{-1} in pig slurry were on average 17, 33 and 67 m^3 ha^{-1}, respectively. Corresponding rates for cattle slurry were 28, 56 and 111 m^3 ha^{-1}, respectively.

Response of spring barley to pig slurry

In practical farming most of the slurry is applied during the autumn and winter. The effect of N may then vary from almost nil to 100% depending on the losses of N through surface run-off, ammonia volatilization and denitrification and leaching of nitrate [1, 2, 7, 10, 22]. In these treatments the slurry was applied just before sowing, using injection techniques at Askov and spreading in bands on the soil surface followed by harrowing on the same day at Uppsala. The effect of pig slurry on the yield of barley in these two experiments for a two-year period at Askov and one-year at Uppsala is

Table 3. Dry matter and nutrient content of cattle and pig slurries used in the experiments

		DM %	Total N %	NH_4^+-N %	P %	K %	Mg %
Cattle slurry	Mean	4.8	0.31	0.18	0.06	0.26	0.04
	Range	1.7–9.7	0.10–0.96	0.06–0.31	0.02–0.13	0.13–0.40	0.01–0.08
Pig slurry	Mean	4.6	0.44	0.30	0.11	0.26	0.03
	Range	1.6–8.5	0.28–0.55	0.25–0.37	0.03–0.17	0.17–0.34	0.01–0.06

given in Fig. 2. The yield curves were calculated using a second degree function.

The unfertilized treatment at Askov gave an average yield of 2210 kg grain ha^{-1}. Pig slurry at rates corresponding to 50, 100 and 200 kg NH_4^+-N ha^{-1} yr^{-1} improved the yield by 101, 132 and 164%, respectively. Fig. 2 illustrates that slurry gave large yield increases throughout.

At Uppsala the barley yield in the untreated plot was 1240 kg ha^{-1}. Slurry NH_4^+-N at rates of 50, 100 and 200 kg ha^{-1} yr^{-1} gave yield increases of 58, 119 and 103%, respectively. The N effect at Uppsala was lower than at Askov throughout. The highest slurry rate at Uppsala also resulted in yield reductions which probably depended on the somewhat unfavourable climatic condition during 1985.

Yield response to NH_4^+-N in slurry compared to CAN-N is also apparent from Fig. 2. At Askov the overall yield response to NH_4^+-N in slurry in comparison with CAN-N was +6% and at Uppsala −22%. The difference depends probably on the method of applying the slurry, which was injected at Askov and spread on the soil surface at Uppsala. The difference in response at different rates of N application was fairly constant at Askov but increased with increasing amounts of applied N at Uppsala.

Response of spring barley to cattle slurry

In this experiment, with a three-year crop rotation, the amount of cattle slurry applied was divided so that the nurse crop, barley, undersown by ley and the other two crops were given 60% and 120%, respectively, of the average rates of 50, 100 and 200 kg NH_4^+-N ha^{-1} yr^{-1}.

Fig. 3 shows the effect of cattle slurry on the yield of barley grain in three trials. The result from Askov is the mean of two experimental years. At Askov the cattle slurry was injected prior to sowing and at Uppsala and Jokioinen it was surface spread. Grain yields in the control treatments at Askov, Jokioinen and Uppsala were 2440, 950 and 2140 kg ha^{-1}, respectively. Cattle slurry in rates of 30, 60 and 120 kg NH_4^+-N ha^{-1} yr^{-1} gave yield increases of 37, 69 and 105%, respectively at Askov, 72, 141 and 208%, respectively at Jokioinen and 22, 36 and 101%, respectively at Uppsala.

Fig. 3 also shows the extra yield of grain obtained with increasing rates of CAN-N. The curves show that CAN-N had a slightly better yield effect than cattle slurry. The relative effect of NH_4^+-N in cattle slurry in comparison with CAN-N at a rate of 100 kg N ha^{-1} yr^{-1} was −11, −20 and −19% at Askov, Jokioinen and Uppsala, respectively.

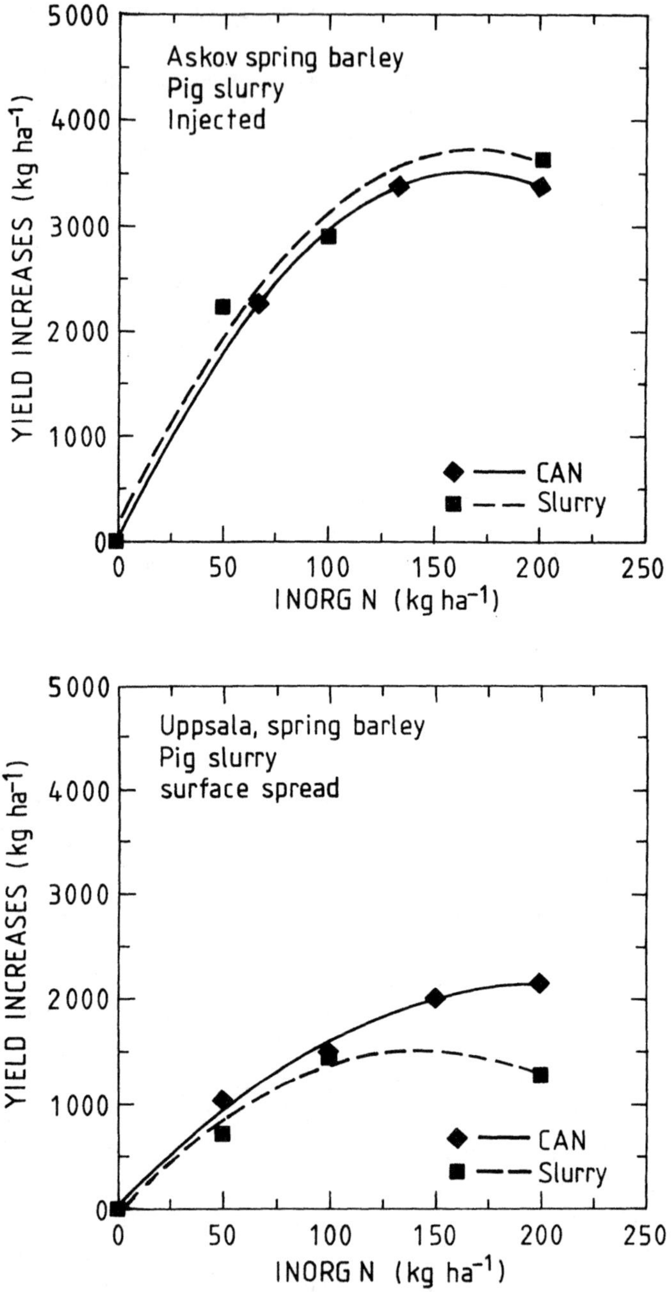

Figure 2. Effect of different rates of NH_4^+-N in pig slurry and N in CAN on the grain yields of spring barley at two experimental sites: Askov and Uppsala.

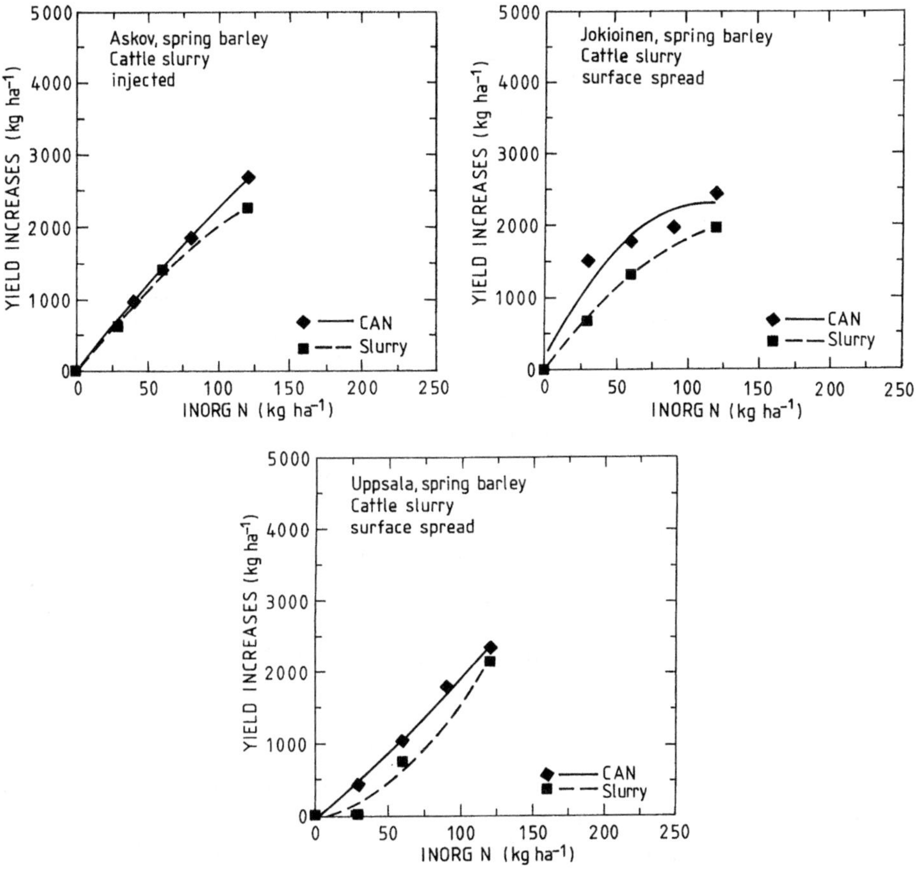

Figure 3. Effect of different rates of NH_4^+-N in cattle slurry and N in CAN on the grain yields of spring barley at three experimental sites: Askov, Jokioinen and Uppsala.

Response of ley and grass to cattle slurry

The seed mixture used in the ley consisted of 20% red clover (*Trifolium pratense*), 30% timothy (*Phleum pratense*), 25% meadow fescue (*Festuca pratensis*) and 25% perennial ryegrass (*Lolium perenne*). At harvest the percentage of clover in the yield varied from 6% in the highest fertilized treatments to 75% in the control. At Askov the grass used was Italian ryegrass (*Lolium multiflorum*).

Fig. 4 shows the effect of N on grass yield as the sum of three cuts at Askov. The yield in the control was 2500 kg DM $ha^{-1} yr^{-1}$. Increased rates of NH_4^+-N in cattle slurry amounting to 60, 120, and 240 kg $ha^{-1} yr^{-1}$, according to the response curve gave increased yields of 67, 116 and 180%,

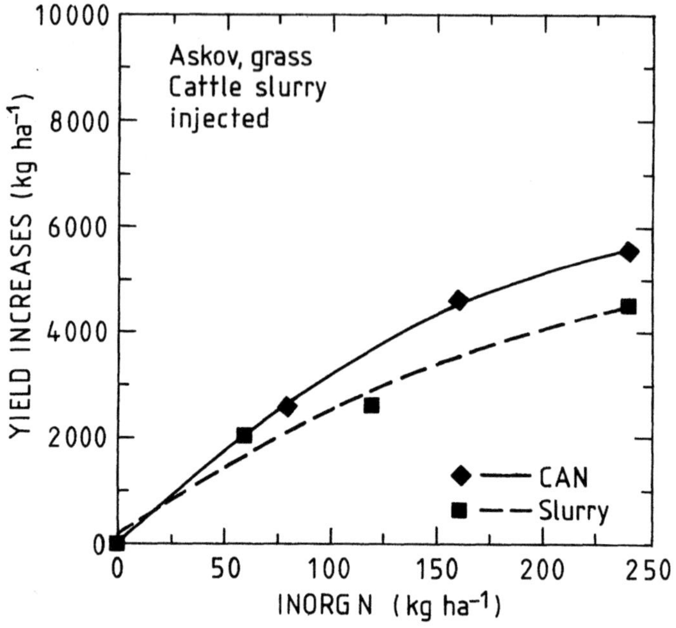

Figure 4. Effect of different rates of NH_4^+-N in cattle slurry and N in CAN on the DM yield of grass at Askov.

respectively. The Figure illustrates that the yield increases were lower after cattle slurry than after CAN at the same level of inorganic N. The relative effect of slurry on yield was −19, −21 and −19% compared to CAN at levels of 60, 120 and 240 kg NH_4^+-N $ha^{-1} yr^{-1}$, respectively. Injection of slurry during the growing season caused some mechanical damage in the form of cut-off roots and open furrows behind the injector coulters. Furthermore, injection at any time at a row spacing of 40 or 50 cm caused different-coloured strips in the stand with dark-green colour along the injected rows and pale green in between. Even though it levelled out with time, this unevenness in the N supply had a negative effect on the yield. Probably these findings are the explanation of the average 20% lower efficiency of injected slurry NH_4^+-N.

The yield increase from the ley after application of increasing rates of cattle slurry and CAN in three trials is shown in Fig. 5. The yield increases are given as the sum of two or three cuts at the individual site. The annual yields of DM in the unfertilized treatments at Jokioinen, Uppsala and Everöd were 4400, 6160 and 5860 kg ha^{-1}, respectively. Slurry, particularly at low rates, had a tendency to give negative effects. However, higher slurry rates increased the yield. The average yield increases at the three experimental sites at rates of 60, 120 and 240 kg NH_4^+-N $ha^{-1} yr^{-1}$ in slurry were only

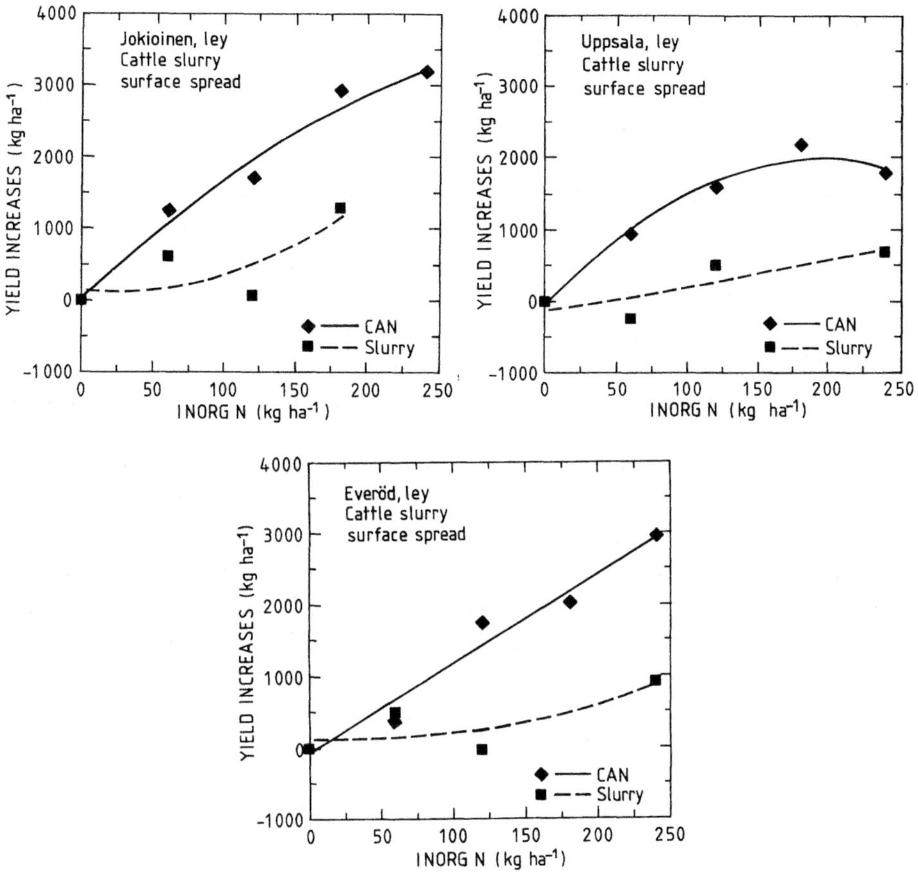

Figure 5. Effect of different rates of NH_4^+-N in cattle slurry and N in CAN on the DM yields of the leys at Jokioinen, Uppsala and Everöd.

2, 5 and 22%, respectively. The explanation of these low responses can be sought in the relatively high proportion of clover in the swards, large N losses due to volatilization of ammonia and the fact that some part of the crop was covered by slurry. In addition, it is possible that the cattle slurry also caused chemical damage, and particularly to the leguminous plants [11, 15]. The large difference between the yield curves for cattle slurry and CAN illustrates that the low N effect only partly depends on the clover content in the sward. The relative effect of cattle slurry in comparison with CAN at 60, 120 and 240 kg inorganic $N\,ha^{-1}\,yr^{-1}$ was -85, -83 and -66%, respectively. There were no particularly large differences in reaction of the ley to slurry or CAN at the three experimental sites.

Response of fodder beet to cattle slurry

Fodder beet has a long growing period and requires relatively large amounts of plant nutrients to give large yields. Consequently, fodder beet is usually considered to be a good crop when animal manure is available [8, 9]. Fodder beet is included in the crop rotation at Askov. Increased yields as DM of roots and tops are given in Fig. 6. All rates of cattle slurry and CAN, from 60 to 240 kg inorganic N ha^{-1} yr^{-1}, have given increased yields, both in roots and tops. In the unfertilized treatment the DM yield was 7750 kg ha^{-1} of which 5660 kg were roots and 2090 kg were tops. The root yield after supply of 60, 120 and 240 kg NH_4^+-N ha^{-1} yr^{-1} in cattle slurry increased, according to the response curve, by 49, 85 and 106%, respectively. The yield increase of tops with corresponding rates of cattle slurry was 47, 89 and 141%, respectively. It can be seen that the increased yield was divided proportionally between roots and tops apart from the highest rate of slurry, which favoured the yield of tops.

The response to NH_4^+-N in cattle slurry was largely comparable with the response to CAN-N. The relative effect of NH_4^+-N in cattle slurry on the root yield compared to CAN-N at 60, 120 and 240 kg inorganic N ha^{-1} yr^{-1} was +1, +3 and −15%, respectively. The corresponding values for tops were −4, −11 and −30%, respectively. At high N rates the cattle slurry thus had a lower effect than CAN.

Comparison between injected and surface-spread slurry

In some of the experiments a comparison was also made between injected and surface-spread slurry. The slurry was applied to spring barley and fodder beet just before sowing, to the ley at the start of the growing season, and after the first cut, and to the grass at the start of the growing season and after first and second cuts. The result of this comparison is summarized in Fig. 7. As shown in the Figure, in spring barley it was better to inject both cattle and pig slurry instead of applying it on the surface. Injection gave on average 14% increases in the yield of grain in comparison with surface spreading. The yield increases may be assumed to depend on the suppressive effect of injection on losses of ammonia. No particularly large differences between pig slurry and cattle slurry could be noticed.

In the trial with fodder beet the injection of slurry gave yields of roots and tops which were 20% and 47% larger than those given by surface-spread slurry. In trials with grass the effect of injection was only 12% better. The poorer response of grass to injected slurry probably depends on the root damage caused by the injector tines [11]. Damage by injector tines to roots of the ley was even greater than that caused to the grass. In particular, roots

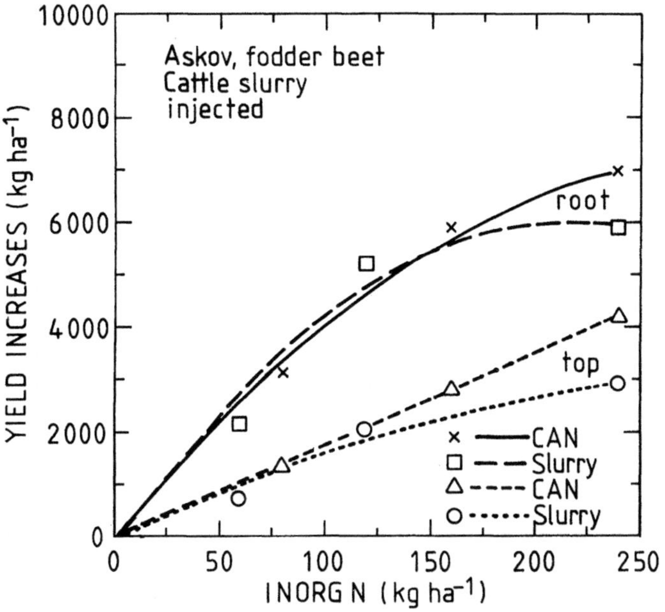

Figure 6. Effect of different rates of NH_4^+-N in cattle slurry and N in CAN on the DM yield of fodder beet tops and roots as Askov.

of clover were severely damaged since they are unable to recover in the same way as grass once they have been cut off. Nonetheless, the injection of cattle slurry still gave an increased DM yield of 18% at Jokioinen and 13% at Everöd. At Uppsala, on the other hand, the injection led to a 20% yield decrease. The damage caused by the injection was particularly evident after application in the spring. In comparison with surface-spread slurry at three experimental sites, injected slurry on average gave only 2% yield increases.

Conclusions

Animal manure represents a large plant nutrient resource and at the same time a potential risk of environmental pollution. Its value depends largely on the degree of utilization in crop production.

Slurry contains both directly available and slow-acting organically bound nutrients. Among the various constituents of slurry, particular attention must be paid to the NH_4^+-N. It is exposed to risks of being lost through volatilization as ammonia or through denitrification and leaching after conversion to nitrate [1, 2, 4, 10, 20]. In addition, the inorganic N has the dominating yield increasing effect among the plant nutrients [8, 9, 10]. Consequently, it is possible for a short period to say that good slurry

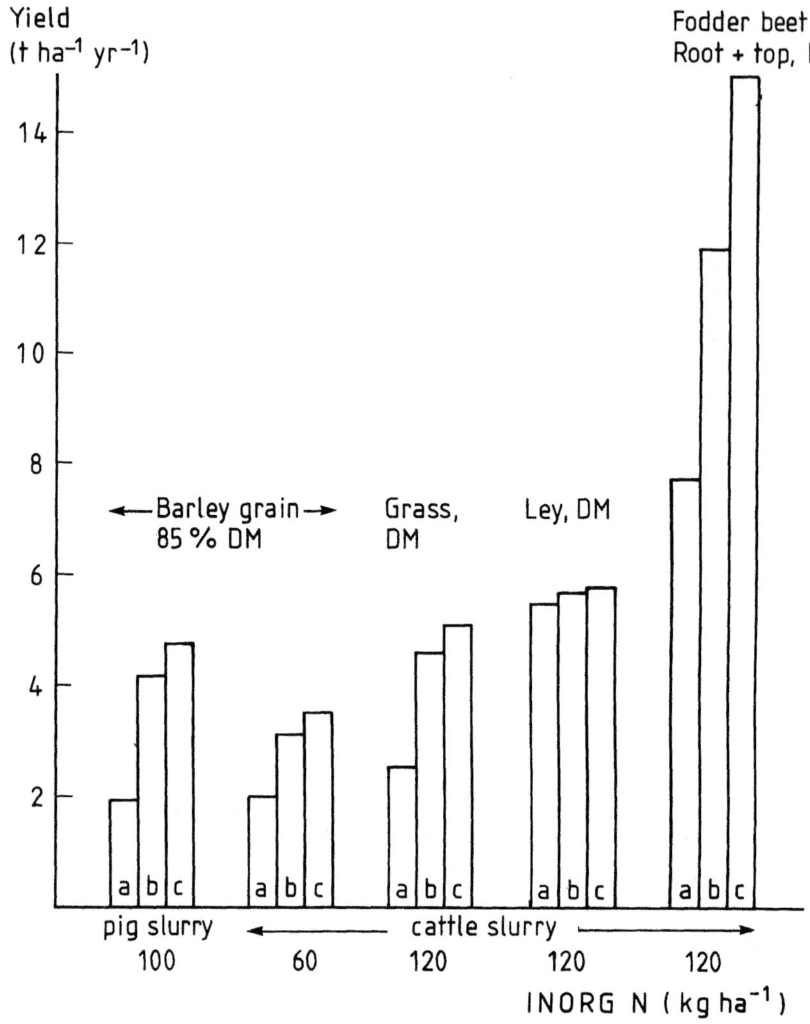

Figure 7. Effect of surface-spread and injected slurries on yield of different crops. Treatments: a = control, b = surface-spread slurry, c = injected slurry.

management is the same as effective utilization of NH_4^+-N. Regarding soil fertility in a longer perspective, the mineralization of organically bound N and the balance between supplied and removed nutrients is of great importance. In this project it has only been possible to study the short-term effect of slurry so far.

In these experiments, as in many other slurry investigations, the Nitrogen Meter for rapid determination of NH_4^+-N in slurry was found to be an important aid in applying the correct amount according to the different N requirements of the crops. In combination with other factors it is possible to achieve increased degrees of utilization of the applied N. The time of application and the method of application are of great importance in this respect. With regard to NH_4^+-N in slurry the application should coincide with the time considered suitable for applying fertilizer. This is confirmed by studies on the most suitable time to spread slurry [7]. Slurry which is applied during the autumn usually has a very minor yield effect. The soil temperature is still sufficiently high for rapid nitrification of NH_4^+-N to NO_3^--N which disappears through denitrification or leaching [1, 10]. Spreading during the winter on frozen soil may also lead to large losses of N, not least through surface run-off, particularly if the land is undulating [22]. Inorganic N in slurry applied in the spring for spring-sown crops has just as good an effect as fertilizer N [8, 9, 18]. In this project, when slurry was injected in the spring for barley and fodder beet, the results are in good agreement with these observations.

In addition to the correct time of application, injection of slurry into the soil to protect ammonia from volatilization is an important factor for increasing the N efficiency. In barley, injection always resulted in yield increases in our trials in comparison with surface spreading. On grassland, surface spreading is considered to be unsuitable on account of the risk of polluting the crop with slurry which may lead to faulty fermentation during ensiling and to the spread of different diseases [12, 19]. Injection, on the other hand, results in mechanical damage to the sward [11]. Danish investigations have demonstrated that the yield loss in grassland caused by injection tines amounted to 15–19% after one run and 20–30% after three runs with an injector on loam soil, and slightly less on sandy soil [11]. In trials with grass at Askov, injected slurry gave about 20% lower yield increase than surface-spread CAN. This difference in effect probably depends on mechanical damage caused by the tines. In comparison with surface-spread slurry the injected slurry was still 12% more effective.

Ley was found to be a poor recipient of slurry. This is partly because of the small N requirement when there is a high clover content and partly since clover in particular is damaged by the slurry and disappears from the sward.

The results show that spreading in the spring combined with injection into the soil gives a high effect of slurry in all crops apart from ley. The risk of environmental pollution probably decreases to a corresponding extent. Obtaining this effect leads, however, to extra costs because of the greater investment required in storage space, tankers, machinery for injection and a greater labour requirement. The extra costs must be covered by the increase in yield.

References

1. Christensen B T (1985) Loss of plant nutrients from manure (Dk). In Husdyrgødning og dens anvendelse. Tidskrift for Planteavls Specialserie, Beretning nr. S 1809, 30–55
2. Christensen B T (1986) Ammonia volatilization loss from surface applied animal manure. In Kofoed A Dam, Williams J H and Hermite P l', eds. Efficient land use of sludge and manure. CEC seminar (Askov), pp 193–203. London: Elsevier Applied Science Publishers
3. Egnér H, Riehm H and Domingo W R (1960) Untersuchungen über die chemische Bodenanalyse als Grundlage für die Beurteilung des Nährstoffzustandes der Böden. II. Chemische Extraktionsmethoden zur Phosphor- und Kaliumbestimmung 26, 199–215
4. Kemppainen E and Heimo M (1983) Better use of plant nutrients in animal manure (Sv). Litteraturöversikt 1983. Jokioinen ISSN 0359–7652
5. Kjellerup V (1986) Agros Nitrogen Meter for estimation of ammonium nitrogen in slurry and liquid manure. In Kofoed A Dam, Williams J H and Hermite P l', eds. Efficient land use of sludge and manure. CEC seminar (Askov), pp 216–223. London: Elsevier Applied Science Publishers
6. Klausen P S (1985) Plant nutrients (Dk). In Husdyrgødning og dens anvendelse. Tidskrift for Planteavls Specialserie, Beretning nr. S 1809, 30–55
7. Klausen P S and Nemming O (1982) Time of application of pig manure to barley (Dk). Tidskrift for Planteavl 86, 189–192
8. Kofoed A Dam (1981) Crop uptake of nitrogen. In Brogan J C, ed. Nitrogen losses and surface run-off from landspreading of manures, pp 125–160. The Hague: Martinus Nijhoff/Dr Junk
9. Kofoed A Dam and Nemming O (1979) Experiments on heavy applications of animal manure to land. In Gasser J K R, ed. Effluents from livestock, pp 184–217. London: Applied Science Publishers
10. Kolenbrander G J (1973) Fertilizers, farming practice and water quality. The Fertiliser Society 135, 1–28
11. Larsen K E and Keller P (1985) Injection of slurry to grass (Dk). Tidskrift for Planteavl 89, 19–24
12. Lund E (1980) Survival of virus in slurry (N). Ugeskrift Jordbrug 125, 27–49
13. Miljöskydd och djurhållning – allmänna råd. Statens Naturvårdsverk. Råd och Riktlinjer (1983) ASBW 91-38-06833-8, Stockholm
14. Miljøstyrelsen (1984) NPO-Redegørelsen, pp 1–218. Copenhagen
15. Schechtner G, Tunney H, Arnold G H and Keuning J A (1980) Positive and negative effects of manure on grassland with special reference to high rates of application. In Prins W H and Arnold G H, eds. The role of nitrogen in intensive grassland production. Proceedings of an international Symposium of the European Grassland Federation (Wageningen), pp 77–93. Wageningen, The Netherlands: PUDOC
16. Simán G (1981) Production and utilization of animal manure in Sweden (Sw). Nordisk Jordbruksforskning 63, 356–357
17. Steineck S (1974) Plant nutrients in manure (Sw). Nordisk Jordbruksforskning 56, 58–59
18. Steineck S (1981) Slurry-nitrogen efficiency (Sw). Medd. från södra jordbruksförsöksdistriktet. ISBN 91-576-1056-8
19. Thunegard E (1975) On the persistence of bacteria in manure. A field and experimental study with special reference to Salmonella in liquid manure. Acta Veterinaria Scandinavia, suppl. 56, pp 1–87. Stockholm
20. Tunney H (1981) Nitrogen uptake by crop-grassland. In Brogan J C, ed. Nitrogen losses and surface run-off from landspreading of manures, pp 161–165. The Hague: Martinus Nijhoff/Dr Junk

21. Tveitnes S (1980) Effect of animal manure on crop production and pollution (N). Slutrapport nr. 349. NLVH ISBN 82-7290-060-2
22. Uhlen G (1981) Surface run-off and the use of farm manure. In Brogan J C, ed. Nitrogen losses and surface run-off from landspreading of manures. The Hague: Martinus Nijhoff/Dr Junk

Extent, effects and tackling of a regional manure surplus; a case-study for a Dutch region

P J M VAN BOHEEMEN
Advisory Service for Soil, Water and Fertilizer Aspects in Stock Farming, P.O. Box 12, 6700 AA Wageningen, The Netherlands

Key words: manure surplus, slurry application, environmental aspects, manure legislation, manure boards, manure transport

Abstract. In the Peel Region about 111 t manure per ha of agricultural land is produced. Only a small part is disposed of elsewhere.

Crop growth and quality are not yet significantly adversely affected. Nearly all the nitrate leached from the root zone is denitrified in the upper zone of the groundwater. At several places the phosphate sorption capacity of the upper layer of the soil profile has already been filled up. The vigour of forests is remarkably low and natural areas are influenced by ammonia deposition and changing groundwater quality.

Recently every farm has been restricted in manure production and application. Over about 15 years the restrictions will be in such a way that the desired environmental qualities will be ensured. In the meantime the infrastructure for manure transport to areas with a manure shortage should be strengthened and plants for manure processing introduced. A reduction in nutrient excretion by changing animal nutrition is considered to be possible. Results obtained up to the present are discussed.

Introduction

Before the Second World War Dutch livestock was practically fed with farm grown crops. All the manure was applied on own land. There was a cycling of nutrients and a maintenance of soil fertility could be nearly reached. The rather small losses of nutrients were balanced by some inputs of fertilizers and concentrates.

In the 1960's an intensification started. The development was attended by an increased use of imported concentrates and of fertilizers, especially nitrogen (N). Associated with this was an increased production of manure. In some regions an enormous surplus has been arised. The Peel Region is one of them as illustrated in the next paragraph.

The situation in the Peel Region is discussed in detail as an example. Thereby, attention is paid to agricultural and environmental aspects as well as to the ways taken to solve the surplus problem.

H.G. v.d. Meer, et al. (eds.), Animal Manure on Grassland and Fodder Crops.
 ISBN 978-94-010-8137-5.

Spatial variation in manure production

Table 1 gives the number of animals in The Netherlands as well as the amount of manure, nitrogen (N), phosphorus (P) and potassium (K) produced by them. For N and P, figures of the same magnitude have been reported previously [26]. The N lost in stable and from storage, and therefore not available for spreading on land, is not included. This is approximately $59 \times 10^6\,kg\,yr^{-1}$ in total.

About one third of the poultry manure is mixed with an absorbing material such as wooden shavings or is dried, and therefore can be typed as solid manure. Nearly all the other manure is produced as slurry.

If the total production is spread equally on all the agricultural land, each hectare would receive yearly 49 t manure with 292 kg N, 52 kg P and 245 kg K. However, the livestock density and thereby the manure production per ha agricultural land varies within The Netherlands. This is illustrated by Fig. 1, which gives the average P production per ha for different agricultural regions. Figures of the average manure, the N or the K production show a comparable pattern [7].

In the south-east there is a rather large area with high P production. The Peel Region forms the heart of it. Fig. 2 gives the P production in the individual communities of these region.

Land use in the peel region

The area of the Peel Region is 85 000 ha of which 50 000 ha is agricultural. It belongs to the catchment area of the river De Aa and consists of slightly undulating, sandy soils. Originally the sand was covered by upland peat which has been dug.

The average excess winter rainfall is 280 mm. In summer the mean rainfall deficit is 90 mm. Excess water drains through brooks to the river De Aa. In dry periods groundwater discharge is limited by means of weirs. Most of the brooks can be supplied with water through inlets constructed at the crossings with canals.

Near the brooks the groundwater level varies between 10 and 150, on the higher ground between 30 and > 200 cm below the surface. At several places groundwater is extracted for industrial and public water supply ($14 \times 10^6\,m^3\,yr^{-1}$). This extraction takes place in the aquifer between 30 and 60 m depth, which is recharged mainly from outside the Peel Region. There are also a lot of extraction points for agricultural irrigation (sprinkling).

Table 1. Manure production in The Netherlands in 1984 and its contents of N, P and K (calculated partly from [8 and 9])

Animal group		Number in May 1984 ($\times 10^3$)	Manure ($t \times 10^6$)	N* ($kg \times 10^6$)	P ($kg \times 10^6$)	K ($kg \times 10^6$)
Cattle	– veal calves	638	2	5	1	3
	– steers	262	3	8	2	15
	– others	3 440**	76	435	58	382
Pigs	– fattening	5 730	10	71	19	53
	– breeding	1 472***	6	33	11	21
Poultry		84 480	2	37	13	20
Total			99	589	104	494
Average production in t or kg per ha agricultural land			49	292	52	245

* exclusive of losses in stable and from storage, estimated at about 59×10^6 kg N yr^{-1}
** in mature cattle units
*** exclusive of piglets

The waste water of the municipalities and industries is treated in sewage plants. The effluents are discharged to the river De Aa. Only a small proportion of the sewage sludge (about 30% in 1984) is still applied to agricultural, mainly arable, land [35].

In the 1950's the farms were small (10 ha) and had both grass and arable crops. Besides cattle there were also pigs and poultry. The shortage of land encouraged an intensification of stock farming. Nowadays fodder maize has a near monopoly on the arable land (Table 2). This is because it tolerates a large amount of slurry, it can be grown every year and the work can be put out to contractors.

Livestock density data are given in Table 3. The numbers of steers, veal calves, pigs and poultry have been calculated on the area of arable land, because little manure from these groups is applied to grassland. The pigs and poultry densities are very high; 58 and 571 animals per ha, respectively.

Surplus of manure and nutrients

As a consequence of the high livestock density, there is a very high availability of manure nutrients in the Peel Region. Per hectare of agricultural land there is about two and a half times the average for the whole country (Table 4).

On grassland the use of mineral fertilizers is at the same level as elsewhere. On the arable land (mainly maize land) much less mineral fertilizers are

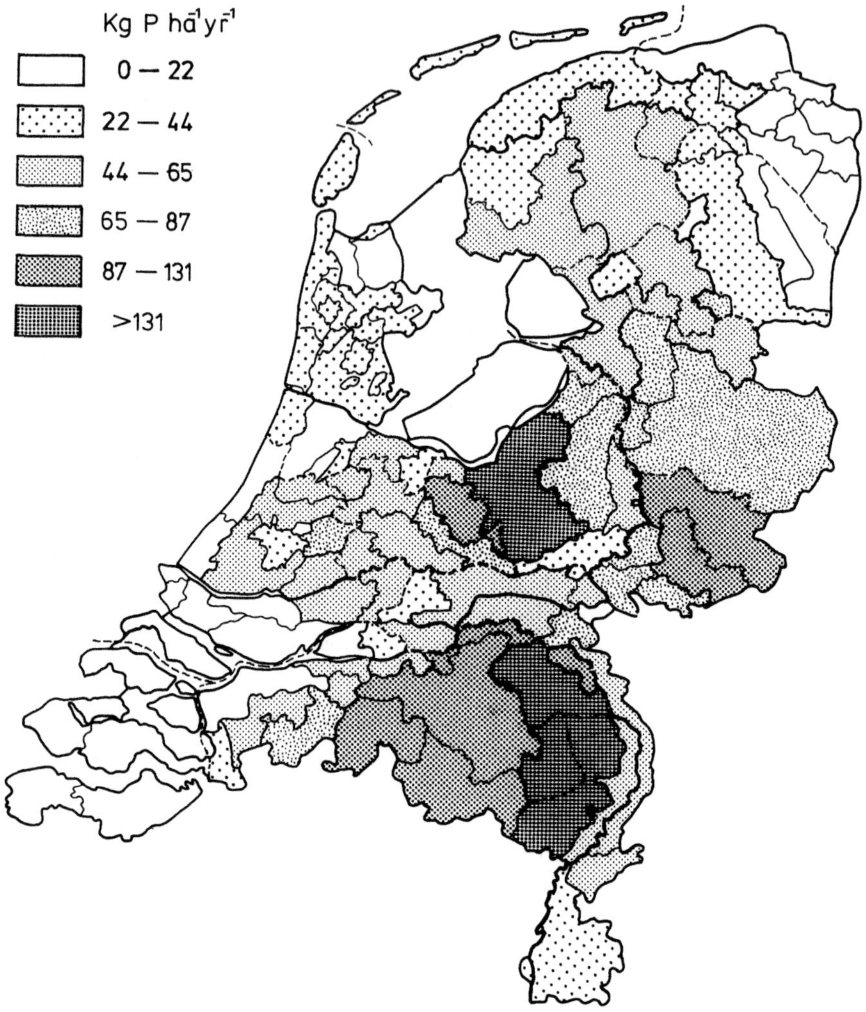

Figure 1. The annual P production in manure in different agricultural regions in The Netherlands, expressed in kg P per ha agricultural land [9].

applied. The data support the statement that the pig and poultry manure is mainly applied to arable land.

The effectiveness of N from manure, compared to that from inorganic fertilizers, depends on the time of application (autumn or spring) and on the method of application (to the surface or by injection). For spring application the effectiveness for the whole growing season is 30–60%. The long-term effectiveness of P and K is 100%.

The manure surplus can be spoken of in different ways. Here it is dealt with in agricultural and environmental terms. From the agricultural point

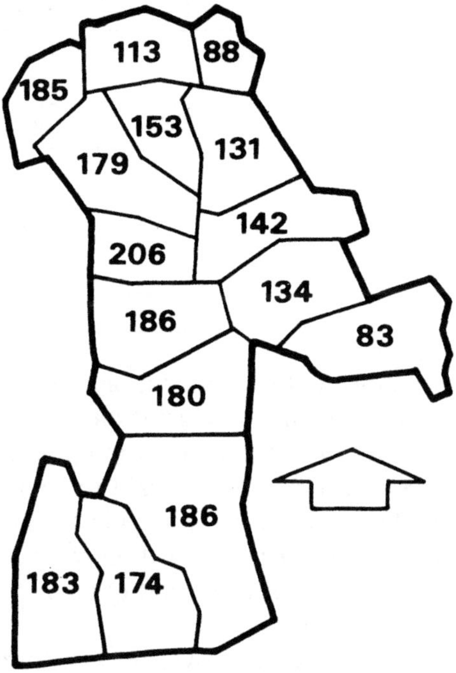

Figure 2. The annual P production in manure in the 15 municipalities of the Peel Region, expressed in kg P per ha agricultural land [9].

of view, a surplus exists if the application of the available nutrients endangers crop growth or quality [6, 11, 17]. There can be distinguished risks for the following growing season such as lodging of cereals and hypomagnesaemia of cows, and long-term risks such as heavy metal effects.

Application of manure (and fertilizers) can lead to considerable losses of nutrients. Losses to groundwater, surface waters and to the atmosphere can be injurious to the environment. Eventual consequences are, for example, a

Table 2. Use of agricultural land in The Netherlands and in the Peel Region in May 1984 [9]

Type of crop	The Netherlands		Peel Region	
	ha	%	ha	%
Grass	1 178 500	(59)	30 395	(61)
Arable crops – fodder maize	166 200	(8)	13 832	(28)
– others	549 700	(27)	3 688	(7)
Horticultural crops	117 900	(6)	2 041	(4)
Fallow	3 800	(0)	74	(0)
Total	2 016 100	(100)	50 030	(100)

Table 3. Livestock density in The Netherlands and in the Peel Region in May 1984 [9]

Group of animals	Number per ha grassland		Number per ha arable land	
	Neth.	Peel	Neth.	Peel
Dairy and in-calf cows	2.2	3.2	–	–
Calves and yearlings	1.7	2.4	–	–
Steers	–	–	0.4	0.9
Veal calves	–	–	0.9	2.3
Pigs (excl. piglets)	–	–	10	58
Poultry	–	–	118	571

lower suitability of groundwater for public supply, a lower quality of surface water for recreation and a change of vegetation in natural ecosystems. If these effects arise and cannot be prevented by modifying the method or the time of manure application, it can also be said to be a manure surplus.

In the next paragraphs more attention is paid to the harmful effects of manure application in the Peel Region. The following points are discussed:

- manure application and crop growth and quality,
- manure application and water quality,
- manure application and natural ecosystems and forests.

Manure application and crop growth and quality

At the present time the grass and the fodder maize grown in the Peel Region are not damaged significantly by the excess application of nutrients. Apply-

Table 4. Nutrients in the manure produced in 1984 (calculated partly from [8 and 9]) and in mineral fertilizers applied in 1980 [36], given for The Netherlands and for the Peel Region

	Manure (t)		N* (kg)		P (kg)		K (kg)	
	Neth.	Peel	Neth.	Peel	Neth.	Peel	Neth.	Peel
Per ha grassland								
– manure of cattle except veal calves and steers	64	93	369	537	50	72	324	472
– mineral fertilizers	–	–	257	286	7	2	9	9
Per ha arable land								
– manure of veal calves, steers, pigs and poultry	32	155	215	1109	55	253	134	773
– mineral fertilizers	–	–	140	44	28	10	71	14
Per ha agricultural land								
– manure	49	111	292	714	52	133	245	558
– mineral fertilizers	–	–	238	204	19	5	48	13
– total	–	–	530	916	71	138	293	571

* exclusive of losses in stable and from storage.

ing all the cattle manure (excluding that of steers and veal calves) to grassland leads to an annual rate of 537 kg N ha^{-1}, 72 kg P ha^{-1} and 472 kg K ha^{-1}. The K figure is worrying, but in practice hardly any case of hypomagnesaemia is noticed. This is attributed to the fact that a part of the cattle slurry is also applied to maize land and that a part of the K applied to the grassland is leached. Furthermore, besides grass the cows are also fed concentrates containing magnesium and silage maize with a rather low K content. A fourth reason is that much grassland is reseeded frequently (every four years). Then the land is ploughed and nutrients are placed below the sod. This practice explains greatly why soil fertility of grassland in respect to P and K declined in the period 1973–1983 (Table 5) [30].

A proportion of the pig slurry is applied to grassland, especially on dairy farms with pigs. In general, the copper content of the sod is still low. Values of 10 mg per kg dry soil (extracted in 0.43 M HNO_3) are found, but values of 20 mg kg^{-1} which are potentially dangerous for sheep [25], are rare. On the majority of the grassland the content is less than 5 mg kg^{-1} as a result of reseeding, which is potentially deficient. The same situation was reported in a comparable region in the centre of The Netherlands [31].

The slurry produced by the pigs, poultry, veal calves and steers goes mainly to the maize land. Many of the farms with these animals grow only maize. If the total available amount was applied to the fodder maize, this crop would receive 1109 kg N, 253 kg P and 773 kg K ha^{-1} yr^{-1}. This is much more than the uptake by the crop. Experience has shown that excessive slurry applications do not damage the growth or the quality, but on the contrary they improve yield [34].

Undoubtedly, an accumulation of the harmful metals copper, cadmium and zinc is taking place. In the long run this will result in a lower growth rate due to copper and lower quality due to cadmium. For the Peel Region data about the contents of heavy metals in the top soil of arable land are not

Table 5. Availability of P and K, measured in about 850 soil samples (0–5 cm depth) from the Peel Region during the winters of 1972–1973 and 1982–1983 (G. Lommerse, unpublished)

Nutrient	Degree of availability	Percentage of analyses	
		1972–1973	1982–1983
Phosphorus	low	10	27
	sufficient	45	51
	high	45	22
Potassium	low	12	21
	sufficient	50	56
	high	38	23

available. In a comparable region values of 20–30 mg copper per kg dry soil were found at 0–20 cm below surface [31]. The limit for forage crops is 50 mg kg^{-1} [25]. If the present application rates were to be continued, this limit will be reached in 50–100 years in several areas.

Manure application and water quality

Nitrogen. The high application rates explain why leaching of nitrate occurs. In the upper metre of the groundwater, contents of 50 g NO_3^--N m^{-3} are found locally (Fig. 3). At greater depth there is hardly any nitrate [13, 23]. However, at this greater depth rather high contents of chloride and sulphate are measured. The disappearance of nitrate is ascribed to denitrification which is possible because of the presence of peat in the soil profile (0.4%

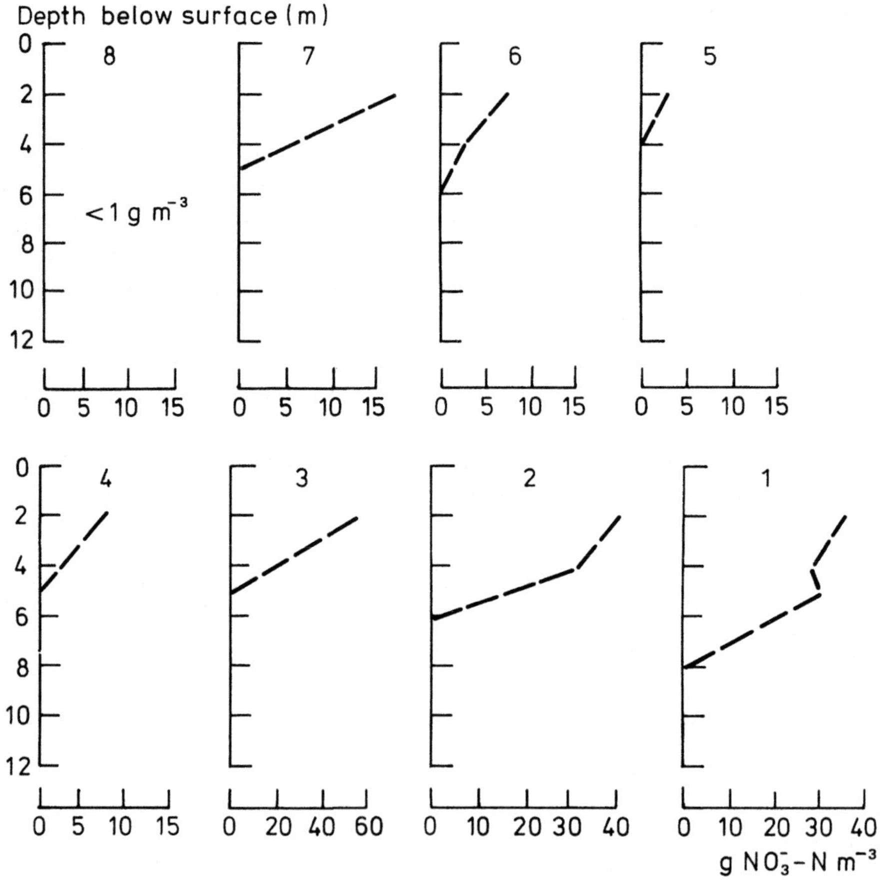

Figure 3. Nitrate concentrations in groundwater (g NO_3^--N m^{-3}), measured at eight locations in the Peel Region [23].

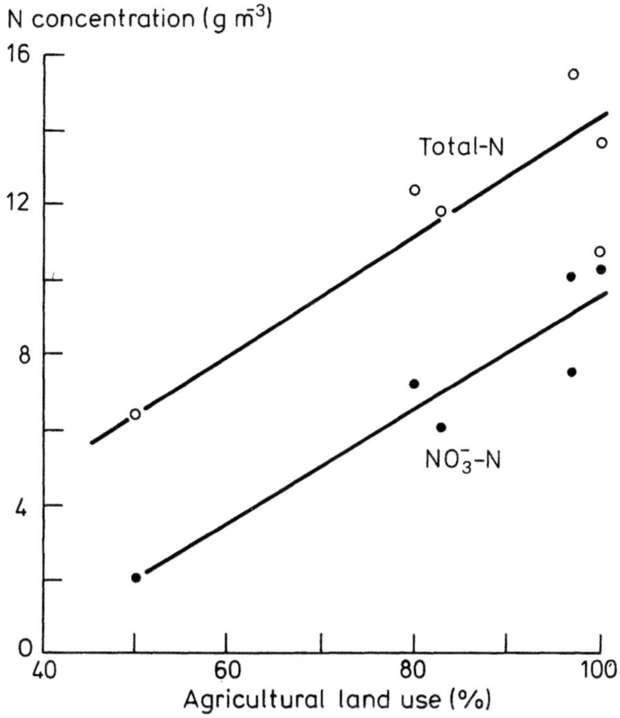

Figure 4. Concentrations of NO_3^--N and total N in water discharged from six small catchment areas in relation to the percentage of land used for agriculture, measured in October 1981–May 1982 [19].

organic matter) [32]. In other sandy regions contents of more than $10\,g\,NO_3^-$-$N\,m^{-3}$ have been measured at great depths [2, 3, 15].

The nitrate content of the surface water is also low. Contents are higher in areas with proportionally more agricultural land (Fig. 4). In two small catchment areas water is discharged with 8–10 $g\,NO_3^-$-$N\,m^{-3}$ (Table 6). Here the yearly loading of the surface water is estimated at 45–50 $kg\,N\,ha^{-1}$. Ninety percent is considered to originate from agricultural activities. The ammonium and ammonia concentrations are 2–4 $g\,NH_4^+$-$N\,m^{-3}$. To ensure an adequate oxygen content the ammonium concentration should be lower than 1 $g\,NH_4^+$-$N\,m^{-3}$ [27].

Table 6. Nitrogen and P content of surface water discharged from two catchment areas within the Peel Region [18]

Catchment area	Nitrate-N ($g\,N\,m^{-3}$)	Ammonium-N ($g\,N\,m^{-3}$)	Organic-N ($g\,N\,m^{-3}$)	P ($g\,P\,m^{-3}$)
De Vlier	9.8	2.4	5.3	0.75
Karweische Loop	7.9	3.5	2.8	0.67

Table 7. Quality of the water discharged outside the Peel Region by the river De Aa [35]

Parameter		1982	1983	1984
O_2	$g\,m^{-3}$	7.8	7.6	8.6
pH	–	7.5	7.5	7.5
Cl^-	$g\,m^{-3}$	62	55	56
SO_4^{2-}	$g\,m^{-3}$	60	58	63
NO_2^--N and NO_3^--N	$g\,m^{-3}$	4.2	4.5	5.7
NH_4^+-N	$g\,m^{-3}$	1.8*	1.4*	1.6*
NH_3-N	$g\,m^{-3}$	0.02	0.01	0.02
BOD	$g\,m^{-3}$	2.0	1.6	2.1
$P_2O_4^{2-}$-P	$g\,m^{-3}$	0.8*	0.8*	0.9*

* average values for the summer period

The N content of the water discharged outside the Peel Region by the river De Aa is given in Table 7.

Phosphorus. In the 0–100 cm layer, if it remains unsaturated with water, a large amount of P from manure can be absorbed. The sorption capacity depends on the contents of reactive iron and aluminum. In the Peel Region there are soils with a sorption capacity of $5\,t\,P\,ha^{-1}$ which would be saturated after 20 years at $250\,kg\,P\,ha^{-1}\,yr^{-1}$, but there are also soils with a sorption capacity of $20\,t\,P\,ha^{-1}$ [24].

Calculations indicate that at several places the upper part of the soil has already been filled up. If the present manure application rates continue, a rapid increase in the problem has been forecasted (Table 8). In such areas the P content of the upper groundwater will be raised to $50–100\,g\,P\,m^{-3}$ [15]. At this moment the deep groundwater (10 m below surface) nowhere contains more than 0.04 g ortho-$P\,m^{-3}$. There the concentration of total-P is always lower than $1\,g\,m^{-3}$, and often even lower than $0.3\,g\,m^{-3}$ [13].

On average, it takes a long time before water reaching the groundwater reservoir is discharged to a water course. However, on small strips beside a ditch or trench and on areas with underdrainage the transport takes little time. Here any P saturation of the upper soil profile will lead rapidly to a P loading of the surface water. A second possiblity for a P loading of the surface water is run-off from areas with poor drainage.

Table 8. Increase of the area (ha) with a top soil saturated with phosphate, if the present manure application rates were to be continued [5]

Soil layer considered	Year		
	1985	2000	2025
0–50 cm below surface	30	6 000	12 000
A + B horizon	220	1 600	5 600
Soil layer not saturated with water in winter	2 600	5 600	11 000

In the two small catchment areas mentioned earlier, the surface water contains $0.7\,g\,P\,m^{-3}$. The yearly loading amounts to $2\,kg\,P\,ha^{-1}$ of which three quarters is assumed to originate from agricultural activities [18]. In summer the discharge of the river De Aa contains $0.9\,g\,P\,m^{-3}$ (Table 7). To prevent eutrophication the concentration should be lower than $0.2–0.3\,g\,P\,m^{-3}$ [27].

Potassium. On grassland the application of K should be limited because of the danger of hypomagnesaemia. However, for fodder maize there is no constraint. Several experiments have proved that the application rates occurring in practice lead to significant leaching losses [12, 29]. This explains why at several places in the Peel Region contents of $20–60\,g\,K\,m^{-3}$ have been measured in the upper groundwater (Pomper, private communication). At a depth of 10 m below surface the K concentration everywhere is below $10\,g\,m^{-3}$ [13].

Organic matter. Leaching of organic matter cannot be ruled out, but the magnitude will be small. Loading of the surface water by run-off is considered to be a significant problem. Quantitative information for the Peel Region is not available.

Manure application and natural ecosystems and forests

The production, storage and application of manure are accompanied by volatilization of ammonia. Deposition of this ammonia can take place within a short distance (several hundreds of metres). Therefore, the Peel Region has not only a high ammonia emission, but also a high deposition.

The ammonia deposition is undesirable. Firstly because N fertilization takes place. The increased N input causes a strong disturbance of the nutrient supply to plants and trees in natural areas and forests. This is related to the fact that in the Peel Region the soils have a low inherent fertility. Secondly, the ammonia deposition accelerates the acidification of the soil. It has been calculated that in the Peel Region the ammonia deposition causes 40% of the acidification (also 40% from SO_2 and the other 20% from NO_x).

Both effects are considered to be in part responsible for the very low vigour of the forests in the Peel Region. A survey undertaken in 1986 found that 30% of the trees were dying or had already died, and that 67% were of low vigour [4]. Recently, in an attempt to reverse the trend, fertilization experiments have been set up [10, 28].

Plants and trees are not only influenced via the atmosphere, but also by the groundwater system. The grey zone in the Piper-diagram of Fig. 5a illustrates the water quality spectrum occurring originally in the Pleistocene sand areas of The Netherlands. The right part of the zone corresponds with areas at watersheds. There infiltration takes place and the groundwater is similar to rain water (*i.e.* Ca^{2+} + Mg^{2+} less than 50% of the secondary cations and HCO_3^- about 0% of the primary anions). During transport through the underground the water is enriched with Ca^{2+} and HCO_3^-. Therefore, the groundwater in the valleys was of the Ca^{2+}/HCO_3^--type (left part of the grey zone).

As a result of human activities, the groundwater in the Peel Region is changing into a new, unnatural and spatially non-differentiated type, *i.e.* the Ca^{2+}/SO_4^{2-}-type (Fig. 5b/c). Besides acids from the atmosphere, manure is considered to be causing this change [21, 22]. To back up this statement some data follow for 1 ha fodder maize receiving 50 t slurry from fattening pigs. In the concerning year the soil water is enriched with 120 kg Ca because of the acid input from the atmosphere, 180 kg Ca because of its presence in the slurry and −35 kg Ca because of the acid neutralizing effect of the pig slurry. The enrichment with sulphate amounts to 40 kg SO_4^{2-}-S from the atmosphere and 30 kg SO_4^{2-}-S from the slurry. On the other hand, the maize will extract 40 kg Ca and 20 kg S.

The new water type leads to a different degree of base saturation and a lower pH in the root zone. This affects the availability of N and P to plants.

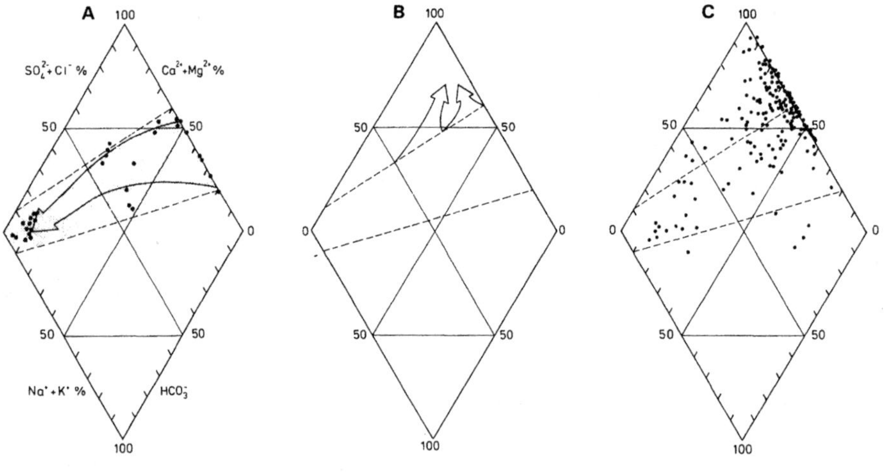

Figure 5. Piper diagrams characterizing groundwater composition [22]. a. Grey zone gives the spectrum originally found in the Pleistocene sand areas. Arrow indicates the change in quality from watershed to valley. b. Arrows indicate recent trends in the Pleistocene sand areas. c. Results of recent groundwater analyses in the Peel Region.

In the long run the vegetation of natural areas will become less valuable and less varied [21].

The chemical composition of the surface water is also changing as a result of the above development. The CO_2 concentration is falling, which influences the quality of aquatic life [1, 21, 22].

Solving the surplus problem

General

The negative effects of the excessive production and application of manure can be avoided or reduced by a policy based on one or both of the following principles:

- Overcoming or eliminating the effects caused by an emission (curative policy). Such a policy will not always have consequences for an individual polluter, at least if the quality desired is not endangered.
- Limiting of actions causing emission (preventive policy). In this framework individual polluters will be affected directly.

In The Netherlands the problem of manure surplus will be tackled mainly by the second approach. In principle each farmer, including those in the Peel Region, will be confronted with the same measures. In the long term the rules will be set so that the qualities desired in the different environmental compartments (soil, water and air) are ensured. In the short term this cannot be realized, because it is not possible to find another destination for a large amount of manure.

In regions with valuable natural areas or groundwater extraction for public water supply more severe rules can be imposed. No attention is paid to this possibility here.

For solving the general problem not one, but several paths will be taken. The most important ones are described beneath.

Restrictions on increasing livestock numbers

Increases of livestock numbers will be only allowed if the farm area is large enough for application of the manure in an acceptable way. This occurs if the yearly production of manure P is lower than 44 kg P ha^{-1}, which corresponds to 3.0 milking cows per ha or 17 fattening pigs per ha. The restriction brings on that on many farms in the Peel Region increases of livestock numbers will only be possible if a large area of land is bought.

Restrictions to the manure application rate

The final goal of the present policy is the introduction of a manure application rate whereby the input of nutrients to the soil equals the uptake by the crop. In practice this goal can only be reached by a policy whereby the maximum permitted application rate is reduced gradually in time. The maximum rates recently enacted are given in Table 9.

Phosphorus has been chosen as the limiting parameter, because it is stable as compared to N, it is easily applied in excess, and, if applied in excess, it endangers the environment by eutrophication.

Urine and filtrate from a manure separator contain relatively high quantities of N. Application of these liquids has been limited on grassland to 50 and on arable land to $25\,m^3\,ha^{-1}\,yr^{-1}$.

Calculations [14] showed that as a consequence of the new restrictions one third of the total manure production can not be applied on the producing farm in the initial years. About 1×10^6 t manure will be disposed outside the Peel Region. This amount will be transported to the northern and southwestern parts of the country and will only involve manure from fattening pigs and poultry.

Restrictions to time and method of application

As well as the application rate, the time and the method of application influence the leaching of nutrients to groundwater, the run-off of nutrients to surface water and the volatilization of ammonia to the atmosphere.

At this time insufficient storage is available to enable the introduction of severe restrictions on the time of manure application. In the first phase it will be forbidden to spread manure on grassland in the period 15 September–1 December, and when the soil is covered with snow also during 1 January–15 February. For arable land the restrictions depend on soil type. On the sandy soils of the Peel Region application will be forbidden between harvest and 1 November. If a second crop is grown only October will be closed.

Table 9. Maximum allowed rates for the application of P in the form of manure ($kg\,P\,ha^{-1}\,yr^{-1}$). In brackets $kg\,P_2O_5\,ha^{-1}\,yr^{-1}$

Type of land use	Period			
	1987–1991	1991–1995	1995–1999	after 1999
Grassland	109 (250)	87 (200)	76 (175)	–
Arable land — maize	153 (350)	109 (250)	76 (175)	–
— other crops } Horticultural land }	55 (125)	55 (125)	55 (125)	–

As for the method of application the following rules have been introduced. On arable land manure has to be incorporated within two days, for example by ploughing. A comparable measure for grassland would be an obligation to inject the slurry into the soil. A decision on such a directive has been postponed till more information is available about the effectiveness of injected P and K.

Transport to areas with a manure shortage

In many parts of the country manure production is rather low (Fig. 1) and manure from surplus areas can be used. In the past, so-called manure boards have been set up to promote this practice. They were prepared to mediate between producers and users.

The Peel Region forms the south-eastern part of the province North-Brabant in which a provincial manure board is working. This board undertook the following activities [33]:
- subsidizing of manure transport (Table 10);
- standardizing the conditions and prices of private companies;
- analyzing and controlling manure quality;
- promoting and supporting the construction of manure storages.

The transport, the acquisition of new manure users, and contact with present users was left to private companies. The board set out to create a good working climate.

Subsidizing of manure transport was possible if:
- the manure was liquid and had a certain minimum quality (DM content);
- the manure was transported from a region with a surplus to one with a shortage and the transport distance was a minimum of 75 km for poultry slurry and 50 km for any other type of slurry;
- a certain amount was payed for collecting the manure (Table 10).

In 1985 the North-Brabant manure board subsidized the transport of 160 000 m^3 slurry. About half came from the Peel Region. The average costs

Table 10. Some tariffs, in guilders (*f*) per m^3, used by the manure board of North-Brabant in 1985 [33]

	Type of slurry	
	Poultry	Other
Collecting charge	*f*1.75	*f* 3.00
Subsidy on transport		
distance 50–60 km	*f*0.00	*f* 3.50
distance 100–110 km	*f*4.00	*f* 9.00
distance 150–160 km	*f*7.25	*f*12.25

of transport were f7.87 per m³ higher than the payments by the slurry users. About one third of the deficit was covered by the collecting charges (f2.54 per m³).

The mean transport distance was 124 km. Fig. 6 shows in which directions the manure was transported. It also illustrates the journeys over distances of 8 to 75 (poultry slurry) and 8 to 50 km (other slurry). Subsidizing such journeys, which in 1985 accounted for 250 000 m³, is considered not to be necessary and stopped some years ago. The same can be said of the yearly transport of 200 000 m³ solid, mainly poultry, manure.

The transport subsidized nowadays by the manure board is distributed very unequally over the year. The months of August–October form a peak period. This is related to the fact that the slurry is transported to arable land where traditionally manure is applied at the end of the growing season and where no capacity is available to store slurry. The manure board is now promoting the construction of portable storage capacity, especially on farms where the use of slurry is ensured for several years by contracts.

The unsubsidized transport over 8 to 50 and 8 to 75 km, respectively, is mainly undertaken in spring and presumably to land cultivated with maize.

In April 1987 the manure boards got legal status. At the same time they were required to take on some new tasks like the obligation to remove

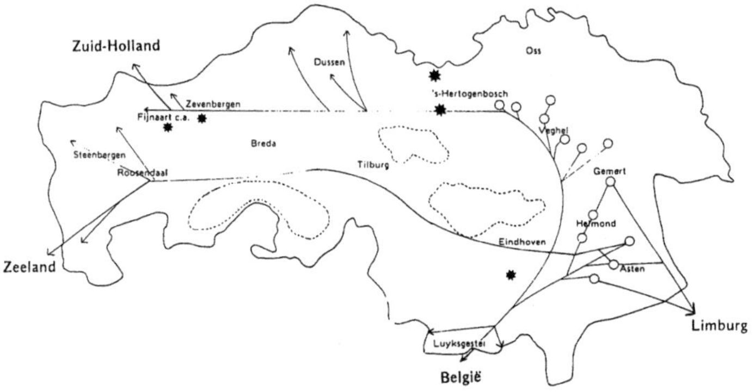

Figure 6. Directions of manure transports over more than 8 km in the working area of the North-Brabant manure board (Gemert and Asten are lying within the Peel Region).

Type of transport	Production area	User area
a. Subsidized transport: > 50 or 75 km (see text)	○	→
b. Unsubsidized transport: 8–50 km (pig slurry) 8–75 km (poultry slurry)	---------	*

manure upon payment of the costs involved and to register all manure transport.

Processing of manure

At different places in The Netherlands research into slurry processing is going on. Until now only small experimental plants with only one or a few steps of the total process have been tested.

Recently, plans have been developed for building large plants for complete processing of pig slurry. One plant will be built in the Peel Region with a capacity of $1 \times 10^5\,t\,yr^{-1}$ initially (at an investment of 21 million guilders), increasing after some years to $1 \times 10^6\,t\,yr^{-1}$ [16]. In the first stage the slurry will be digested anaerobically to produce energy as biogas, in the second the slurry will be separated into a cake and a filtrate. The cake will be dried. The N in the filtrate will be removed by nitrification/denitrification and the volume of the filtrate reduced by evaporation. The processing costs will be at least 15 guilders per ton and the costs to the farmer will be at least 10 guilders per ton.

Reducing the mineral and heavy metal production via animal nutrition

The mineral and heavy metal problem can be reduced by changing the diet of the animals. The following possibilities exist [20]:

- Use of concentrates and roughages which are better attuned to animal requirements. In theory this gives a wide assortment of special feeds for each animal category and stage of life. Some possibilities will presumably be exploited in practice because of economic benefits, others require a directive such as the recent lowering of the maximum copper content of feed for fattening pigs.
- Use of concentrates and roughages with a better availability of minerals. For producing concentrates other raw materials should be chosen. The addition of enzymes like phytase is another possibility.

Conclusion

Manure application rates are so high that a continuation would lead to unacceptable and irreversible pollution. In the next few years the new rules will lighten the most burdened and most vulnerable areas. At the same time new technical developments are being introduced whereby a final solution may be reached.

References

1. Aart P J M van der, Kemmers R H en Verhoeven J T A (1987) Eutrofiëring in het landelijk gebied: Problemen en mogelijke oplossingen vanuit een landschapsoecologisch kader. Verslag Symposium 'Integraal waterbeheer, een nieuwe aanpak' (in press)
2. Appelo C A J, Krajenbrink G J W, Ree C C D F van en Vasak L (1982) Beïnvloeding van de grondwaterkwaliteit in het infiltratiegebied van de noordwestelijke Veluwe. Bodembescherming 11, Ministerie van Volkshuisvesting, Ruimtelijke Ordening en Milieubeheer. Den Haag: Staatsuitgeverij
3. Bleuten W en Cerutti M (1984) De huidige en toekomstige nitraat- en sulfaatbelasting van grond- en drinkwater van de Nederlandse pleistocene zandgebieden. H_2O 17, 208–212
4. Boer W M J den (1986) Vitaliteit van naaldbossen in de Peel. ROM 12, 3–9
5. Breeuwsma A en Schoumans O F (1986) Fosfaatophoping en -uitspoeling in de bodem van mestoverschotgebieden. Rapport 1866. Wageningen: Stichting voor Bodemkartering
6. Breimer T en Smilde K W (1986) De effecten van organische mestdoseringen op de zware metaalgehalten in de bouwvoor van akkerbouwgronden. In Titulaer H H H, ed. Themadag 'Organische stof in de akkerbouw'. Themaboekje 7. Lelystad: Proefstation voor de Akkerbouw en Groenteteelt in de Vollegrond
7. Centraal Bureau voor de Statistiek (1982) Produktie van dierlijke mest 1980. Den Haag: Staatsuitgeverij
8. Centraal Bureau voor de Statistiek (1986) Produktie van dierlijke mest 1984. Kwartaalbericht Milieustatistieken 2 (4), 7–11
9. Centraal Bureau voor de Statistiek (1986) Several detailed statistics on land use, livestock density and manure production in 1984. Voorburg: Centraal Bureau voor de Statistiek
10. Diest A van (1986) Aanvullende bemesting met mineralen in Nederlandse bossen. Meststoffen 2/3, 4–8
11. Dijk T A van (1979) Maximaal toelaatbare hoeveelheden dierlijke mest op bouw- en grasland. Stikstof 8, 290–296
12. Dijk T A van (1985) De uitspoeling van mineralen op bouwland waaraan jaarlijks drijfmest wordt toegediend. Rapport 2–85. Haren: Instituut voor Bodemvruchtbaarheid
13. Duyvenboden W van, Gast L F L en Taat J (1985) Landelijk Meetnet Grondwaterkwaliteit. Bodembescherming 46A en 46B. Ministerie van Volkshuisvesting, Ruimtelijke Ordening en Milieubeheer. Den Haag: Staatsuitgeverij
14. Gorissen A (1986) Berekening, transport en verwerking van de mestoverschotten voor twee fasen. Intern rapport. Den Haag: Landbouw Economisch Instituut
15. Haan F A M de, Lexmond Th M en Riemsdijk W H van (1986) Belasting van bodem en water door mineralenoverschotten uit de intensieve veehouderij. Milieu 2, 34–43
16. Hengel H van den (1986) Van drijfmest naar Promest? Boer en Tuinder 40 (2004), 20–21
17. Instituut voor Bodemvruchtbaarheid (1978) De mest- en gierverspreiding op landbouwgrond in de EG. I. Wetenschappelijke basis voor het beperken van de verspreiding en criteria voor regulerende maatregelen. Informatie over landbouw. Rapport 47. Brussel: Commissie van de Europese Gemeenschappen
18. Instituut voor Cultuurtechniek en Waterhuishouding (1985) Jaarverslag 1984, pp 71. Wageningen: Instituut voor Cultuurtechniek en Waterhuishouding
19. Instituut voor Cultuurtechniek en Waterhuishouding (1986) Jaarverslag 1985, pp 32. Wageningen: Instituut voor Cultuurtechniek en Waterhuishouding
20. Jongbloed A W, Steg A, Simons P C M, Janssen W M M A, Lenis N P, Meijs J A C en Vreman K (1985) Berekeningen over de mogelijke vermindering van de uitscheiding aan N, P, Cu, Zn en Cd via de voeding door landbouwhuisdieren in Nederland. Mededeling 3. Lelystad: Instituut voor Veevoedings Onderzoek

21. Kemmers R H (1986) Perspectives in modeling of processes in the root zone of spontaneous vegetation at wet and damp sites in relation to regional water management. In: Water management in relation to nature, forestry and landscape management. Proceedings and Information 27. Den Haag: Commission for Hydrological Research TNO
22. Kemmers R H en Jansen P C (1985) Hydrologie in relatie tot de beschikbaarheid van vocht en voedingsstoffen voor natuurlijke begroeiingen. Cultuurtechnisch Tijdschrift 24 (4), 195–211
23. Koppers R G M (1984) Onderzoek naar de verplaatsing van stikstof in de ondergrond van een veehouderijbedrijf ten oosten van Deurne (N-B). Nota 1519. Wageningen: Instituut voor Cultuurtechniek en Waterhuishouding.
24. Korzelius E en Breeuwsma A (1983) Het fosfaatbindend vermogen van zandgronden. Rapport 1745. Wageningen: Stichting voor Bodemkartering.
25. Landbouwadviescommissie Milieukritische Stoffen (1986) Signaalwaarden voor de gehalten van milieukritische stoffen in grond met het oog op landbouwkundige gebruiksmogelijkheden van verontreinigde bodems. Den Haag: Ministerie van Landbouw en Visserij
26. Meeuwissen P C en Meer H G van der (1987) Emissie van stikstof en fosfaat uit landbouwgronden in relatie tot bemesting en bedrijfsvoering. Wageningen: Centrum voor Agrobiologisch Onderzoek (in press)
27. Ministerie van Verkeer en Waterstaat (1986) Indicatief Meerjarenprogramma Milieubeheer 1985–1989. Den Haag: Staatsuitgeverij
28. Mohren G M J, Burg J van den, Burger F W en Oterdoorn J H (1986) Fosforgebrek veroorzaakt door hoge stikstoftoevoer in Douglas-opstanden. Nederlands Bosbouw Tijdschrift 58 (9), 239–245
29. Oosterom H P en Steenvoorden J H A M (1984) Drijfmestgiften op snijmaïspercelen (zandgrond) en de uitspoelingsverliezen naar het grondwater. Nota 1499. Wageningen: Instituut voor Cultuurtechniek en Waterhuishouding
30. Prins W H en Boer D J den (1985) Aspects of nitrogen and phosphate fertilization on grassland. Fertilizers and Agriculture 90, 27–36
31. Riemsdijk W H van, Lexmond Th M en Haan F A M de (1983) Fosfaat- en kopertoestand van de cultuurgrond in de provincie Gelderland. Wageningen: Landbouwhogeschool
32. Ruiken M J en Steenvoorden J H A M (1986) Organische stofgehalten in de ondergrond van het Zuidelijk Peelgebied. Nota 1709. Wageningen: Instituut voor Cultuurtechniek en Waterhuishouding
33. Stichting Brabantse Mestbank (1986) Jaarverslag 1985. Tilburg: Stichting Brabantse Mestbank
34. Schröder J (1985) Effecten van grote drijfmestgiften bij de teelt van snijmaïs. Publikatie 30. Lelystad: Proefstation voor Akkerbouw en Groenteteelt in de Vollegrond
35. Waterschap De Aa (1986) Jaarverslag 1985. Den Bosch Waterschap De Aa
36. Wijnands J H M, Kruif F F de, Lodder K en Luesink H H (1983) Het kunstmestgebruik in de land- en tuinbouw in 1979/1980. Rapport 3. 125. Den Haag: Landbouw Economisch Instituut

Nitrogen losses from animal manures: from grazed pastures and from applied slurry

S C JARVIS[a], M SHERWOOD[b] and J H A M STEENVOORDEN[c]
[a] *Institute for Grassland and Animal Production, Hurley, Maidenhead, Berkshire SL6 5LR, UK*
[b] *An Foras Taluntais, Johnstown Castle Research Centre, Wexford, Ireland*
[c] *Institute for Land and Water Management Research (ICW), P.O. Box 35, 6700 AA Wageningen, The Netherlands*

Key words: ammonia volatilization, denitrification, excreta, fertilizer, grazing, leaching, nitrate, nitrogen, slurry, surface run-off

Abstract. Less than 20% of inorganic fertilizer nitrogen (N) applied to grass swards is recovered in animal products. In this paper, the extent of N transformations and losses either from excreta returned directly to grazed pastures or from slurry applied to grassland is reviewed. The three major pathways of N loss after return of excreta, namely gaseous emission (*i.e.* through ammonia (NH_3) volatilization and denitrification), surface run-off and leaching are discussed. The impact of the grazing ruminant or the application of slurry on potential N loss is considerable. For example, NH_3 volatilization can result in losses of from 60–80% of the urinary N applied to pastures. Under grazing, significant NH_3 losses have been recorded which are greatest (i) during or immediately after the grazing period, (ii) with high stocking densities and (iii) in cycles of wetting and drying. NH_3 losses from applied slurry are variable, but are always greatest immediately after application and are reduced by injection. There are strong interactions between the denitrification and leaching processes which have yet to be quantified; both processes are greater under grazed than under cut swards. Denitrification rates are dependent upon temperature, moisture and nitrate status. Slurry application, through increased soluble carbon contents and anaerobic conditions, increases the potential for denitrification; this is greatest when slurry is injected rather than surface-spread.

There have been few studies of run-off from grazed swards but, although greater than from cut swards, the rates of loss by this means are low. Major proportions of the run-off loss that occur from slurry are associated with events immediately after application. Relationships between rates of slurry application and N in run-off waters have been established. The N accumulated from animal returns to grazed swards results in leaching losses of nitrate-N that are from 3.5 to 6 times greater than those from cut swards. There is little information on the extent of nitrate loss following slurry addition, although it is known that losses depend upon rate and time of application. Injection of slurry does not increase the risk of leaching, provided that rates of fertilizer addition are modified to take account of the mineral N in the slurry.

Introduction

Concern over the losses of nitrogen (N) from grassland has grown considerably with respect to both economic and possible environmental consequen-

H.G. v.d. Meer, et al. (eds.), Animal Manure on Grassland and Fodder Crops.
 ISBN 978-94-010-8137-5.

ces [34, 35]. For example, from an annual input of more than 500 kg N ha^{-1} from mineral fertilizers and purchased feeds, less than 20% is exported from the farm in milk and meat [27]. Nitrogen has a well-defined cycle of transformations in soil-plant systems. In grassland, this cycle is much influenced by the presence of grazing animals and the return of the major part of the consumed N, either directly in dung and urine, or in slurries from housed animals. The extent of the losses associated with these returns have been difficult to quantify, especially under grazing. Recent techniques, developed for use on both plot and field-scale trials [9, 40, 43] have enabled balances of N inputs and outputs to be better established. Such information is essential before management strategies can be devised which will permit more efficient use of both fertilizer N and N returned in manures and slurries. In this review we indicate how work on the three major pathways of N loss from excreta, *i.e.* through gaseous emission, surface run-off and leaching has progressed over recent years.

Gaseous losses

The losses of N in gaseous forms after the return of excreta to grassland are considerable. Two major processes are involved (i) the volatilization of ammonia (NH_3) after the hydrolysis of urea and (ii) denitrification, *i.e.* the reduction of nitrate to the gases nitrous oxide (N_2O) and nitrogen (N_2).

Volatilization of ammonia

Of the N ingested by ruminants, some 75–95% is excreted, and the major proportion (up to 80%) is contained within urine. Losses of NH_3 largely result from the high urea or ammonium (NH_4)-ion contents of urine and slurry.

Losses under grazing. The deposition of urine is very uneven, and urine affected areas can contain the equivalent of 300–1000 kg N ha^{-1} [36], some of which can be lost as NH_3. Studies of losses from urine-treated pasture [2, 3] indicate that the retention of urinary N in the soil-plant system in improved pasture is only 20–40% of that added.

Urea hydrolysis is rapid and obeys first order kinetics during the first 24 h after application [45] but volatilization is then very dependent upon environmental conditions. In New Zealand, the greatest losses occurred during warm, dry conditions and were much reduced during cooler and/or wetter conditions [1]; 16, 66 and 5% of the urinary N applied was subsequently

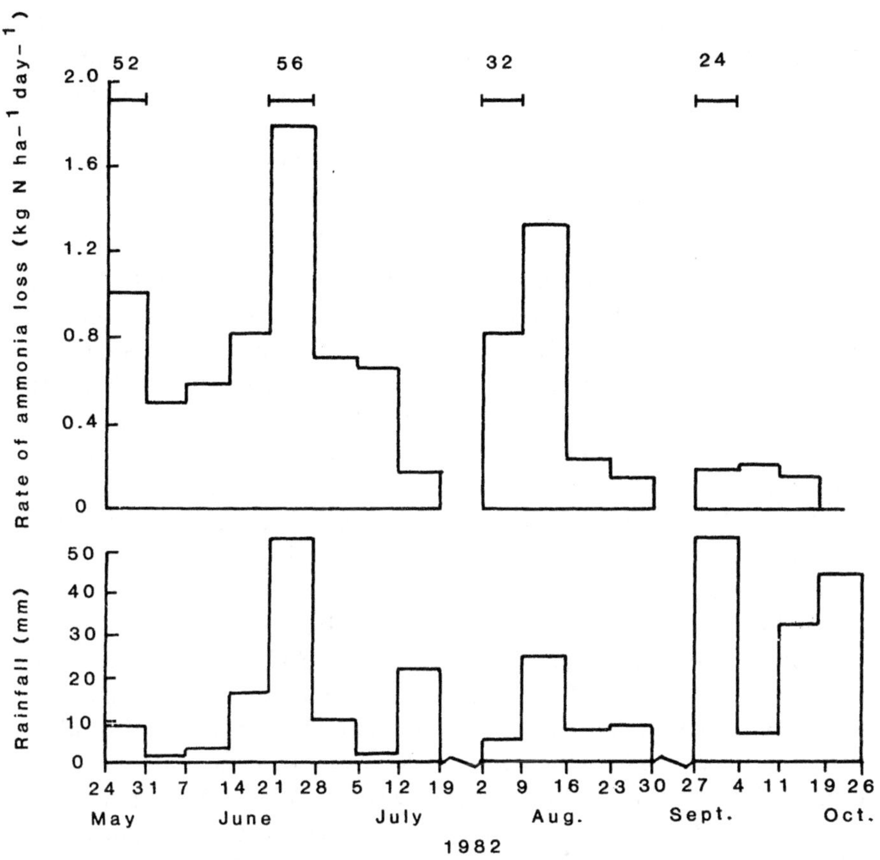

Figure 1. Mean losses of NH_3 from grazed swards (yearling steers in a 28-day rotation) receiving 420 kg N ha^{-1} yr^{-1} and corresponding values for accumulated rainfall. Bars show grazing period and stocking rates (steers per ha) (from Ryden [36]).

volatilized under warm moist, warm dry and cool moist conditions, respectively. The higher rate of loss under New Zealand conditions was also approached in the UK where 40% of the urinary N applied to grass plots was volatilized over a 17-day period [41]. Other recent results from rotationally grazed pasture at Hurley (Fig. 1) indicate that average NH_3 losses in excess of 1 kg N ha^{-1} day^{-1} may occur [36]. The following observations were made from the study at Hurley, (i) the highest rates of loss occur during and immediately after grazing, (ii) the highest rates of loss are associated with the high stocking densities, (iii) cycles of wetting and drying promote NH_3 loss and (iv) losses continue at a reduced rate but are still in excess of 0.2 kg N ha^{-1} day^{-1} for at least three weeks after grazing stops. The continued release from grazed swards contrasts with that from urine-treated swards from which NH_3 losses are almost complete within 7 days after application [35]. Although losses from a mixed ryegrass/clover sward were only one-tenth of those from a ryegrass sward [35], Australian studies on a sheep-grazed lucerne pasture [10] found losses of 0.2–0.7 kg N ha^{-1} day^{-1}. On a yearly basis, this represented a loss of approximately 50% of the 200 kg N ha^{-1} fixed by the lucerne.

Losses from applied slurry. Losses of NH_3 will occur directly from excreta deposited in the buildings that house animals and also during slurry collection and storage. There have been only a few measurements [36] of the magnitude of these losses. It has, however, long been known that further losses occur when slurries are spread on to land surfaces, but the extent of these have usually been estimated indirectly from shortfalls in the total N budget. The variability in the extent of the loss is indicated by the wide-ranging effects that slurry addition can have on subsequent crop growth, *i.e.* through a variable supply of available N. Heck [15] found that 37 to 50% of the N in previously anaerobically stored manure was lost by exposure to dry conditions. Lauer *et al.* [23] showed that NH_3 loss from slurry followed a first order rate process and concluded that general evaporative conditions and precipitation were the major controlling factors. In a survey of a number of studies [22], it was also shown that (i) environmental conditions are critical in determining the release of NH_3 and (ii) 80% or more of the loss occurs within 10–15 days. Initial high rates of NH_3 loss after application are a common finding; thus when cattle slurry was applied to grassland at 80 t ha^{-1} to give 248 kg total N ha^{-1}, 1.65 kg NH_3-N ha^{-1} h^{-1} were lost during the first 12 h after application, but this fell to 0.10 kg ha^{-1} h^{-1} within four days [56]. Research in Canada [5] found that the 24–33% losses of the NH_3-N occurring within 6–7 days from the application of dairy shed slurry containing 325 kg N ha^{-1}, resulted in hourly rates of losses as high as 2.9 kg N ha^{-1}. Furthermore, up to 0.3 kg N ha^{-1} h^{-1} continued to be lost in cold wet weather when conditions are presumed to minimize the potential for loss.

A series of recent measurements with wind tunnels has also shown similar rates and timings of losses from cattle slurry applied to ryegrass swards on a freely-drained loam (Table 1). During the first 24 h, from 9–38% of the applied inorganic N was lost as NH_3 from applications of 40–80 t slurry ha^{-1}. The losses of mineral N through NH_3 volatilization from a 80 t ha^{-1} application ranged from 27 to 75 kg N ha^{-1} over 10 days; the losses from spring applications were only 40% of those made in winter. In one of the studies [56], total losses of N through NH_3 volatilization represented 77 and

Table 1. Losses of NH_3-N from surface-applied cattle slurry to a perennial ryegrass sward

Application			Losses of NH_3-N, kg N ha^{-1}		% of inorganic N lost as NH_3	
Time	Rate		Range on a daily basis	Cumulative loss over 10 days	After 24 h	After 10 days
	t ha^{-1}	kg inorganic N ha^{-1}				
November[1]	40	56	nd	28	nd	50
	80	112	0–30.0	75	27	67
March[1]	40	56	nd	16	nd	34
	80	112	0.7–12.7	27	11	28
December[2]	80	105	0–39.6	67	38	64
April[2]	80	110	1–9.6	29	9	26

[1] Pain *et al.* [32]
[2] Thompson *et al.* [56]

50 kg N ha^{-1} from winter and spring applications, respectively, *i.e.* 73 and 45% of the inorganic N contents.

Denitrification

Losses under grazing. Denitrification loss from grassland is likely to be greater than that from arable cropping because of the greater amounts of oxidizable carbon present under grass. Although the losses of N through denitrification in grassland have received more attention than NH_3 losses, until recently measurements have been confined largely to cut swards. Inorganic N originating from urine and not lost through NH_3 volatilization may be nitrified, resulting in nitrate-N concentrations equivalent to 200–1710 kg ha^{-1} [4]. Under appropriate conditions, part of this nitrate-N may be lost as N_2 or N_2O through denitrification, otherwise it may leach to ground or surface waters.

Measured rates of denitrification under a grazed sward are from 2 to 20 times greater than those from a cut sward receiving an otherwise similar treatment [38]. Table 2 summarizes some of the rates of loss found under grazing and cutting and also indicates the immediate influence of excreta deposition. Higher rates of denitrification occur towards the end of the grazing season (Table 2) when the nitrate derived from excreta or fertilizer N and not utilized by the sward accumulates, and the soil returns to field capacity. There is then a decline in activity with declining soil temperature. In spring the rapid uptake by the growing sward of any nitrate that has not been lost by leaching prevents major losses by denitrification even though environmental conditions may be as favourable as those at the end of the previous season.

Table 2. Denitrification losses from (a) irrigated grassland during summer after grazing or cutting, and adding 60 kg N ha^{-1} [38]; (b) unirrigated swards during autumn after 420 kg N ha^{-1} added during the growing season [42]

	Rate of denitrification kg N ha^{-1} day^{-1}		
	Cut sward	Grazed sward	
		Areas not affected by excreta	Areas affected by excreta
a. *Irrigated*			
Day 3	0.040	0.058	0.199
Range over 20 days	0.004–0.040	0.013–0.064	0.032–0.4
b. *Unirrigated*			
21st October (soil temp. 11 °C)	< 0.001	0.164	0.318 (urine) 0.386 (dung)
30th November (soil temp. 6 °C)	0.009	0.016	0.027 (urine) 0.073 (dung)

Present evidence indicates that high rates of denitrification occur when nitrate concentrations are at or above 5–10 mg N kg^{-1} and when the upper part of the horizon is at or close to field capacity. The strong interaction between nitrate content, temperature and moisture status and losses either by denitrification and/or leaching have been well demonstrated but have yet to be quantified.

Losses from applied slurry. Effects of slurry application on denitrification losses have been the subject of an extensive recent review [11]. The addition of large quantities of organic matter and the resultant increase in anaerobic conditions increase the potential for denitrification loss.

There is a soil zone immediately beneath the applied slurry which is rich in anaerobic microsites and this prevents nitrate accumulating in subsoils [6, 60]. Continuous detection of significant rates of denitrification at 40 cm for several months after slurry application during winter [11] also indicates that nitrate accumulation at depth is reduced by this process. Comparison of the effects of cattle slurry or fertilizer N addition show that slurry greatly extends the period in which N_2O is detected in the soil atmosphere.

A number of recent studies at Hurley have directly measured losses after winter and spring applications of slurry [32, 56]. On the well-drained Hurley soil, variation in denitrification was largely a reflection of the variation in soil temperature. Surprisingly, significant denitrification continued at temperatures below 5 °C, an effect which is not usually recorded with inorganic

N addition. However, in laboratory studies [46] the presence of a suitable carbon source promoted significant rates of denitrification between 2 and 5 °C. The form of the carbon in slurry may encourage such effects in the field. In the Hurley studies [56], application of 105 kg NH_4^+-N ha^{-1} (80 t slurry ha^{-1}) during December, resulted in rates of denitrification equivalent to losses of 0.03 to 0.47 kg N ha^{-1} day^{-1} during the first 10 weeks after application and continued over the next 2 months at less than 0.12 kg ha^{-1} day^{-1}. Total losses over the whole period were equivalent to 30 kg N ha^{-1} (*cf.* 53 kg N ha^{-1} from the same rate of injected slurry). Application of slurry during April (110 kg N ha^{-1}) resulted in a maximum daily rate of denitrification of only 0.2 kg N ha^{-1} and total losses of 5 kg N ha^{-1} over an eight week period (*cf.* 18 kg N ha^{-1} for injected slurry). Denitrification losses from the surface-spread slurry were equivalent to 12 and 2% of the total N added (248 and 262 kg N ha^{-1}, respectively, for winter and spring applications).

Run-off

Losses from grazed pastures. Studies on N losses through run-off from grazed pastures have mostly been carried out in the USA. Some of these studies were carried out in watersheds or on small catchments so that the collected water contained N derived from sub-surface lateral flow as well as surface run-off. Nitrogen fertilizer had also been applied to some of these pastures and may have contributed directly to the N content of the run-off water.

Greater losses of N were reported [29] from continuously grazed than from rotationally grazed pastures in Oklahoma but both types of pasture lost less in run-off than they received in precipitation. In a 4-year study in Ohio [7], mean losses of < 1.0 and 10.4 kg N ha^{-1} yr^{-1} were found in surface and in sub-surface lateral flow, respectively, from summer grazed pastures, compared with 6.8 and 4.5 kg N ha^{-1} yr^{-1} from winter grazed pastures where animals were fed supplementary hay. Run-off losses of total Kjeldahl-N and nitrate from pastures in Nebraska [44] were decreased by 19%, and increased by 45%, respectively, in the presence of grazing animals. In another study in Idaho, also involving extensive management, losses of < 1 to 4 kg N ha^{-1} yr^{-1} occurred from grazed pastures, compared with < 0.5 kg ha^{-1} yr^{-1} from non-grazed areas [18].

Nutrient run-off from a grazed unfertilized pasture was compared with that from an adjacent woodland in Ohio [30]; N losses were found to be similarly low from both. Owens *et al.* [31] studied surface and sub-surface runoff from rotationally grazed pastures receiving 224 kg N ha^{-1} yr^{-1}. They found very little N transported in surface run-off but found mean annual transport of 46.6 and 26.2 kg N ha^{-1} in sub-surface lateral flow from winter grazed and summer grazed pastures, respectively.

In summary therefore, the N losses through surface run-off from manures deposited on grazed pastures are low and would account for only a very small fraction of the total annual N losses in intensive systems. Nitrogen transport in sub-surface run-off was high in one study [31], but part of this N may have been derived directly from the applied fertilizer rather than from animal excreta *per se*.

Losses from land-applied slurry. Research on run-off has intensified in the past 10–15 years in an effort to provide a clearer understanding of the factors which contribute to non-point source pollution. The results of this research have incidentally provided a good general picture of the magnitude of N losses from land-applied slurry and the factors which govern it.

It is notable that in many studies big losses (10–20%) of applied nutrients (including N) were mostly associated with run-off events which occurred soon after slurry application. These run-off events sometimes resulted from snowmelt while others resulted from heavy rainfall soon after spreading on frozen or poorly drained soils. Thus, N losses of up to 20% were found [28] when dairy manure was spread on frozen ground and this was followed by conditions which favoured early spring run-off. Hensler *et al.* [17] reported a loss of 10% of the applied N when 19 mm rain fell within 24 hours of spreading dairy manure on frozen ground. Similarly, N losses of up to 20% from manure spread on frozen lucerne plots in spring have been reported [62], and increased losses resulting from adverse weather conditions following winter spreading in the USA have also been reported [8, 20]. Uhlen [57] found N losses of 2–25% from manure spread in winter or autumn in Norway, with the highest losses occurring when slurry was spread on 50 cm of packed snow. Following laboratory and field investigations at Cornell University, it was estimated [51] that where infiltration is limited, as much as 50% of manurial N could be lost in run-off. It was also pointed out, however, that frozen soil is not necessarily impermeable with movement of slurry occurring down cracks. At Johnstown Castle, cumulative total N losses of up to 17% of the applied N were recorded where the first run-off event occurred one day after spreading on poorly drained soils which were not frozen [48].

As the time interval between spreading and the first run-off event increases, the loss of N in the run-off water decreases rapidly [33, 54, 61]. This may be partly due to a reduction in the N present through volatilization of NH_3 during that interval, but it also reflects increased adsorption of NH_4-ions by the soil colloids. Cumulative losses under these circumstances are therefore reduced and may vary from 0–10% depending on the interval to first run-off, and the total rainfall [24, 48].

Good contact between the NH_4-ions and the soil colloids is also a probable reason why N run-off losses from ploughed ground are lower than from sodded areas [16, 62]. Injection or incorporation of the manure into the soil results in negligible N losses through run-off (and also NH_3 volatilization) [25, 33, 57].

There are several reports that run-off volume from manured areas decreases with time compared with untreated control areas [16, 26, 62] due to improved infiltration. At Johnstown Castle this trend has been observed on a manured, well-drained soil but not on a poorly-drained heavy clay soil (M. Sherwood, unpublished results).

Effect of rate of application. There is generally a good relationship between the rate of manure application and the concentration of N and the total content of N in run-off water. Khaleel *et al.* [19] examined data from six major studies in the USA, and concluded that the total N lost in run-off (y) could be described as a function of the N applied (x) by simple equations (data expressed in $\mathrm{kg\,N\,ha^{-1}\,yr^{-1}}$)

(i) for areas which are subject to snowmelt

$$y = 0.105x - 6.39$$

(ii) for areas which are not subject to snowmelt

$$y = 0.00161x + 5.6$$

Although such equations may be useful indicators of trends, equation (ii) takes insufficient account of the high concentrations of N in run-off water which can result from rainfall soon after manure application.

At Johnstown Castle a very good relationship between total N concentration in run-off water and the load applied at two different experimental sites has been found (Table 3). It can also be seen that N concentrations from treatments at the Castlebridge site, where four days elapsed before the occurrence of the first run-off event, were very much lower than at the Burrel's Field site where the first run-off event occurred within two days of spreading. This is in agreement with previous findings [33, 61].

The N concentration in the water from each consecutive run-off event is always much lower than that of the previous event. This has been found in a number of studies [52] and is also evident in Table 3.

Table 3. Kjeldahl-N (TKN) added in pig slurry at two rates of application at two sites in October 1980, and concentrations in run-off water at first and second run-off events after application

Treatment	1st run-off event					2nd run-off event			
	TKN applied (kg ha^{-1})	Days after application	Vol. run-off (mm)	TKN conc. (mg l^{-1})	TKN in run-off (kg ha^{-1})	Days after application	Vol. run-off (mm)	TKN conc. (mg l^{-1})	TKN in run-off (kg ha^{-1})
Burrel's Field (imperfectly drained loam)									
High rate	432	2	10.7	377	40	8	11.1	70	7.8
Low rate	212	2	15.7	192	30	8	12.5	30	3.8
Castlebridge (impermeable gley)									
High rate	490	4	26.5	101	27	6	9.1	66	6.0
Low rate	263	4	17.5	63	11	6	7.0	43	3.0

Leaching

Leaching of N is a natural process which takes place when rainfall exceeds water loss by evapotranspiration. The movement of nitrate into water systems in this way is one of the major causes of concern over losses of N. The amounts of N leached will be dependent not only on the amount of nitrate present but also on climatic factors (e.g. amount and timing of rainfall, evapotranspiration), soil type and depth of groundwater table, as well as interactions with the growing crop. The inefficiency in N use in grassland, especially under grazing systems, may lead to high nitrate losses through leaching [13, 55].

Losses under grazing. Most direct measurements of leaching from grassland have been carried out in lysimeter based experiments under cut swards. Thus, Garwood *et al.* [14] reported an average annual loss of 7-9 kg N ha^{-1} over a 8-year period from a sward on a freely drained loam in S.E. England which received 250 kg N ha^{-1} yr^{-1}. Research of this nature has led to the generally held view that leaching losses from grassland are small. However, Kolenbrander [21] indicated that grazing cattle increase N leaching losses by factors of 3.5 to 5. Recently, it has been shown [39] that when 420 kg N ha^{-1} yr^{-1} were applied to cut or grazed swards, 29 and 162 kg N ha^{-1} yr^{-1}, respectively, were lost by leaching. Work in New Zealand, has indicated that from 110 to 190 kg N ha^{-1} yr^{-1} are also lost from intensively managed grass/clover swards [50]. Much of the increased loss under grazed systems results from the accumulation of inorganic N from animal returns during the grazing season.

Losses from applied slurry. As in the case of grazed grassland, nitrification has to occur before there is significant movement of N down the soil profile; there is little leaching of either ammonium ions or forms of N other than nitrate. There have been only few attempts to measure N lost by leaching after slurry addition. In general, on arable land there is a linear increase in the N content of groundwaters with increasing slurry addition [53, 58]; approximately 25% of the applied slurry N is lost by leaching. With additions of slurry equivalent to 540 kg N ha^{-1}, it was found [58] that between 20 and 30% of the N applied in autumn was leached. Time of application has an important effect, the amount leached from grassland decreasing as the date of application approaches the start of crop growth [58]. Losses approaching those found by the Dutch and German workers have also been found under Irish conditions [47]. However, in the Irish study [47], a striking effect was that of soil type; whereas 15% of the applied N was leached from

Table 4. Nitrogen application rates (kg N $ha^{-1} yr^{-1}$) as mineral fertilizer and injected or surface-applied cattle slurry averaged for 5 experimental years. Each slurry rate was also combined with each mineral fertilizer rate

Nitrogen source	Nitrogen applied (kg $ha^{-1} yr^{-1}$)			
	Rate 1	Rate 2	Rate 3	Rate 4
Mineral fertilizer	0	190	380	580
Cattle slurry (surface)	0	50	100	190
Cattle slurry (injected)	0	100	190	380

a well-drained loam, only 2% was leached from an impermeable gley. This contrasts with findings from a French study [12] in which soil type had only a minor effect on amounts of N leached.

An experiment has recently been conducted to compare leaching after injection of slurry with that from mineral fertilizers and slurry spread on the surface of a loamy sand at Ruurlo in The Netherlands (Steenvoorden, unpublished) (Table 4). Slurry application took place in early spring and mineral fertilizer N addition was split into 5 to 7 dressings depending on the number of cuts in a growing season. Over the 5-year period of study, the average yearly values for precipitation and evaporation (from open water) were 754 and 640 mm, respectively, and the precipitation surplus discharged to groundwater was 265 mm per year. Nitrate leaching was determined by collecting soil water samples at 1 m depth at regular time intervals in autumn and winter with ceramic cup samplers [55], and was calculated as the product of water fluxes and nitrate concentrations. The influence of surface-applied or injected slurry on nitrate leaching was quite different (Fig. 2). Leaching was not affected by surface-spread slurry and the same relationship was found for all three rates of slurry application. Injection of 100 kg N $ha^{-1} yr^{-1}$ as slurry did not lead to higher leaching losses, despite the greater availability of inorganic N. However, at the highest rate of injection, leaching increased sharply at all rates of application of mineral fertilizer N.

The relationship between 'available mineral N' (*i.e.* the sum of mineral N from fertilizers and slurry) and nitrate leaching losses (Fig. 3) was different for mineral fertilizers, injected or surface-spread slurries. Available mineral N from the slurry was estimated by multiplying total slurry N input and relative apparent recovery of slurry N (equivalent to the ratio of apparent recovery of slurry N/apparent recovery of fertilizer N). The relative apparent recovery of slurry N was 24% when surface-spread and 58% when injected. Leaching was greatest from the treatment receiving mineral N alone, and the use of slurry in combination with mineral N eventually resulted in lower rates of leaching (Fig. 3). This may, in part, be explained by an increase in denitrification at the end of the season through inputs of

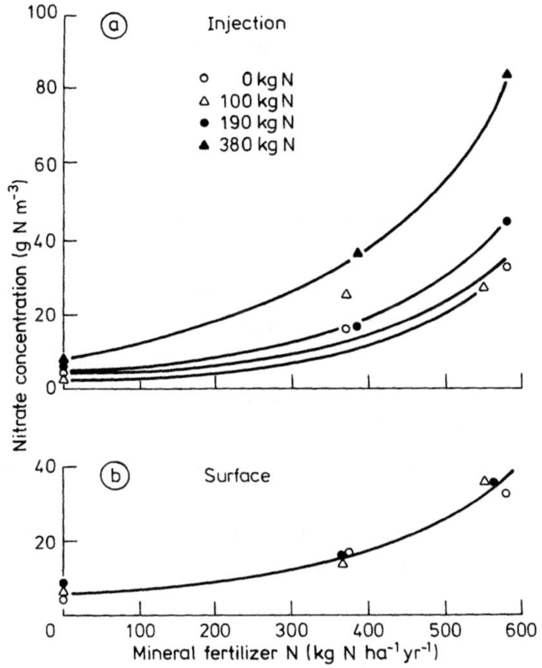

Figure 2. Nitrate concentrations ($g\,N\,m^{-3}$) in leachates at 1 m depth as influenced by N application rates ($kg\,ha^{-1}\,yr^{-1}$) from fertilizer and from cattle slurry applied by injection (a) and on soil surface (b).

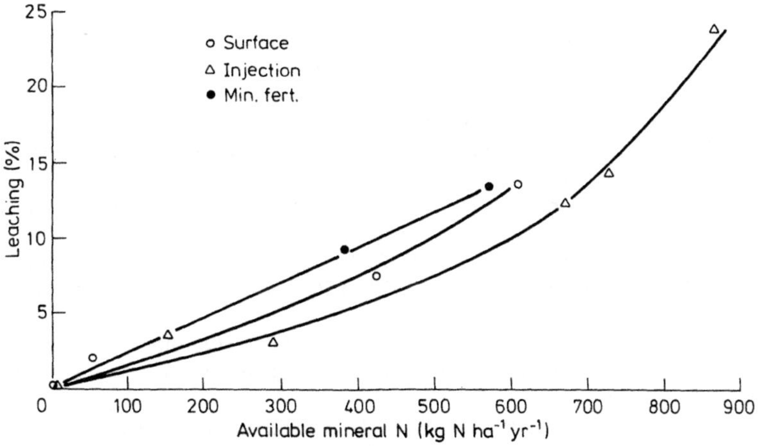

Figure 3. Nitrate leaching losses, as a percentage of 'available mineral N' (mineral N from fertilizers plus that from slurry) as influenced by available inorganic N from slurry and fertilizer.

organic carbon. The effects during the growing season are included in the calculation of apparent N recovery. Extra denitrification losses for surface-applied slurry will be small since the aeration status of only a narrow zone of the top soil is affected; in contrast, injection of slurry at approximately 15 cm results in anaerobic conditions in a much larger zone of soil and thus increases denitrification. The increased rate of denitrification, and a probable greater N uptake by the sward result in lower nitrate leaching from slurry at higher N levels. From an environmental point of view, slurry injection in spring is a promising approach since it results in lower rates of NH_3 volatilization, higher apparent N recovery, lower needs for mineral fertilizer N, lower nitrate leaching losses and reduced surface run-off.

The conclusion from this and other studies [54] is that slurry injection in grassland does not increase the risk of leaching provided that the rate of fertilizer N application is modified to take account of the higher availability of the mineral N from slurry. Efficiency of N use in grassland systems can therefore be improved by slurry injection instead of surface spreading. However, the consequences of this for nitrate leaching on different soil types have not been examined to any extent.

References

1. Ball P R (1982) Nitrogen balances in intensively managed pasture systems. In Gandar P W, ed. Nitrogen balances in New Zealand ecosystems, pp 47–66. Palmerston North, New Zealand: DSIR
2. Ball P R and Keeney D R (1981) Nitrogen losses from urine-affected areas of a New Zealand pasture under contrasting seasonal conditions. In Smith J A and Hays V W, eds. Proceedings of the XIV International Grassland Congress (Lexington), pp 342–344. Boulder, Colorado: Westview Press
3. Ball P R, Keeney D R, Theobald P W and Nes P (1979) Nitrogen balances in urine-affected areas of a New Zealand pasture. Agronomy Journal 71, 309–314
4. Ball P R and Ryden J C (1984) Nitrogen relationships in intensively managed temperate grasslands. Plant and Soil 76, 23–33
5. Beauchamp E G, Kidd G E and Thurtell G (1982) Ammonia volatilization from liquid dairy cattle manure in the field. Canadian Journal of Soil Science 62, 11–19
6. Burford J R, Dowdell R J, Crees R and Hall K C (1979) Soil aeration and denitrification. ARC Letcombe Laboratory Annual Report 1978, p 26. Wantage, UK: Letcombe Laboratory
7. Chichester F W, Van Keuren R W and McGuinness J L (1979) Hydrology and chemical quality of flow from small pastured watersheds: II. Chemical quality. Journal of Environmental Quality 8, 167–171
8. Converse J C, Bubenzer G D and Paulson W H (1976) Nutrient losses in surface runoff from winter spread manure. Transactions of the American Society of Agricultural Engineers 19, 517–519
9. Denmead O T (1983) Micrometeorological methods for measuring gaseous losses of

nitrogen in the field. In Freney J R and Simpson J R, eds. Gaseous loss of nitrogen from plant-soil systems, pp 131–157. The Hague: Martinus Nijhoff/Dr Junk

10. Denmead O T, Simpson J R and Freney J R (1974) Ammonia flux into the atmosphere from a grazed pasture. Science 185, 609–610
11. Dowdell R J (1981) Denitrification in soils treated with animal slurry. In Brogan J C, ed. Nitrogen losses and surface run-off from landspreading of manures, pp 397–408. The Hague: Martinus Nijhoff/Dr Junk
12. Duthion C (1981) Nitrogen leaching after spreading pig manure. In Brogan J C, ed. Nitrogen losses and surface run-off from landspreading of manures, pp 274–283. The Hague: Martinus Nijhoff/Dr Junk
13. Garwood E A and Ryden J C (1986) Nitrate loss through leaching and surface runoff from grassland: effects of water supply, soil type and management. In Meer H G van der, Ryden J C and Ennik G C, eds. Nitrogen fluxes in intensive grassland systems, pp 99–113. Dordrecht, The Netherlands: Martinus Nijhoff Publishers
14. Garwood E A, Salette J and Lemaire G (1980) The influence of water supply to grass on the response to fertiliser nitrogen and nitrogen recovery. In Prins W H and Arnold G H, eds. The role of nitrogen in intensive grassland production, pp 59–65. Wageningen, The Netherlands: PUDOC
15. Heck A F (1931) Conservation and availability of the nitrogen in farm manure. Soil Science 31, 335–363
16. Hensler R F, Erhardt W H and Walsh L M (1971) Effects of manure handling systems on plant nutrient cycling. In Livestock waste management and pollution abatement. Proceedings of an international symposium on livestock wastes, pp 254–257. St Joseph, Michigan: ASAE
17. Hensler R F, Olsen R J, Witzel S A, Altoe O J, Paulson W H and Johannes R F (1970) Effects of method of manure handling on crop yields, nutrient recovery and runoff losses. Transactions of the American Society of Agricultural Engineers 13, 726–731
18. Jawson M D, Elliott L F, Sazton K E and Fortier D H (1982) The effect of cattle grazing on nutrient losses in a Pacific Northwestern setting. Journal of Environmental Quality 11, 628–631
19. Khaleel R, Reddy K R, Overcash M R and Westerman P W (1978) Transport of potential pollutants in runoff water from land areas receiving animal wastes: A review. American Society of Agricultural Engineers, Technical Paper No. 78–2058. St Joseph, Michigan: ASAE
20. Klausner S D, Zwerman P J and Ellis D F (1976) Nitrogen and phosphorus losses from winter disposal of dairy manure. Journal of Environmental Quality 5, 47–49
21. Kolenbrander G J (1981) Leaching of nitrogen in agriculture. In Brogan J C, ed. Nitrogen losses and surface run-off from landspreading of manures, pp 199–216. The Hague: Martinus Nijhoff/Dr Junk
22. Kolenbrander G J (1981) Effect of injection of animal waste on ammonia losses by volatilisation on arable land and grassland. In Brogan J C, ed. Nitrogen losses and surface run-off from landspreading of manures, pp 425–430. The Hague: Martinus Nijhoff/Dr Junk
23. Lauer D A, Bouldin D R and Klausner S D (1976) Ammonia volatilization from dairy manure spread on the soil surface. Journal of Environmental Quality 5, 134–141
24. Long F L (1979) Runoff water quality as affected by surface-applied dairy cattle manure. Journal of Environmental Quality 8, 215–218
25. Long F L, Lund Z F and Hermanson R E (1975) Effect of soil incorporated dairy cattle manure on runoff water quality and soil properties. Journal of Environmental Quality 4, 163–166

26. McCaskey T A, Rollins G H and Little J A (1971) Water quality of runoff from grassland applied with liquid, semi-liquid and 'dry' dairy waste. In Livestock waste management and pollution abatement. Proceedings of an international symposium on livestock wastes, pp 239–242. St Joseph, Michigan: ASAE
27. Meer H G van der and Uum-Van Lohuyzen M G van (1986) The relationship between inputs and outputs of nitrogen in intensive grassland systems. In Meer H G van der, Ryden J C and Ennik G C, eds. Nitrogen fluxes in intensive grassland systems, pp 1–18. Dordrecht, The Netherlands: Martinus Nijhoff Publishers
28. Minshall N E, Witzel S A and Nichols M S (1970) Stream enrichment from farm operations. American Society of Chemical Engineers. Journal of Sanitary Engineers Division, 96 (SA2), 513–524
29. Olness A, Smith S J, Rhoades E D and Menzel R G (1975) Nutrient and sediment discharge from agricultural watersheds in Oklahoma. Journal of Environmental Quality 4, 331–336
30. Owens L B, Edwards W M and Van Keuren R W (1983) Surface runoff water quality comparisons between unimproved pasture and woodland. Journal of Environmental Quality 12, 518–522
31. Owens L B, Van Keuren P W and Edwards W M (1983) Nitrogen loss from a high fertility, rotational pasture program. Journal of Environmental Quality 12, 346–350
32. Pain B F, Thompson R B, Lande Cremer L C N de la and Holte L ten (1987) The use of additives in livestock slurries to improve their flow properties, conserve nitrogen and reduce odours (this volume)
33. Ross I J, Sizemore S, Bowden J P and Haan C T (1979) Quality of runoff from land receiving surface application and injection of liquid dairy manure. Transactions of the American Society of Agricultural Engineers 22, 1058–1062
34. Ryden J C (1984) Fertilisers for grassland. Chemistry and Industry 18, 652–657
35. Ryden J C (1984) The flow of nitrogen in grassland. Proceedings of the Fertiliser Society 229, 44 p
36. Ryden J C (1985) Ammonia loss from grassland systems. In Nielsen V C, Voorburg J H and Hermite P l', eds. Odour prevention and control of organic sludge and livestock farming, pp 33–42. London: Elsevier Applied Science Publishers
37. Ryden J C (1985) Ammonia loss from grazed swards. Grassland Research Institute Final Annual Report 1984–1985, p 23. Hurley, UK: GRI
38. Ryden J C (1985) Denitrification loss from managed grassland. In Gotterman H L, ed. Denitrification in the nitrogen cycle, pp 121–134. London: Plenum Publishing Corporation
39. Ryden J C, Ball P R and Garwood E A (1984) Nitrate leaching in grassland. Nature 311, 50–53
40. Ryden J C and Dawson K P (1982) Evaluation of the acetylene-inhibition technique for the measurement of denitrification in grassland soils. Journal of the Science of Food and Agriculture 33, 1197–1206
41. Ryden J C, Lockyer D R, Ball P R and Bristow A W (1984) Ammonia losses from grazed swards. Grassland Research Institute Annual Report 1983–1984, pp 28–30. Hurley, UK: GRI
42. Ryden J C and Nixon D J (1985) Denitrification loss from grazed and cut swards. Grassland Research Institute Final Annual Report 1984–1985, pp 21–22. Hurley, UK: GRI
43. Ryden J C, Skinner J H and Nixon D J (1987) Evaluation of a soil core incubation system for field measurement of denitrification using the acetylene-inhibition technique. Soil Biology and Biochemistry (in press)
44. Schwepers J S and Francis D D (1982) Chemical quality of runoff from grazing land in

Nebraska: 1. Influence of grazing livestock. Journal of Environmental Quality 11, 351–354

45. Sherlock R S and Goh K M (1984) Dynamics of ammonia volatilization from simulated urine patches and aqueous urea applied to pasture. I. Field experiments. Fertilizer Research 5, 181–195
46. Sherwood M (1980) The effects of land spreading of animal manures on water quality. In Gasser J K R, ed. Effluents from livestock, pp 379–390. London: Applied Science Publishers
47. Sherwood M (1986) Nitrate leaching following application of slurry and urine to field plots. In Dam Kofoed A, Williams J H and Hermite P l', eds. Efficient land use of sludge and manure, pp 150–157. London: Elsevier Applied Science Publishers
48. Sherwood M and Fanning A (1981) Nutrient content of surface run-off water from land treated with animal wastes. In Brogan J C, ed. Nitrogen losses and surface run-off from landspreading of manures, pp 5–17. The Hague: Martinus Nijhoff/Dr Junk
49. Snijders P J M (1986) Personal communication
50. Steele K W and Shannon P (1982) Concepts relating to the nitrogen economy of a northland intensive beef farm. In Gandar P W, ed. Nitrogen balances in New Zealand ecosystems, pp 85–90. Palmerston North, New Zealand: DSIR
51. Steenhuis T S, Bubenzer G D, Converse J C and Walter M F (1981) Winter-spread manure nitrogen loss. Transactions of the American Society of Agricultural Engineers 24, 436–449
52. Steenvoorden J H A M (1981) Landspreading of animal manure and runoff. In Brogan J C, ed. Nitrogen losses and surface run-off from landspreading of manures, pp 26–33. The Hague: Martinus Nijhoff/Dr Junk
53. Steenvoorden J H A M (1986) Nutrient leaching losses following application of farm slurry and water quality considerations in The Netherlands. In Dam Kofoed A, Williams J H and Hermite P l', eds. Efficient land use of sludge and manure, pp 168–177. London: Elsevier Applied Science Publishers
54. Steenvoorden J H A M and Bouma J (1987) Optimizing the use of soils: new agricultural and water management aspects. Proceedings of the EC-Symposium on Soil Protection (Berlin) (in press)
55. Steenvoorden J H A M, Fonck H and Oosterom H P (1986) Losses of nitrogen from intensive grassland systems by leaching and surface runoff. In Meer H G van der, Ryden J C and Ennik G C, eds. Nitrogen fluxes in intensive grassland systems, pp 85–97. Dordrecht, The Netherlands: Martinus Nijhoff Publishers
56. Thompson R B, Ryden J C and Lockyer D R (1987) Fate of nitrogen in cattle slurry following surface application or injection to grassland. Journal of Soil Science (submitted for publication)
57. Uhlen G (1978) Nutrient leaching and surface runoff in field lysimeters on a cultivated soil. II. Effects of farmyard manure spread on a frozen ground and mixed in the soil on water pollution. Scientific Reports of the Agricultural University of Norway 57 No. 28, pp 1–23
58. Vetter H and Steffens G (1981) Leaching of nitrogen after spreading of slurry. In Brogan J C, ed. Nitrogen losses and surface run-off from landspreading of manures, pp 251–269. The Hague: Martinus Nijhoff/Dr Junk
59. Vetter H and Steffens G (1981) Gaseous N-losses. In Brogan J C, ed. Nitrogen losses and surface run-off from landspreading of manures, pp 409–415. The Hague: Martinus Nijhoff/Dr Junk
60. Wallingford G W, Murphy L S, Powers W L and Manges H L (1975) Denitrification in soil treated with beef-feedlot manure. Communications in Soil Science and Plant Analysis 6, 147–161

61. Westerman P W and Overcash M R (1980) Short term attenuation of runoff pollution potential for land applied swine and poultry manure. In Livestock waste: A renewable resource. Proceedings of the 4th international symposium on livestock wastes, pp 289–292. St Joseph, Michigan: ASAE (Publ.)
62. Young R A and Mutchler C K (1976) Pollution potential of manure spread on frozen ground. Journal of Environmental Quality 5, 174–179

Phosphorus and heavy metals: accumulation and consequences

W H VAN RIEMSDIJK[a], Th M LEXMOND[a], C G ENFIELD[b] and S E A T M VAN DER ZEE[a]

[a] *Department of Soil Science and Plant Nutrition, Agricultural University, De Dreijen 3, 6703 BC Wageningen, The Netherlands*

[b] *Robert S. Kerr Environmental Research Laboratory, United States Environmental Protection Agency, PO Box 1198, Ada, Oklahoma 74820, USA*

Key words: manure, phosphate, heavy metals

Abstract. Land-spreading of animal manure in quantities exceeding plant uptake of phosphorus and heavy metals results in an accumulation of these elements in soil. When the cumulative excess becomes large compared to the buffering capacity of the soil adverse effects may arise. Phosphorus may leach to ground- and surface water causing eutrophication. Heavy metals may bring about a reduction in crop yield and livestock performance or affect the quality of the produce. Generally these negative effects will develop only after a relatively long period of large manure applications, but soils differ widely in their buffering capacities. For a proper risk-assessment it is essential to have a thorough understanding of the behaviour of the elements in the soil system and also to take into account the spatial variability of the physical and chemical properties of the soil as well as the spatial variability of the cumulative manure input.

Introduction

General statements related to the appreciation of animal manure are incompatible with the huge diversity in agricultural activities. This statement still holds when attention is confined to animal husbandry in Europe, which ranges from extensive hill sheep farming to the "bio-industrial" production of pigs and poultry.

Whether animal manure is considered as fertilizer or as waste depends strongly on the assessor's interest. Extreme positions may be taken by the farmer on the one hand and the conservationist on the other. The farmer might not be inclined to consider manure as waste unless he is unable to dispose of the amount produced without noticeable damage to the productivity of his land or the quality of its produce. The conservationist might consider it as waste whenever its application to soil induces either emissions exceeding natural levels to the atmosphere or groundwater or to an increase in the content of any element in the soil. Some of these differences may be explained by the different time scales considered. As an entrepreneur, the

H.G. v.d. Meer, et al. (eds.), Animal Manure on Grassland and Fodder Crops.
 ISBN 978-94-010-8137-5.

farmer counts costs and returns over a relatively short term whereas the conservationist is concerned with all changes which eventually may turn into undesirable effects upon natural values considered worthy of protection.

The elements to be discussed in this contribution are characterized by a relatively low mobility in soil under normal conditions. They are strongly retained by the soil and undesirable effects induced by these elements, both on agricultural production and the environment, are not likely to appear in the short term. This does not imply, however, that such undesirable effects do not have to be considered in relation to land application of manure. It merely indicates that present day practices have to be evaluated on a long term basis. It seems appropriate to start such an analysis by a general discussion of element behaviour in agro-ecosystems. In the second part of this paper attention is focused on phosphorus (P). Then some consideration is given to the heavy metals copper (Cu), zinc (Zn), and cadmium (Cd), which appear to be of prime concern in this context.

Element behaviour in agro-ecosystems

The behaviour of elements in agro-ecosystems can be evaluated from rates of transfer across the system boundaries (input and output) and between the main compartments, *viz.* soil, crop, livestock and manure. Agro-ecosystems are open to the atmosphere and they may border on surface water. Their lower boundary is determined by the depth to which the soil is involved in supplying nutrients and water to the crop, thus largely excluding non-rooted subsoil and groundwater. A final boundary is the farm gate via which supplies may enter the system and produce leaves.

Within the system, elements pass from the soil to the crop which includes all plants, grown either to be harvested or to be grazed. Elements in dead plant material remaining in the field reenter the soil compartment. Livestock takes up elements from the crop compartment by consumption of harvested crops or by grazing of forage; in addition, uptake directly from the soil occurs as a result of ingestion of soil particles. Faeces and urine return elements straight to the soil during grazing; if they are collected and stored a separate manure compartment can be discerned from which elements eventually return to the soil.

Starting from the principle that agro-ecosystems are managed with production of food as one of their main objectives [9], it is obvious that some kind of output, be it animals and animal products or marketable crops, has to be realized. This entails an output of elements from the system. Moreover, unintentional losses may occur from several compartments, in particular of nitrogen (N) and sulphur (S) which can be found in volatile forms (causing losses to the atmosphere) and as anions (NO_3^-, SO_4^{2-}) of considerable mobility in soil (leading to losses to subsoil and groundwater). For P and heavy metals such losses may be considered less likely, since they are not volatile and, under normal conditions, practically immobile in soil. Under certain circumstances, losses to surface water via surface run-off carrying suspended soil and manure as well as dissolved matter may be of considerable importance, as may also losses via wind blown dust. These unintentional losses may directly affect the environment.

On the opposite side of the balance-sheet a similar distinction between deliberate input and unintentional gains can be made. Examples of the latter are deposition from the atmosphere and, occasionally, from surface water during flooding. Deliberate input of elements takes place via the application of purchased materials, such as mineral fertilizers and manure, soil conditioning materials (e.g. sewage sludge, town refuse compost), irrigation water, seed, pesticides, animal feed (both concentrates and roughage), bedding materials, and veterinary medicine. These materials may, however, also contain significant amounts of elements whose application is not intended. The relative importance of these contributions may vary for different elements.

The balance of a particular element in agro-ecosystems can be characterized by comparing the inflow and outflow of that element. When these flows are equal the system is in a steady state, when they are unequal the system is declining (outflow > inflow) or accumulating (inflow > outflow) in respect of that element. The system may, at the same time, be in different states for different elements.

As far as essential elements (nutrients) are concerned, decline of the system's content eventually results in decreasing productivity. The time elapsing before this becomes apparent and the rate at which productivity decreases depend on the deficit of the balance-sheet and on the buffering capacity of the soil.

When an element is naturally abundant and accumulates in a chemical form of similar solubility to its natural compounds, accumulation may proceed without any appreciable effect on the sytem itself, its produce or on the environment. Iron can be mentioned as an example of such behaviour. In many cases however, accumulation involves a steady increase in activity and/or mobility in the soil. The rate of this increase again depends on the soil's buffering capacity and on the actual surplus of the balance-sheet. Such accumulation induces an increase in element flows between the compartments, and from the system to its surrounding environment. Increasing flow rates between the compartments may, depending on the element considered, also result in increasing contents in crops and livestock and consequently in the produce.

An appreciation of such changes depends among other factors on the initial state of the system and on the time scale considered. Phosphorus is a striking example in this respect. Accumulation of P has taken place in many agro-ecosystems for decades with highly beneficial effects on soil fertility and generally without increased losses to the environment. Many Dutch soils have now reached a P level which enables application of P in quantities equal to those removed with the crops [3]. Further accumulation of P is apparently unnecessary from an agricultural point of view. However, in many cases accumulation still proceeds at an appreciable rate and the risk of P leaching can no longer be ignored. The processes involved form the subject of the next section.

Phosphorus

Introduction

Phosphorus is one of the three major plant nutrients. The behaviour in the soil of P present in animal manures and slurries, is thus of importance when they are applied as sources of P fertilizer. Where slurry is a waste-product of intensive animal husbandry and is applied to soil in quantities that far exceed the requirements for efficient plant production, the behaviour of P in soil is of importance with respect to the risk of P leaching to groundwater and to surface water where it may contribute to eutrophication. A thorough understanding of the dynamics of P behaviour in soil is of prime importance, both for the fertility aspect and for the environmental aspect. Although the behaviour of P in soil has been studied for several decades it is still difficult to predict quantitatively. In part this is due to the many different chemical forms in which P may be present in soil. Also, the rate of transformation of P from one chemical form to another which is more stable, often controls P behaviour. One generally studies only the net effect of many different processes that occur simultaneously. To understand and predict P behaviour with respect to plant uptake or with respect to leaching, it is often necessary to describe not only the net effect of a series of reactions but also the individual processes that occur in soil. All these processes are influenced by the soil mineral composition and by the composition of the soil solution. In this chapter the main chemical processes occurring when slurry P reacts with soil are discussed. A distinction is made between calcareous soils and noncalcareous (acid) soils. A quantitative insight into the different reactions is used to develop models that can predict P behaviour on a field- or regional-scale. Many new problems have to be tackled when one wants to apply the detailed understanding of P transformations in soil to field-scale models. Before the behaviour of P in soil is discussed, the chemical forms of P in animal slurry are considered.

Phosphate in animal slurries

Phosphate in animal slurries is mainly present as solid-phase inorganic P. The fraction present in the solid fraction may differ considerably for different slurries. This variability may be ascribed partly to the variation in dry matter content of slurries but also to the composition of the animal feed, where mineral P is often added. According to Fordham and Schwertmann [7, 8] roughly half of total P in cattle slurry is present in the solid fraction, whereas Gerritse [11] reports a fraction of 0.9 for a pig slurry. Both in the solid phase as well as in the liquid phase P is mainly present as inorganic P (80%), the remainder being organic P. In the solid fraction the mineral struvite ($MgNH_4PO_4 \cdot 2H_2O(s)$) has been identified [8] and the presence of octocalcium-phosphate ($Ca_4H(PO_4)_3 \cdot 3H_2O(s)$) [8] and dicalcium-phosphate ($CaHPO_4 \cdot 2H_2O(s)$) [8, 14] have been suggested. That dicalcium-phosphate is a likely constituent in animal slurries is corroborated by measurement of the composition of the soil solution in the topsoil of a field that had received excessive amounts of pig slurries [14]. Of the total P present in solution of slurry, only a small fraction is present as dissolved organic P (1–2%) [12]. The inorganic P solids in slurry are relatively soluble. When these slurries are applied to soil the minerals gradually dissolve in the water percolating through the soil. The dissolved inorganic P then starts to react with soil components and/or with other ions in the soil solution. This leads to the formation of adsorbed P present at the soil colloid-soil solution interface and/or to formation of P containing solids (where the largest part of the P is not in direct contact with the soil solution). This last fraction is referred to by the term precipitated P. The organic P fraction in the slurry is subject to mineralization to inorganic P when applied to soil. When the soil is in steady state with respect to its organic matter content, the behaviour of P from slurries is dominated by the behaviour of inorganic P [11]. In the following paragraphs emphasis will be on the behaviour of inorganic P in soil.

Adsorption/desorption

Phosphate may adsorb on the surfaces of metal oxides and metal hydroxides e.g. $Fe_2O_3(s)$ and $Al(OH)_3(s)$, on the edges of clay minerals, on the surface of $CaCO_3(s)$ and on iron or aluminum ions associated with soil organic matter. The main factors that determine the extent of adsorption are the surface area of the reactive minerals in the soil and the composition of the soil solution. With respect to the composition of the soil solution, the concentration of inorganic P and the pH are the main factors that influence the adsorption. Near the root surface exudates like oxalate or citrate may also considerably influence the adsorption behaviour of P, due to competition by these ions for the same reactive sites The adsorption/desorption of P in soil is in itself complicated if one wants to understand the dynamics of the distribution of P over the soil solution and the different solid surfaces. For pure oxide minerals several detailed electrochemical models have been presented that are able to describe adsorption as a function of solution composition [2, 13, 27, 28]. The relationship between P concentration in solution (at constant temperature, pH, ionic strength, etc.) and the amount of P adsorbed per unit amount of soil or sorbent is called the adsorption isotherm. Determining the shape of this isotherm for a soil is very difficult in practice, especially at high P concentrations in solution. This is because the amount of P removed from solution during the measurement may be adsorbed or precipitated. One also has to establish the adsorbed P in the sample as taken from the field. A relevant concentration range for adsorption studies would be from very low up to the maximum that may occur in the soil solution directly after addition of fertilizer or slurry. For pig slurry, the P concentration in the soil solution of the top soil may be as high as $3\,\mathrm{mmol\,P\,l^{-1}}$ [14] due to buffering by dissolution of the relatively soluble precipitates present in the slurry.

In Fig. 1 the shape of the adsorption isotherm as measured with iron-oxide (e.g. goethite) as the sorbent is compared with a curve obtained by measuring the "removal" of P from solution in a slightly acid soil. The isotherm for goethite shows a very flat portion for solution concentrations higher than approximately $0.2\,\mathrm{mmol\,l^{-1}}$, whereas the relationship for the soil in this range maintains a significant slope. In the horizontal portion of the iron oxide isotherm, the solid phase does not release, or buffer, P to the solution when the P concentration in solution is diminished due to removal by plant uptake or leaching, as long as one is in the region of the plateau. The question thus arises whether the solid phase of the soil of Fig. 1 buffers phosphate in the concentration range above $0.2\,\mathrm{mmol\,l^{-1}}$. Desorption experiments using soil packed columns presaturated with P [30], have shown that no buffering takes place above $0.2\,\mathrm{mmol\,l^{-1}}$. This soil is thus characterized by a reversible isotherm of a shape that is comparable with that for P adsorbed on the metal oxide surface. The soil curve of Fig. 1 is thus equal to a reversible adsorption isotherm plus a certain amount of precipitated P. Precipitated P should be able to buffer the P concentration in solution by

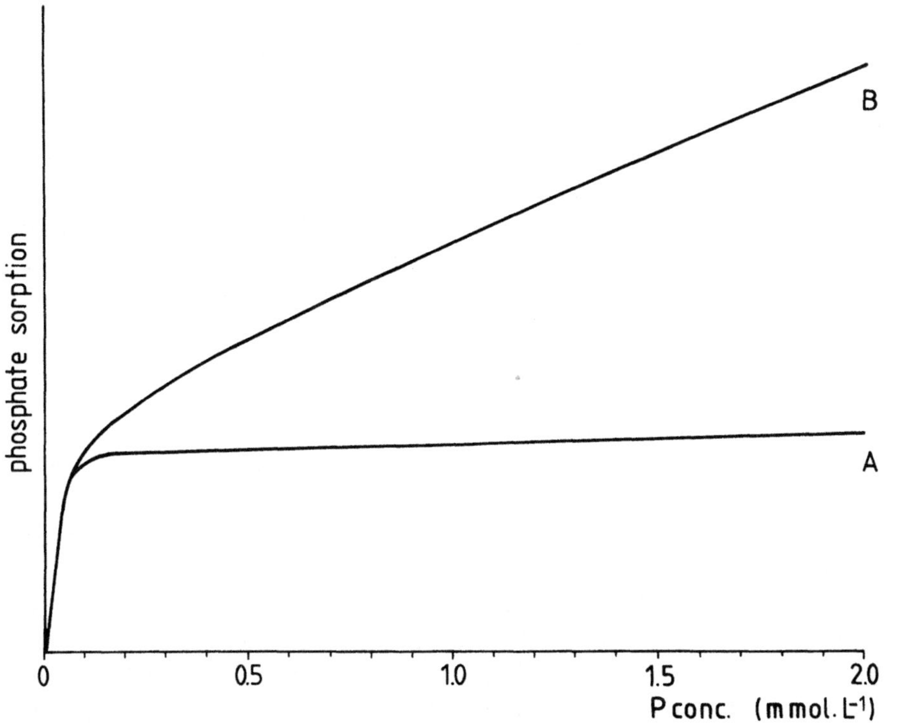

Figure 1. The shape of the phosphate adsorption isotherm (A) measured for goethite at a constant pH (< 7) and the shape of the amount of phosphate reacted with an acid soil for long reaction times (ca 1 year) as a function of the concentration in solution (B).

dissolution. However, if the precipitate formed is relatively insoluble, the buffering may take place at P concentrations far below the region of the adsorption plateau. Both for fertility aspects and out of environmental concern, the distinction between the different processes in a quantitative manner is quite important. In addition to the shape of the adsorption isotherm, the units on the vertical axis are of importance. A soil that has a large reactive surface area per unit amount of soil will show, for a given P concentration, a much higher amount of P adsorbed per unit soil than the soil that has a lower reactive surface area (assuming the same type of reactive surfaces are present in both soils). This can be further illustrated by considering a given soil and comparing this with a mixture of the same soil on a 1:1 basis with pure quartz sand (which does not adsorb P). In this case the adsorption process is not affected by the mixing, but the amount adsorbed per kg of "soil" at a given P concentration is reduced by 50% due to the mixing. This effect disappears when the amount adsorbed is expressed per unit amount of reactive surface area. In practice, expressing adsorption in terms of reactive surface area is not done since the composition and the specific surface area ($m^2 kg^{-1}$) of the different reactive surfaces for a given soil are seldom determined. For noncalcareous sandy soils it seems that the adsorption data can be normalized reasonably well by considering the amount of P adsorbed per unit amount of reactive iron plus aluminum [1, 29, 31]. A relatively low content of reactive surface area is beneficial from the point of view of soil fertility but unfavourable with respect to leaching of phosphate to groundwater. The higher the content of reactive surfaces the smaller the penetration of P per unit amount of P applied in slurry.

Precipitation/dissolution

In the preceding paragraph, it was mentioned that in addition to adsorption, precipitation of P may occur. In calcareous soil one of the most stable (least soluble) phosphate minerals will be the calcium phosphate hydroxyapatite ($Ca_5(PO_4)_3OH(s)$). A whole range of different calcium phosphate minerals exists which vary quite significantly in their solubility. When readily soluble forms of P are applied to a calcareous soil, these will gradually dissolve. Dissolved phosphate will partly be transformed to adsorbed P and partly be transformed to less soluble calcium phosphate precipitates. The kinetics of this transformation from more soluble to much less soluble calcium phosphates are complicated and the reaction rate is relatively slow. This means that relatively high P concentrations may exist temporarily in calcareous soil. The equilibrium situation would be a soil solution in equilibrium with apatite, which is characterized by a very low concentration of P. From the point of view of soil fertility it is beneficial that these reaction rates are rather low, from the point of view of leaching it is a disadvantage since the slower these reactions are the more leaching may occur [5, 30].

In slightly acid soils, phosphate precipitates may be formed with iron and aluminum. However, when a soil is in the pH range 5–6 the solubilities of iron and aluminum oxides/hydroxides in soil are very low. Thus, the phosphate-ions can not directly combine in solution with dissolved metal ions to a significant degree, due to the imbalance between the concentration of P and of dissolved metal ions. Instead of direct precipitation, penetration of P to the interior of the poorly crystalline metal oxides may take place, leading to the formation of coatings of metal phosphate on the surface of the amorphous metal oxide particles. The kinetics of this process are slow and decrease rapidly with the amount of P that has reacted with the metal oxide [1, 23–26]. After a certain degree of conversion the rate will become so slow that the conversion of metal oxide to metal phosphate virtually stops. The degree of conversion that may be reached depends on the particle size, on the P concentration prevailing in solution, and on the reaction time [23, 24]. It has been found that the overall average conversion of the reactive metal oxide fraction of slightly acid soil (at 3 mmol $P\,l^{-1}$) may reach values as high as 0.6. This value is fairly constant for different soils varying quite markedly in their absolute amount of reactive iron and aluminum [32]. An average conversion of 0.6 indicates that the very small and highly reactive metal oxide particles are completely converted into metal phosphate, whereas the more crystalline larger particles are only partially converted. Large crystalline iron oxide particles like goethite, are probably not transformed at all at these P concentrations. This conversion process may be compared with the oxidation of pure iron and aluminum metal in the presence of oxygen. The rate and extent of this reaction also depends on particle size. Very small iron particles (pyrophoric iron) will burn spontaneously when brought in contact with air (oxygen), and be transformed into iron oxide. Whereas only a few molecular layers of a sheet of aluminum metal in contact with air will be converted to aluminum oxide, effectively protecting the underlying pure metal from further corrosion.

The metal phosphate coating formed in soil (pH 5–6) is probably relatively insoluble with an equilibrium phosphate concentration well below $10^{-5}\,M$. The occurrence of this diffusion precipitation process may contribute significantly to the total sorption capacity (see also Fig. 1). The availability of these precipitated phosphates for plant growth is expected to be less than that of the adsorbed phosphate. One reason for this is that the buffering of P adsorbed near the adsorption maximum occurs at much higher P concentrations than buffering due to dissolution of these slightly soluble precipitates. In practice, special techniques are required to discriminate between adsorbed P and precipitated P. The use of column techniques or desorption of soil P using iron oxide coated filter paper as a sink for phoshate [29, 30] seems promising in this respect. Whatever technique will be used a sharp and unequivocal distinction between the two processes is probably impossible due to the gradual transition from adsorption to precipitation.

The qualitative nature of P behaviour in soil has been understood for quite some time. In general, when soluble phosphate is applied to soil most of the added P will rapidly adsorb on the surface of soil minerals. The slow transformation to precipitated P will further lower the P concentration in solution which induces a decrease of the amount of adsorbed P. Both the P concentration in solution and the amount of reversibly adsorbed P will thus decrease as a function of time, even in the absence of leaching or plant uptake. This phenomenon occurs both in acid and alkaline soils. Apart from a mechanistic insight there is the need for a quantitative description of the dynamics of P behaviour in soil.

Recently some progress has been made in this field both for calcareous and noncalcareous soils.

Application of laboratory results to P behaviour in the field

An insight into the mechanisms, and a quantitative description of the different P transformations, taking place in the soil for a relevant set of conditions on the laboratory scale is a prerequisite for understanding and describing the behaviour of P in the field. However, additional problems arise when one wants to apply data obtained in the laboratory to the field. The main problem is heterogeneity in the field both in a physical and a chemical sense. The distribution of P in the soil depends amongst others on the amount of P applied, both recently and in the past, the content of reactive surfaces and reactive particle size distributions. These factors may differ considerably depending on spatial coordinates. The physical properties governing water transport may also vary greatly in the field. They are of special importance to P transport when the P is applied with large quantities of water (e.g. land-treatment of sewage water). In this case, those places with the highest permeability for water receive the highest input of P, leading to a very high variability of P input. To obtain all the parameters necessary to describe the dynamic transport for a homogeneous soil column requires a considerable amount of time and effort. It is thus impractical to characterize a whole heterogeneous field in such a way as would be necessary for a totally deterministic description of transport of a reactive constituent such as phosphate.

From a fertility point of view the heterogeneity also plays an important role with respect to P fertility. For example, P fertilizer is often purposely distributed unevenly over the field by concentrating it near the rows where the crop is growing. The buffering of P will take place on a higher concentration level in the case of band or row placement of phosphate (due to a higher fractional saturation of the adsorption capacity) than when the same amount of fertilizer is evenly distributed. This practice causes a strong heterogeneity with respect to P content in the soil. After ploughing this induced heterogeneity will still partly remain. Apart from heterogeneity which arises due to an uneven distribution of applied P, the content of reactive material may differ considerably on the field scale. These two aspects (heterogeneity of input and of sorption capacity) strongly determine transport of P under field conditions. It is obvious that those places where there is a combination of a high input of P and a low sorption capacity will start to release P to the groundwater much earlier than one would calculate based on averages of input and sorption capacity for the whole field.

Knowledge of distribution functions for sorption capacity and P input may be used to describe P transport using stochastic transport models [31].

In the case of annual applications of high amounts of slurries as in regions with a high density of animals, the soil will slowly become saturated with P with respect to the maximum P concentration (= $3\,\mathrm{mmol\,P\,l^{-1}}$) that may occur in the soil solution for such conditions. The rate of movement of this saturation front for a homogeneous soil column can be simply calculated when it is assumed that the saturation front moves piston like (blockfront) or that its shape is not altered upon transport to greater depth due to the kinetic nature of the chemical reactions, using the following formula:

$$y = \frac{A - U}{BS} 141$$

where A is the amount of P applied per year ($\mathrm{kg\,P_2O_5\,ha^{-1}\,yr^{-1}}$), U is plant uptake (same units), B is dry bulk density of the soil ($\mathrm{kg\,m^{-3}}$), S is the sorption capacity ($\mathrm{mmol\,P\,kg^{-1}}$) and y is the rate of movement of the saturation front in $\mathrm{cm\,yr^{-1}}$. In Fig. 2 y is plotted as a function of (A − U) for a constant value of B ($1500\,\mathrm{kg\,m^{-3}}$) and some values of S as they are found for sandy soils in The Netherlands. There are situations where y may be estimated to be several cm per year. When it is considered that such high quantities of slurries have already been applied for some 15 years and that the groundwater table is often not more than 1 m below the soil surface, it can be estimated that within the coming 20–30 years breakthrough of very high P concentrations to the groundwater (and eventually also to surface waters) will take place if these high rates of slurry application continue. Calculations based on average inputs and average sorption capacity suggest of course a much less urgent problem, because there are also situations where it would take several hundred years before leaching to the groundwater occurs. For a correct evaluation of the risk of P leaching from high inputs of slurries it is thus essential to take the effect of heterogeneity into account.

Heavy metals: copper, zinc, and cadmium

The group of elements indicated as heavy metals comprises over half the elements in the periodic system. Although they have a number of physical characteristics in common, their chemical behaviour shows considerable diversity. Molybdenum usually occurs as an anion (molybdate) with a behaviour resembling that of phosphate in many respects. Copper, Zn and

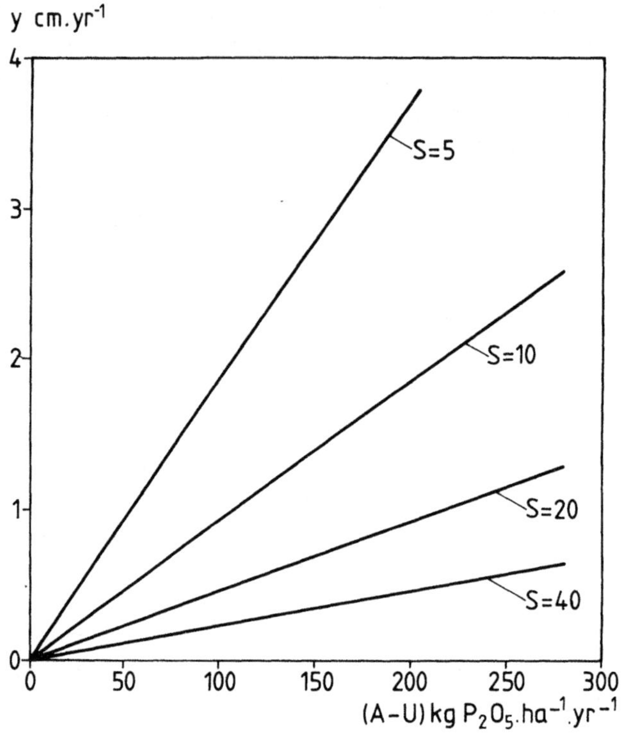

Figure 2. The rate of penetration of the phosphate saturation front as a function of the yearly application of phosphate (minus plant uptake) for four different sorption capacities, S, indicated in the figure.

Cd have a cationic chemistry in common, but differences in their behaviour in soil-plant-animal systems are still largely due to their individual chemical nature. This would justify individual treatment [15, 18, 20–22] but available space precludes it.

The chemistry of Cu, Zn, and Cd is dominated by their ability to form complexes by exchange of one or more coordinated water molecules for either functional groups on the surface of solid soil constituents or ligands in the soil solution. The number and nature of coordinating functional groups on the surfaces depend on the composition of the soil. Organic matter and (hydrous) oxides of Fe, Mn, and Al are generally considered to provide the bulk of the binding or sorption sites. Sorption by purely electrostatic forces, as in simple cation exchange, is a process of limited importance, since the heavy metal cations have to compete with other cations (e.g. Ca) in much higher concentrations. A striking feature of heavy metal sorption is its high sensitivity to changes in soil pH, sorption increasing rapidly with pH. Since soil composition and pH are highly variable, the buffering capacity for heavy metal ions may attain widely divergent values. Low values are to be expected in acid sandy soils poor in humus and high values in calcareous organic soils.

From an agricultural point of view, heavy metals can be divided into those that are essential to crops and livestock (micronutrients, e.g. Cu and Zn) and those that are not. Copper and Zn can be present at levels too low for an adequate supply to crops and livestock and a (limited) increase in the content of the agro-ecosystem can have a beneficial effect on its productivity. Since Cd apparently is not required for agricultural production, accumulation cannot be expected to have beneficial effects.

The three elements may all give rise to undesirable effects on agriculture when present at high levels. Zinc is mainly known for its phytotoxicity, Cu for its potential zootoxicity particularly to sheep. Cadmium is considered primarily as a threat to human health. Several countries already have defined maximum allowable concentrations for Cd in agricultural produce, others are preparing such limits in order to protect consumers. It is becoming the direct interest of the farmer to safeguard produce against Cd concentrations exceeding such limits.

Next to effects on growth, health and composition of crops and livestock, other effects on agro-ecosystems have to be considered. Heavy metals may damage or kill soil organisms (fauna and microflora) involved in the mineralization of organic residues, thus affecting the liberation of nutrients from these materials. As an example, the depressive effect of Cu on earthworm reproduction can be mentioned [19]. In orchards with a history of application of Cu containing fungicides, litter decomposition was shown to be inhibited because of adverse effects on earthworm populations [16]. Hoogerkamp *et al.* [17] demonstrated the importance of earthworms for grassland productivity.

Apart from effects on agro-ecosystems emission of heavy metals into the environment may give rise to pollution of surface- and groundwater. It is to be expected that maximum allowable heavy metal concentrations in surface- and groundwater will be established in the near future. Such standards already exist for drinking water. Then, the total allowable output will in fact be defined and, consequently, also the total allowable input, since in the long term input should balance output for essential elements and input should not exceed output for non-essential elements [10].

In contrast to the situation with P, agriculture cannot always be held fully responsible for the input of heavy metals in agro-ecosystems. In The Netherlands deposition from the atmosphere forms a significant contribution to the total input of several elements (for example Cd, Zn, and lead (Pb)). In more specific areas, *viz.* the forelands of rivers like Rhine and Meuse, deposition during flooding is the major factor determining the loading of soil with heavy metals. In many cases, however, agricultural practices have a decisive influence on heavy metal input. A clear example in this respect is the use of Cu as a feed additive in pig fattening and the inclusion of Zn in the rations to prevent adverse effects of Cu on the animals. Cadmium is present in phosphorites and passes into fertilizers and phosphates used as mineral supplements in compounded feeds. Application of sewage sludge and town refuse compost adds heavy metals to the agro-ecosystem, as is the case with several pesticides. All these routes have to be evaluated with an eye to the possibility of reducing inputs. With respect to animal manure, it is especially important to limit the presence of heavy metals in purchased animal feed. Application rates of animal manure corresponding to crop requirements for the macronutrient present in the highest relative concentration with respect to crop uptake, should prevent heavy metal additions exceeding acceptable output levels.

References

1. Beek J, Riemsdijk W H van and Koenders K (1980) Aluminum and iron fractions affecting phosphate bonding in a sandy soil treated with sewage water. In Banin A and Kafkafi U, eds. Agrochemicals in soils, pp 369–378. Oxford, UK: Pergamon Press
2. Bowden J W, Nagarajah S, Barrow N J, Posner A M and Quirk J P (1980) Describing the adsorption of phosphate, citrate and selenite on a variable-charge mineral surface. Australian Journal of Soil Research 18, 49–60
3. Diest A van (1984) Characteristics of nutrient cycling and nutrient balance sheets in low-input and high-input agriculture. Proceedings of the 18th Colloquium of the International Potash Institute, pp 13–37. Bern, Switzerland: IPI
4. Enfield C G, Harlin C C and Bledsoe B E (1976) Comparison of five kinetic models for orthophosphate reactions in mineral soils. Soil Science Society of America Proceedings 40, 243–249
5. Enfield C G, Phan T and Walters D G (1981) Kinetic model for phosphate transport and transformation in calcareous soils. II. Laboratory and field transport. Soil Science Society of America Journal 45, 1064–1070
6. Enfield C G, Phan T, Walters D M and Ellis R (1981) Kinetic model for phosphate transport and transformation in calcareous soils. I. Kinetics of transformation. Soil Science Society of America Journal 45, 1059–1064
7. Fordham A W and Schwertmann U (1977) Composition and reaction of liquid manure (Gülle), with particular reference to phosphate: I. Analytical composition and reaction with poorly crystalline iron oxide (ferrihydrite). Journal of Environmental Quality 6, 133–136
8. Fordham A W and Schwertmann U (1977) Composition and reaction of liquid manure (Gülle), with particular reference to phosphate: II. Solid phase components. Journal of Environmental Quality 6, 136–140
9. Frissel M J, ed. (1977) Cycling of mineral nutrients in agricultural ecosystems. Agro-Ecosystems 4, 1–354
10. Furrer O J (1984) Konzept zur Festlegung von Grenzwerten für Schwermetallimmisionen in den Boden. Schweizerische Landwirtschaftliche Forschung/La Recherche agronomique en Suisse 23, 195–199

11. Gerritse R G (1976) Phosphorus compounds in pig slurry and their retention in the soil. In Voorburg J H, ed. Utilisation of manure by land spreading (EUR 5672e), pp 257–277. Luxembourg: The Commission of the European Commities.
12. Gerritse R G and Eksteen R (1978) Dissolved organic and inorganic phosphorus compounds in pig slurry. Journal of Agricultural Science (Cambridge) 90, 39–45
13. Goldberg S and Sposito G (1984) A chemical model of phosphate adsorption by soils: I. Reference oxide minerals. Soil Science Society of America Journal 48, 772–778
14. Haan F A M de and Riemsdijk W H van (1986) Behaviour of inorganic contaminants in soil. In Assink J W and Brink W J van den, eds. Contaminated soil, pp 19–32. Dordrecht, The Netherlands: Martinus Nijhoff
15. Hermite P l' and Dehandtschutter J, eds. (1981) Copper in animal wastes and sewage sludge. Dordrecht, The Netherlands: Reidel
16. Hirst J M, Le Riche H H and Bascomb C L (1961) Copper accumulation in the soils of apple orchards near Wisbech. Plant Pathology 10, 105–108
17. Hoogerkamp M, Rogaar H and Eijsackers H J P (1983) Effect of earthworms on grassland on recently reclaimed polders in the Netherlands. In Satchell J E, ed. Earthworm ecology, pp 85–104. London, UK: Chapman and Hall
18. Loneragan J F, Robson A D and Graham R D, eds. (1981) Copper in soils and plants. Sydney, Australia: Academic Press
19. Ma W (1983) Regenwormen als bio-indicators van bodemverontreiniging. Reeks Bodembescherming 15. 's-Gravenhage, The Netherlands: Staatsuitgeverij
20. Nriagu J O, ed. (1979) Copper in the environment. 2 volumes. New York, USA: Wiley
21. Nriagu J O, ed. (1980) Zinc in the environment. 2 volumes. New York, USA: Wiley
22. Nriagu J O, ed. (1980–81) Cadmium in the environment. 2 volumes. New York, USA: Wiley
23. Riemsdijk W H van, Boumans L J M and Haan F A M de (1984) Phosphate sorption by soils. I. A diffusion precipitation model for the reaction of phosphate with metal-oxides in soil. Soil Science Society of America Journal 48, 537–540
24. Riemsdijk W H van, Linden A M A van der and Boumans L J M (1984) Phosphate sorption by soils. III. The diffusion-precipitation model tested for three acid sandy soils. Soil Science Society of America Journal 48, 545–548
25. Riemsdijk W H van and Lyklema J (1980) Reaction of phosphate with gibbsite ($Al(OH)_3$) beyond the adsorption maximum. Journal of Colloid and Interface Science 76, 55–60
26. Riemsdijk W H van and Lyklema J (1980) The reaction of phosphate with aluminum hydroxide in relation with phosphate bonding in soils. Colloids and Surfaces 1, 33–44
27. Riemsdijk W H van and Zee S E A T M van der (in press) Application of ligand exchange, surface precipitation and solid solution models to describe ion adsorption. Proceedings of the 2nd NATO Workshop on soil colloid surfaces (Ghent, 1986)
28. Sigg L and Stumm W (1981) The interaction of anions and weak acids with the hydrous goethite (α-FeOOH) surface. Colloids and Surfaces 2, 101–117
29. Zee S E A T M van der, Fokkink L G J and Riemsdijk W H van (1987) A new technique for assessment of reversibly adsorbed phosphate. Soil Science Society of America Journal (in press).
30. Zee S E A T M van der and Riemsdijk W H van (1986). Sorption kinetics and transport of phosphate in sandy soil. Geoderma 38, 293–309
31. Zee S E A T M van der and Riemsdijk W H van (1986) Transport of phosphate in a heterogeneous field. Transport in Porous Media 1, 339–359
32. Zee S E A T M van der and Riemsdijk W H van (in press) Model for long term phosphate reaction kinetics in soils. Journal of Environmental Quality

The use of additives in livestock slurries to improve their flow properties, conserve nitrogen and reduce odours

B F PAIN[a], R B THOMPSON[a], L C N DE LA LANDE CREMER[b] and L TEN HOLTE[c]

[a] *Institute for Grassland and Animal Production, Hurley, Maidenhead, Berkshire SL6 5LR, UK*

[b] *Institute for Soil Fertility, P.O. Box 30003, 9750 RA Haren, The Netherlands*

[c] *Centre for Agrobiological Research, P.O. Box 14, 6700 AA Wageningen, The Netherlands*

Key words: slurry, additives, pollution, nitrogen, odours

Abstract. In recent years there has been increased interest in using additives in livestock slurries to alleviate management problems, to reduce the risk of air or water pollution and to improve their fertilizer value. The limited amount of independent research undertaken with bacterial or enzymic preparations for accelerating decomposition, so making semi-liquid slurries easier to store and pump, suggests that these products are frequently ineffective or too expensive. More favourable reports have come from some farms. A number of chemicals have been shown to reduce loss of nitrogen (N) by ammonia volatilization but each may have adverse side effects and the economics of their use is questionable.

Nitrification inhibitors have proved to be effective in reducing losses of N by both denitrification and nitrate leaching, especially when used with autumn or winter slurry applications to freely drained soils; agronomic experiments on both grassland and arable crops have given inconsistent results. Chemical additives have proved more effective for reducing odours than bacterial or enzymic preparations. Some success has also been reported for certain re-odorizing or masking agents and for materials with a large sorptive capacity such as bentonites and zeolites. With oxidizing agents, large quantities are normally required to achieve a significant, long-term effect.

Introduction

Faeces and urine collected from housed livestock contain appreciable quantities of plant nutrients which, when recycled to land in the form of slurry, can contribute towards the fertilizer requirements of the farm [26]. However, problems frequently arise with storing, transporting and landspreading large volumes of semi-liquid slurry. The practical problems are often due to the formation of crusts and sediments during storage which make it difficult to withdraw slurry from the store and to make timely, accurate applications to land. There is also increasing concern over pollution of the environment by both the inorganic and organic constituents of slurries. Nitrogen (N), for

H.G. v.d. Meer, et al. (eds.), Animal Manure on Grassland and Fodder Crops.
 ISBN 978-94-010-8137-5.

example, is readily lost from slurry, the loss not only representing a decrease in fertilizer value but also a source of water pollution (e.g. from nitrate leaching) and aerial pollution (e.g. from ammonia emission). In addition, there is an increasing number of public complaints about the emission of offensive odours from livestock buildings, slurry stores and slurry spreading.

Additives for slurry are available which are intended to help prevent or to alleviate the problems outlined above. These materials are normally bacterial or enzymic preparations, chemicals including acids, alkalis, oxidizing agents, bacteria inhibiting or bacteriacidal compounds, odour masking products or products with physical properties such as adsorbents. For many commercially available products, there is a lack of information from objective research and opinion about their effectiveness differs.

Although some products are claimed to provide a range of benefits, this paper discusses the use of additives for 3 main purposes:

1. to improve the flow properties of slurries in order to make handling and storage easier;
2. to reduce losses of plant nutrients, especially N, so reducing the risk of environmental pollution and conserving the value of slurries as fertilizers;
3. to control offensive odours.

To a lesser extent, certain products are claimed to increase the efficiency of treatment processes such as anaerobic digestion, or to have a disinfective effect in the prevention or control of disease.

Improving the physical characteristics of slurries

The main use for additives in this context is to prevent the formation of or to promote break-down of the crusts and sediments in stores so making slurries easier to pump and to apply to land.

The rheological properties of slurries have been shown to be dependent upon total solids (TS) content, the viscosity decreasing with decreasing TS content [5]. The power requirements for pumping should therefore also decrease with decreasing TS; the simplest means of decreasing TS content is dilution by the addition of water. However, it has been demonstrated that the solids in more dilute slurries settle out more rapidly and that this can increase the need for mixing [29]. For cattle slurry, Staley *et al.* [27] concluded that dilution below 8% TS was to be preferred for pumping, but that unnecessary dilution would increase storage, pumping and spreading costs.

A wide range of commercially available bacterial or enzymic preparations are claimed to speed up the rate of breakdown of slurry solids and improve the handling properties. Generally, little information is available about the origin or composition of these products and there is a lack of objective research on their use and efficacy. In a non-sterile environment such as slurry, bacteria which are able to degrade the organic matter are already present in abundance and there would appear little to be gained in reinforcing the population by inoculation. Similarly, since the indigenous bacteria secrete enzymes capable of degrading the external residues, it is questionable whether adding extra enzymes is worthwhile. In an extensive review of research, De la Lande Cremer [13] concluded that bacteria and enzyme preparations for accelerating the breakdown of slurry organic matter were generally ineffective and too expensive though successes, however, have been reported from some farms.

Conserving plant nutrients

Large proportions of the nutrient elements contained in plants and ingested by housed livestock are subsequently voided in faeces and urine. For example, about 80% and 90–95% of the N ingested by dairy cows and beef cattle, respectively, is excreted [3]. As discussed by Jarvis *et al.* [10], this N is readily lost by various pathways resulting in reduction of fertilizer value and environmental pollution. The main pathways for N loss are surface run-off, volatilization of ammonia (NH_3), leaching of nitrate (NO_3^-) and denitrification. The use of additives can provide a means of reducing both gaseous and leaching losses.

Reduction of ammonia volatilization

Very shortly after excretion by livestock, the urea in the urine will be hydrolysed and converted to NH_4^+-N which is subject to loss by NH_3 volatilization. Volatilization will occur from the floors of livestock buildings, from slurry stores and during and following the spreading of slurry on land. Three main methods have been investigated as means of controlling these losses, inhibition of urea hydrolysis, chemical stabilisation and adsorption.

Inhibition of urea hydrolysis. Faeces and urine contain little NH_4^+-N when voided by livestock, but NH_4^+-N is formed rapidly as the urea in the urine is hydrolysed by the commonly occurring enzyme urease. Although specific urease inhibitors exist, it seems unlikely that their use with slurries would be economic. In 1940 it was reported that hydrated lime ($Ca(OH)_2$) prevented the breakdown of urea to NH_4^+-N for several days, but became ineffective as the pH of the lime-manure mixture decreased [19]. More recently, hydrated lime with an 80% calcium oxide (CaO) content has been used in experiments in both the laboratory and buildings. The effect of lime in preventing NH_3 loss was found to be entirely due to the change in pH [20]. Whereas low rates of lime addition increased NH_3 losses, rates sufficient to increase the pH of the mixture to pH 9.6–9.8 conserved both total N and urea in slurry in the passageways of cattle buildings. The optimum rates of lime application ranged between 10 and 16 kg CaO per t slurry, depending upon the rate of deposition by the animals. Further work is needed on the effects of using lime in animal housing on herd health and equipment; assessments are also required on the economics of its use.

Disinfectants have also been used to retard ammonification but with little success.

Chemical stabilization. Additives have been used to convert ammonium carbonate in slurries, from which NH_3 is readily lost, to more stable ammonium compounds. During the early part of this century, gypsum ($CaSO_4 \cdot 2H_2O$), superphosphate ($Ca(H_2PO_4)_2$), phosphoric acid (H_3PO_4) or kainite ($KMg(SO_4)Cl \cdot 3H_2O$) were added to freshly produced manure or spread on land at the time of manure application. Tests were conducted over a 30-year period using these chemicals with manure applied to maize, and superphosphate proved to be the most effective compound in increasing crop yields from amended manures [24]. Both superphosphate and phosphoric acid have the advantage of reacting with both free NH_3 and NH_4-ions to form stable ammonium salts. In laboratory experiments, 94% of the ammonia in fresh dairy cow slurry was lost in 28 days compared to 70 and 53% when either superphosphate or phosphoric acid was added [23]. The use of neither chemical was considered to be economic but additional benefit may derive from the extra phosphorus in deficient soils. Conversely this could be a disadvantage in soils where the phosphorus supply was at a high level.

Sulphuric acid is relatively less expensive and its addition to slurry produces a stable ammonium salt, ammonium sulphate. Fig. 1 illustrates the loss of N from dairy cow slurries by volatilization of NH_3 measured directly using a system of small wind tunnels [15] following application to grassland at Hurley during the autumn. The pH of one batch of slurry was reduced to 5.5 by the addition of sulphuric acid prior to application. The slurry supplied 100 kg NH_4^+-N per ha and, without acidification, approximately 20% of this was lost within 3 days of application compared with less than 5% when acid was added. Subsequently, similar slurry treatments were applied at 3 grassland sites to assess the effect on herbage yields on different soils. Although the results across sites were inconsistent, herbage yields from the acidified slurry were up to 10% higher than those from unamended slurry.

Adsorption. Additives based on clay minerals (e.g. kaolinites, bentonites, zeolites) are claimed to offer a range of benefits including the reduction of NH_3 loss. The effect is through the adsorption of NH_4^+ and depends upon the cation exchange capacity of these materials. Little research appears to have been reported on the use of these materials for preventing NH_3 loss from slurries. However, a natural zeolite (clinoptilolite) applied to soil at a rate equivalent to 10 t per ha reduced NH_3 loss from urea fertilizer by 70–80% [22]. In a review of possible applications for natural zeolites, the use of clinoptilolite in Japan with both poultry and pig slurry was described [21]. In addition to controlling odours and hydrogen sulphide emissions, the material was also claimed to reduce NH_3 loss. Apparently, gaseous NH_3 reacts with the hydrous zeolite to form NH_4-ions which are selectively adsorbed and held in the zeolite structure. The material is either spread on the waste or, in some instances, fed to the livestock.

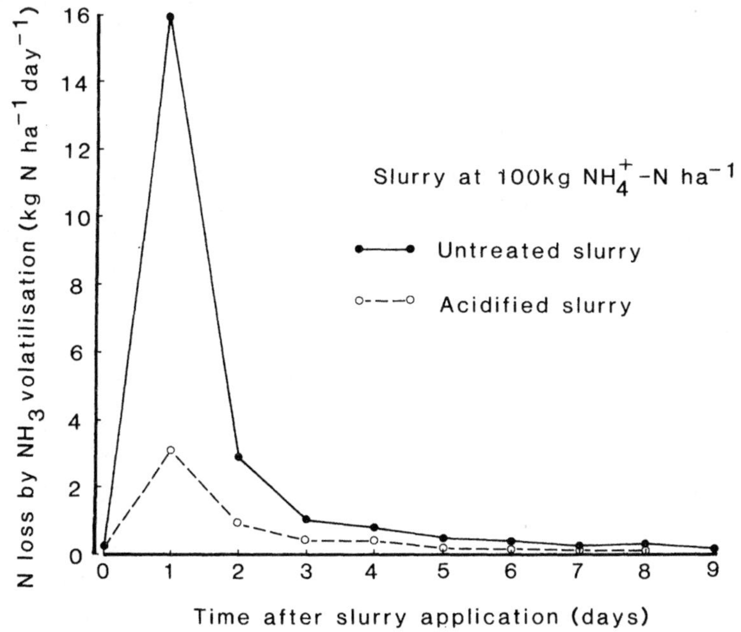

Figure 1. Loss of N by ammonia volatilization from cow slurry and acidified cow slurry applied to grassland.

Inhibition of nitrification

Following the application of slurry to land, a substantial proportion of the NH_4^+-N in slurries can be lost by volatilization of NH_3, unless this is controlled either by injecting or incorporating the slurry into the soil, or by use of an additive. The remaining NH_4^+-N is rapidly nitrified to NO_3^- by soil microorganisms. Although losses from NH_4^+-N within the soil are comparatively minor due to the adsorption of NH_4^+ by the cation exchange complex, NO_3^- is susceptible to loss by both leaching and denitrification under wet conditions as discussed by Jarvis *et al.* [10]. Nitrification inhibitors delay transformation of NH_4^+ to NO_3^- and reduce the loss of slurry N by both denitrification and leaching and so conserving it for uptake by plants.

The characteristics required for an ideal nitrification inhibitor were described by Hauck [8] as follows:

1. Inhibition should be specific to NH_4^+ oxidation. Nitrite (NO_2^-) oxidation should be unaffected since a build-up of NO_2^- in the soil may be phytotoxic. The chemical should have no adverse effects upon other soil microorganisms or other forms of life.
2. Resistance to degradation and physical/chemical properties (e.g. water solubility, vapour pressure and leaching characteristics) should enable an effective concentration to be maintained in the soil for an adequate period.
3. It should be economical to use.

Seven commercially available inhibitors were identified in 1983 [8], only 3 of these having been frequently used in experiments with slurries (Table 1). Nitrapyrin is the most extensively studied compound especially in relation to mineral fertilizer. An increasing amount of work with DCD, particularly with slurries, has been undertaken since the late 1970's in north-west

Table 1. Some nitrification inhibitors used with slurries

Chemical name	Common names	Company
2 chloro-6-(trichloromethyl) pyridine	Nitrapyrin or N-serve	Dow Chemical Co (USA)
Dicyandiamide	Didin or DCD	SKW Trostberg AG (FRG) Shawa Denko (Japan)
5-ethoxy-3-(trichloromethyl) -1,2,4-thiadiazole	Dwell or Terrazole	Olin Corporation (USA)

Europe. Dwell has been used only in a small number of experiments with slurries.

Numerous studies [2, 7] have demonstrated the effectiveness of DCD and nitrapyrin in inhibiting autotrophic nitrification in soils. The mode of action of the two compounds differs, DCD being bacteriostatic inhibits the activity and multiplication of *Nitrosomonas*, whereas nitrapyrin is bactericidal and kills the target organisms. The effectiveness of both inhibitors is dependent upon maintaining an adequate concentration in the soil. As the concentration falls, recovery is likely to be more immediate with DCD as the existing bacterial population regains its activity, but with nitrapyrin a new population has to be established [34]. The concentration of the inhibitor in the soil is determined by the rate of application and the rate of breakdown. Once commenced, the breakdown of DCD proceeds rapidly to form NH_4^+, NO_3^-, H_2O and CO_2 [2]. Since the compound is 66% N, it also contributes to the N required for crop growth. The eventual fate of nitrapyrin is uncertain and concern has been expressed over possible toxic residues [1]. The major factor affecting degradation is soil temperature [11, 33], the rate of degradation of both DCD and nitrapyrin increasing rapidly as soil temperature increases from about 6 °C to 20 °C. Water logging of soil can reduce the rate of DCD degradation [33] and the organic matter of the soil can diminish the effective concentration of nitrapyrin by sorption [11]. The high vapour pressure of nitrapyrin (and of Dwell) compared with DCD limits its use with surface-applied slurry where loss by volatilization reduces effectiveness. Conversely, nitrapyrin appears to leach more slowly than NH_4^+ and DCD more rapidly [4, 11].

Nitrification inhibition in field experiments. The effectiveness of DCD in inhibiting the nitrification of NH_4^+-N in slurry is well illustrated in Fig. 2. These data are from a field experiment where 60 t cattle slurry ha^{-1} containing 276 kg N (114 kg NH_4^+-N) was injected 20 cm deep into a sandy soil in The Netherlands during late November prior to sowing silage maize in the following spring. In the absence of DCD, all the NH_4^+-N applied in November had been nitrified by the following March despite the winter conditions. The addition of DCD at 30 kg ha^{-1} markedly delayed nitrification with only a small proportion of the NH_4^+-N being nitrified by March and approximately 50% by the beginning of May. DCD also influenced the distribution of the added N in the soil profile. In the unamended slurry treatment, a considerably greater proportion of the added mineral N was leached below 40 cm over the winter period. The inhibition of nitrification during the winter resulted in larger amounts of mineral N in the root zone

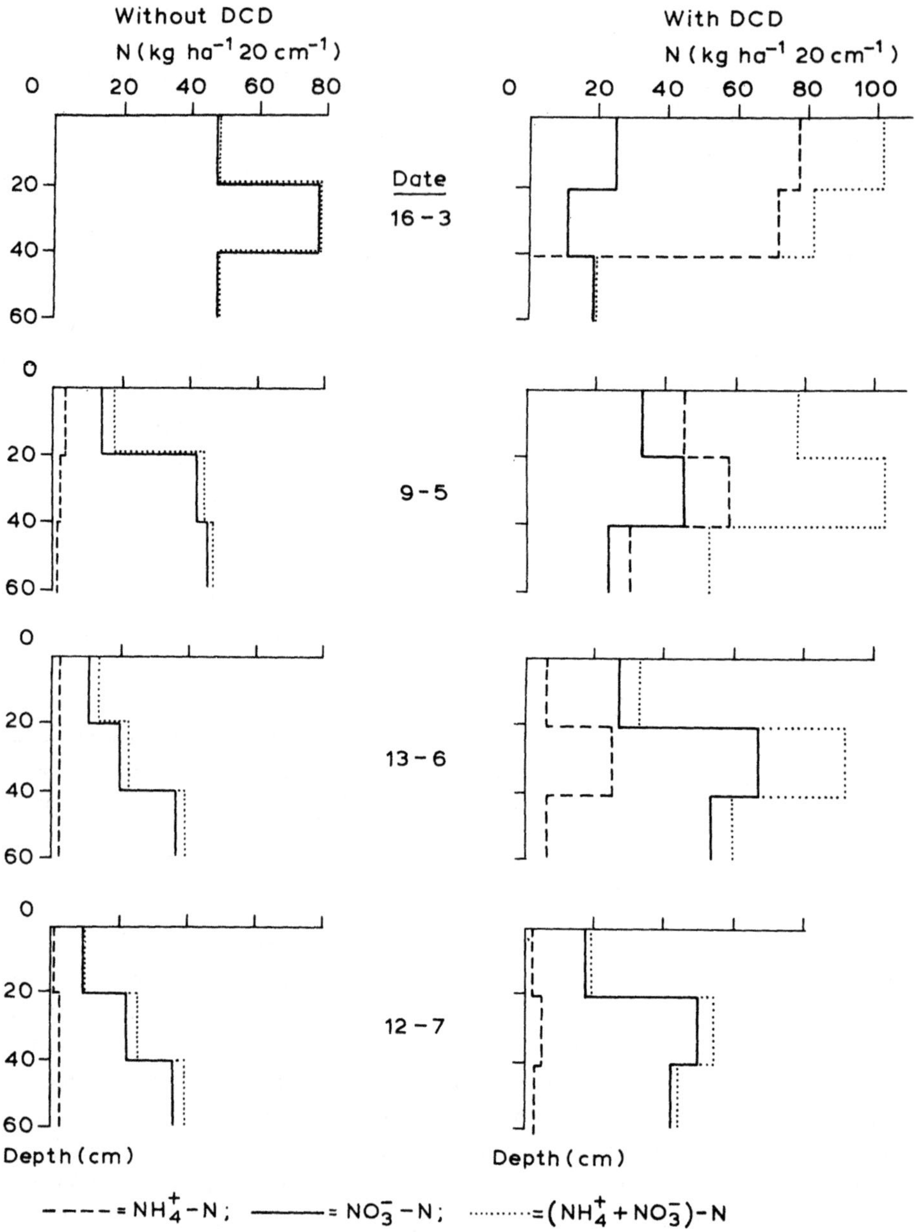

Figure 2. Distribution of inorganic N, NH_4^+-N and NO_3^--N in the soil profile following injection of 60 t cattle slurry ha^{-1} into soil in November with (30 kg ha^{-1}) or without a nitrification inhibitor (DCD).

in mid June. Thus the use of DCD resulted in an additional 120 kg inorganic N ha^{-1} being present in the soil profile to 60 cm.

Similar results were obtained from spring applications of slurry. Without DCD addition, nitrification was complete within 2 months, whereas only 50% of the NH_4^+-N in the slurry was nitrified when DCD was used.

Inhibition of nitrification by nitrapyrin was demonstrated in field experiments in the USA. Complete nitrification of NH_4^+-N in pig slurry injected into a fallow warm soil was delayed from 7 to 15 weeks [17].

Reduction of leaching losses. Loss of slurry N by the leaching of NO_3^- is generally minor in grassland [25], but may be more substantial following application to fallow or arable soils, especially in the autumn [28]. The results of field experiments have demonstrated the effectiveness of DCD in reducing leaching losses from slurries. The mean quantity of NO_3^--N contained in drainage water from a field sown to maize on a sandy soil and receiving either 40 or 80 m^3 ha^{-1} of slurry applied in spring, was reduced by

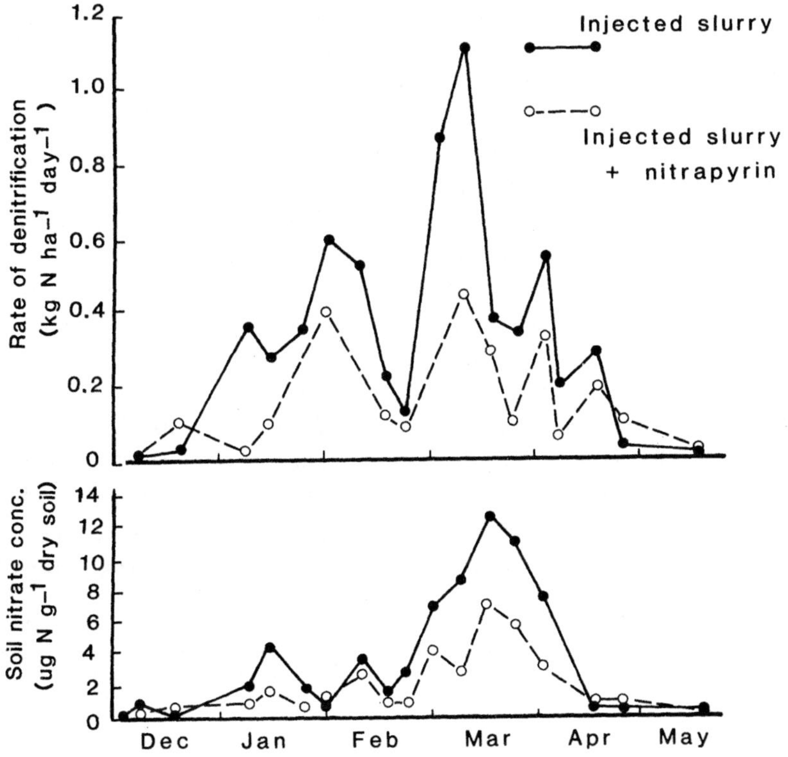

Figure 3. Rate of denitrification and nitrate concentration in soil following injection of cattle slurry with or without a nitrification inhibitor (nitrapyrin).

up to 19% when DCD was used [12]. For autumn applied slurry, the results from experiments in The Netherlands described earlier in this paper (Fig. 2), illustrate the effectiveness of DCD in reducing movement of NO_3^--N down the soil profile.

Reduction of denitrification losses. The potential for high denitrification losses and the apparent dependence upon soil NO_3^- content (Thompson, unpublished) in soils treated with slurry is illustrated in Fig. 3. These data are drawn from an experiment in which cattle slurry was injected into grassland on a freely drained loam soil at the beginning of December. The addition of nitrapyrin resulted in lower soil NO_3^- contents and lower rates of denitrification until the following April and total denitrification losses were reduced from 53 to 23 kg N ha^{-1}. For spring applications, nitrapyrin only slightly reduced denitrification loss (from 18 to 14 kg N ha^{-1}), presumably because warmer soil conditions reduced the effectiveness of the inhibitor. More recent work suggests that DCD is a more effective inhibitor than nitrapyrin for use with injected slurry.

Evidence for the reduction of denitrification by DCD (at 25 kg ha^{-1}) and nitrapyrin (5 litre ha^{-1}) with dairy cow slurry applied to the surface of grassland in early winter is provided in Table 2. The slurry was applied at 80 t ha^{-1} to a freely drained loam (Hurley, Berkshire), and a poorly drained clay soil (North Wyke, Devon), and was acidified prior to application to reduce NH_3 volatilization. The total amount of N lost by denitrification was markedly reduced by DCD addition at Hurley whereas the effect of nitrapyrin was negligible, presumably due to rapid volatilization of this inhibitor. Denitrification losses at North Wyke were small both with and without the inhibitors because poor aeration of the heavy clay soil precluded nitrification for much of the winter. When the experiment was repeated on both soils in the spring, denitrification losses were small.

Although there are few measurements of denitrification loss from slurry applied to arable soils, it is possible that such losses could be substantial especially with slurry incorporated into the soil. Nitrification inhibitors could then be useful, especially with slurry incorporated into well aerated soils in the autumn or, for later maturing crops, such as maize, in the spring.

Table 2. Total denitrification loss (kg N ha^{-1}) from acidified slurry applied at 80 t ha^{-1}

Treatment	Denitrification loss (kg N ha^{-1}) at:	
	Hurley	North Wyke
Acidified slurry	46.3	3.2
Acidified slurry + DCD	6.7	0
Acidified slurry + nitrapyrin	39.1	3.5

Agronomic responses to the use of nitrification inhibitors. The effects of nitrification inhibitors in reducing losses of N from slurry should be reflected in increased crop yields. A selection of data drawn from recent agronomic trials is presented in Tables 3a–c. In most instances, higher crop yields were obtained from slurry applications with an inhibitor than from those without. Generally, the yield increases were relatively greater for autumn than for spring applications and for higher rates of slurry application.

There is evidence (Thompson, unpublished) that although injection reduces loss of slurry N by NH_3 volatilization, it can increase loss by denitrification. This is reflected in the higher grass DM yields given in Table 3a from slurry applied with rather than without an inhibitor. In experiments with surface-applied slurry, DCD was of significant benefit only when used with a relatively high rate of slurry application ($80\,m^3\,ha^{-1}$) to a freely drained soil in the autumn.

A short-term retardation in grass growth from cattle slurry injected into the soil with DCD was reported [14] together with reductions in the yield of vegetable crops when this inhibitor was used at high concentrations [6]. It is not clear to what extent these reductions in yields are due to direct phytotoxicity and to what extent they reflect altered nutritional status, especially NH_4^+/NO_3^- nutrition. Sensitivity appears to vary between different plant species [2].

Although the data presented in Table 3b suggest that DCD augments the yield increase in silage maize obtained from slurry application, other studies have given inconsistent results. Despite clear evidence from soil tests that DCD inhibited nitrification, no increase in the DM yield of maize over 2 years was recorded from either autumn or spring slurry applications [9]. In this instance, it was considered that plant growth was not limited by N supply. In contrast, in a German study [30], significant yield increases were observed from the use of DCD over 3 years. Although the magnitude of the increases varied greatly, it was concluded that the use of DCD gave economic benefits.

Good results have been obtained with root crops (Table 3c) [30, 31]. With sugar beet and potatoes there were significant savings in fertilizer N application from using slurry supplemented with DCD, especially in the autumn [31]. The quality of both crops was equally good and sometimes better when DCD was used.

In reviewing an extensive series of Dutch experiments, De la Lande Cremer [14] noted that the best results from DCD were recorded with root crops, lesser responses were observed with silage maize, and least with grassland. Also, most benefit was obtained on sandy soils and the least on clay soils.

Table 3(a). Grass yields from slurries applied with or without a nitrification inhibitor

Study	Slurry				Inhibitor		Crop yield (t DM ha^{-1})			Notes
	Type	Application rate ($m^3 ha^{-1}$)	Application method	Application time	Name	Application rate	Untreated	Slurry	Slurry + inhibitor	
Thompson, UK (unpublished)	Cattle	80	Injection	Winter	Nitrapyrin	5 l ha^{-1}	4.6	9.3	10.0	
	Cattle	80	Injection	Spring	Nitrapyrin	5 l ha^{-1}	4.3	8.1	8.7	
Enting *et al.*, NL [6]	Cattle	40	Injection	Autumn	DCD	20 kg ha^{-1}	5.9	7.5	8.4	
	Cattle	40	Injection	Spring	DCD	20 kg ha^{-1}	5.9	8.9	9.3	
Pain, UK (unpublished)										
a. Free-drained soil	Cattle	40	Surface	Autumn	DCD	25 kg ha^{-1}	0.9	1.8	1.7	1st silage cut
	Cattle	40	Surface	Spring	DCD	15 kg ha^{-1}		2.2	2.3	
	Cattle	80	Surface	Autumn	DCD	25 kg ha^{-1}		2.2	3.0	
	Cattle	80	Surface	Spring	DCD	15 kg ha^{-1}		2.8	3.0	
b. Poorly drained soil	Cattle	40	Surface	Autumn	DCD	25 kg ha^{-1}	1.9	2.0	2.0	
	Cattle	40	Surface	Spring	DCD	15 kg ha^{-1}		2.9	3.1	
	Cattle	80	Surface	Autumn	DCD	25 kg ha^{-1}		2.8	3.1	
	Cattle	80	Surface	Spring	DCD	15 kg ha^{-1}		4.8	4.8	

Table 3(b). Arable crop yields from slurries applied with or without a nitrification inhibitor

Study	Slurry				Inhibitor		Crop yield (t DM ha^{-1})		
	Type	Application rate ($m^3 ha^{-1}$)	Application method	Application time	Name	Application rate	Untreated	Slurry	Slurry + inhibitor
McCormick *et al.*, USA [18]									
Grain maize	Pig	60	Injection	Autumn	Nitrapyrin	25 mg l^{-1}	3.3	7.3	11.4
Grain maize	Pig	63	Injection	Spring	Nitrapyrin	25 mg l^{-1}	3.3	11.7	11.2
Ten Holte, NL (unpublished)									
Silage maize	Cattle	60	Injection	Autumn	DCD	30 kg ha^{-1}		17.7	17.7
Silage maize	Cattle	60	Injection	Spring	DCD	20 kg ha^{-1}		17.4	17.2
Silage maize	Cattle	90	Injection	Autumn	DCD	30 kg ha^{-1}		16.2	21.5
Silage maize	Cattle	90	Injection	Spring	DCD	20 kg ha^{-1}		17.7	20.3
Vetter & Steffens, FRG [32]									
Winter barley	Pig	30	Surface	Autumn	DCD	15 kg ha^{-1}	3.1	3.4	3.8
Winter barley	Pig	30	Surface	Autumn	DCD	30 kg ha^{-1}	3.1	3.4	3.9
Winter barley	Pig	30	Surface	Winter	DCD	15 kg ha^{-1}	3.1	4.6	5.0
Winter barley	Pig	30	Surface	Winter	DCD	30 kg ha^{-1}	3.1	4.6	5.0

Table 3(c). Root crop yields from slurries applied with or without a nitrification inhibitor

Study	Slurry				Inhibitor		Crop yield (t ha^{-1})		Notes
	Type	Application rate (m^3 ha^{-1})	Application method	Application time	Name	Application rate	Without inhibitor	With inhibitor	
Titulaer, NL [31]									
Potatoes	Pig	50	Surface	Autumn	DCD	25 kg ha^{-1}	57.1	61.2	Mineral N applied at 120 kg N ha^{-1}
Potatoes	Pig	50	Surface	Spring	DCD	25 kg ha^{-1}	61.2	61.4	
Sugar beet	Pig	50	Surface	Autumn	DCD	25 kg ha^{-1}	62.0	75.4	Mineral N applied at 100 kg N ha^{-1}
Sugar beet	Pig	50	Surface	Spring	DCD	25 kg ha^{-1}	71.9	75.7	

The use of nitrification inhibitors with slurries can increase crop yields when the rate of slurry application is matched to crop requirements, and to the soil N status and the anticipated growing conditions. However, when application rate is excessive, yield increases from inhibitors are small or negative, and N losses are delayed rather than prevented. Similarly, the effectiveness of inhibitors is influenced by environmental conditions, especially soil temperature, and so variable effects on crop yields are to be expected.

Odour reduction

Offensive odours from livestock buildings, slurry stores and the land spreading of slurry are the subject of an increasing number of complaints from the public. A range of chemical, bacterial and enzymic additives are claimed to be effective in controlling odours. From a review of recent research on additives [13], it was concluded that proprietary bacterial and enzymic preparations were generally unsuitable for controlling odours but that some chemical additives were more effective. The mode of action of such chemicals is referred to as being either pre- or post-formation control [16], *i.e.* by either preventing the formation of malodorous compounds or by destroying or masking odours, respectively.

Pre-formation chemicals act by killing, or inhibiting the growth of bacteria, or by oxidation (this prevents the formation of odours and also oxidizes any malodorous compounds already present).

The use of superphosphate in animal houses stabilizes NH_4^+-N and is bacteriostatic but cannot be used in areas where soil phosphate levels are already excessively high. Reductions in odour have been obtained by adding Alzogur, a product based on 50% calcium cyanamide in aqueous solution and added directly to slurry [13]. The inhibition or killing of microorganisms can also be achieved with disinfectants such as sodium hydroxide, sodium hypochlorite, chloramines, quaternary ammonium compounds, phenols, formalin and chlorosulphate. The large quantities of these chemicals required for odour control often make their use uneconomic and, furthermore, there are possible side effects on animals, humans, crops and buildings.

Mechanical aeration of slurries is an effective method for reducing odours and the addition of oxidizing agents (permanganates, persulphates, peroxides) is an alternative possibility. Since slurries contain large amounts of organic matter, large quantities of these reagents are necessary to ensure complete oxidation and foam formation can be a problem. Some success has been reported with products based on potassium permanganate and ammonium persulphate [13].

Post-formation control by the use of re-odorization or masking agents has given some success, the agents being added to the slurry or spread by sprinklers. In The Netherlands, positive effects have been obtained with Alamask CPM-NX (Stalfris) [13]. The control of odours with odour neutralizing agents is more difficult to assess because the use of these materials depends upon maintaining a balance between mutually interacting odour components.

Odour reducing properties have also been claimed for products containing clay minerals, usually bentonite (e.g. Gullosan in The Netherlands or Agriben in the Federal Republic of Germany), kaolinite and zeolite (e.g. Biomin) [13]. The effects ascribed to these materials depend upon their large sorptive capacity.

Conclusions

1. A wide range of bacterial or enzymic preparations are available commercially, which claim to accelerate the decomposition of slurries, to control odours etc. Favourable reports have come from some farms but the limited amount of independent research undertaken with these materials suggests that they are generally ineffective and too expensive.
2. A number of chemicals, including hydrated lime, superphosphate and sulphuric acid, reduce NH_3 loss from slurries but each may have undesirable side effects and the economics of their use is questionable.
3. Nitrification inhibitors have proved effective in reducing nitrification of NH_4^+-N in slurries applied to soils and in controlling losses of N by both denitrification and NO_3^- leaching. The best results have been obtained with autumn or winter slurry applications to freely drained soils. Agronomic experiments have produced variable results: substantial yield increases have been recorded in some experiments, inconsistent or nil effects in others.
4. Bacterial and enzymic preparations appear to be unsuitable for controlling odours but some chemical additives are more effective. Some success has also been achieved with certain re-odorizing or masking agents and with materials such as bentonites and zeolites having a large sorptive capacity. With oxidizing agents, large quantities are normally required to achieve a significant, lasting effect.

References

1. Amberger A (1981) Didin as a nitrification inhibitor. Symposium on SKW-Didin nitrogen stabiliser for liquid manure, pp 20–33. Augsburg, Federal Republic of Germany: SKW Trostberg AG
2. Amberger A (1983) Wirkung und Einsatzmöglichkeiten des Nitrifikationshemmstoffes Dicyandiamid. Symposium Nitrifikationshemmstoffe, pp 22–47. Weihenstephan, Federal Republic of Germany: Verband Deutscher Landwirtschaftlicher Untersuchungs- und Forschungsanstalten
3. Blaxter K L (1977) The production of protein. Proceedings of the 2nd international Symposium on Protein metabolism and nutrition. EAAP Publication no. 22, 4–10
4. Bock B R, Lawrence J E and Williams H M (1981) In Hauck R D and Behnke H, eds. Proceedings of a technical workshop on dicyandiamide, pp 25–36. Muscle Shoals, Alabama, USA: SKW Trostberg AG
5. Chen Y R (1986) Rheological properties of sieved beef-cattle manure slurry: rheological model and effects of temperature and solids concentration. Agricultural Wastes 15, 17–33
6. Enting A E, Straatman F A and Lande Cremer L C N de la (1985) DCD, een nitrificatieremmer, Nota 143, 54 pp. Haren, The Netherlands: Instituut voor Bodemvruchtbaarheid
7. Goring C A I (1962) Control of nitrification by 2-chloro-6-(trichloromethyl) pyridine. Soil Science 93, 211–218
8. Hauck R D (1983) Nitrification inhibitors – Potential and limitations. Symposium Nitrifikationshemmstoffe, pp 9–21. Weihenstephan, Federal Republic of Germany: Verband Deutscher Landwirtschaftlicher Untersuchungs- und Forschungsanstalten
9. Holte L ten (1983) Gülleanwendung mit Nitrifikationshemmer: Einfluss auf N-Gehalt des Bodens und Ertrag von Silomais. Symposium Nitrifikationshemmstoffe, pp 290–304. Weihenstephan, Federal Republic of Germany: Verband Deutscher Landwirtschaftlicher Untersuchungs- und Forschungsanstalten
10. Jarvis S C, Sherwood M and Steenvoorden J H A M (1987) Nitrogen losses from animal manures: from grazed pastures and from applied slurry (this volume)
11. Keeney D R (1980) Factors affecting the persistence and bioactivity of nitrification inhibitors. Nitrification inhibitors – potential and limitations. ASA Special Publication No 38, Proceedings of the annual meeting of the American Society of Agronomy, pp 33–46. Chicago, USA: American Society of Agronomy, Soil Science Society of America
12. Kuntze H and Scheffer B (1981) Ways to reduce nitrogen leaching into the water system. Symposium on SKW-Didin nitrogen stabiliser for liquid manure, pp 41–55. Augsburg, Federal Republic of Germany: SKW Trostberg AG
13. Lande Cremer L C N de la (1984) Toevoegmiddelen aan mest en compost. Landbouwmechanisatie 35, 812–817
14. Lande Cremer L C N de la (1986) Preliminary results with the nitrification inhibitor dicyandiamide (DCD) in The Netherlands. In Dam Kofoed A, Williams J H and Hermite P l', eds. Efficient land use of sludge and manure, pp 7–20. London, UK: Elsevier Applied Science Publishers
15. Lockyer D R (1984) A system for the measurement of field losses of ammonia through volatilisation. Journal of the Science of Food and Agriculture 35, 837–848
16. Matthews P J (1979) Odours in sewage and sewage-treatment systems: control and cures. Septic sewage: problems and solutions. Proceedings of symposium, pp 55–60. Bournemouth, UK: Institute of Water Pollution Control
17. McCormick R A, Nelson D W, Sutton A L and Huber D M (1983) Effect of nitrapyrin on nitrogen transformations in soil treated with liquid swine manure. Agronomy Journal 75, 947–950

18. McCormick R A, Nelson D W, Sutton A L and Huber D M (1984) Increased N efficiency from Nitrapyrin added to liquid swine manure used as a fertiliser for corn. Agronomy Journal 76, 1010–1016
19. Midgley A R and Mueller (1940) Effect of lime on the nitrogen content of cow manure. Bulletin 456. Burlington, USA: Vermont Agricultural Experimental Station
20. Muck R E and Herndon F G (1985) Hydrated lime to reduce manurial nitrogen losses in dairy barns. Transactions of the ASAE, Paper no. 83-4063, 201–208
21. Mumpton F A and Fishman P H (1977) The application of natural zeolites in animal science and aquaculture. Journal of Animal Science 45, 1188–1203
22. Ryden J C (1984) Fertilisers for grassland. Chemistry and Industry (17 September 1984), 652–657
23. Safley L M, Nelson D W and Westerman P W (1983) Conserving manurial nitrogen. Transactions of the ASAE, Paper no. 81-2579, 1166–1170
24. Salter R M and Schollenberger C J (1939) Farm manure. Ohio, USA: Ohio Agricultural Experimental Station
25. Sherwood M (1986) Nitrate leaching following application of slurry and urine to field plots. In Dam Kofoed A, Williams J H and Hermite P l', eds. Efficient land use of sludge and manure, pp 150–157. London, UK: Elsevier Applied Science Publishers
26. Smith K A and Unwin R J (1983) Fertiliser value of organic manures in the UK. Proceedings no. 221 of the Fertiliser Society of London, 31 pp
27. Staley L M, Bulley N R and Windt T A (1971) Pumping characteristics, biological and chemical properties of dairy manure studies. In Livestock waste management, pp 142–145. Ohio, USA: American Society of Agricultural Engineers
28. Steenvoorden J H A M (1986) Nutrient leaching losses following application of farm slurry and water quality considerations in The Netherlands. In Dam Kofoed A, Williams J H and Hermite P l', eds. Efficient land use of sludge and manure, pp 168–176. London, UK: Elsevier Applied Science Publishers
29. Sobel A T (1966) Physical properties of animal manures associated with handling. In Management of farm wastes, pp 27–32. St Joseph, USA: American Society of Agricultural Engineers
30. Solansky S (1981) SKW-Didin Nitrogen Stabiliser. Test results and recommendations for use. Symposium on SKW-Didin nitrogen stabiliser for liquid manure, pp 78–96. Augsburg, Federal Republic of Germany: SKW Trostberg AG
31. Titulaer H H H (1983) Erfahrungen mit DIDIN zu Gülle in den Niederlanden. Symposium Nitrifikationshemmstoffe, pp 271–289. Weihenstephan, Federal Republic of Germany: Verband Deutscher Landwirtschaftlicher Untersuchungs- und Forschungsanstalten
32. Vetter H and Steffens G (1986) Economical use of slurry on arable land without causing damages on soil, water and plants and with small odour emission. Animal waste utilisation. Proceedings of 5th Consultation of FAO Cooperative Network, pp 99–117. Burgos, Spain: The National Board of Agriculture Sweden
33. Vilsmeier K (1983) Umsatz ^{15}N-Dicyandiamid im Boden. Symposium Nitrifikationshemmstoffe, pp 64–73. Weihenstephan, Federal Republic of Germany: Verband Deutscher Landwirtschaftlicher Untersuchungs- und Forschungsanstalten
34. Zacherl B and Amberger A (1983) Hemmung der Ammonium oxidation durch *Nitrosomonas europaea* mit verschiedenen Nitrifikationshemmstoffe. Symposium Nitrifikationshemmstoffe, pp 55–63. Weihenstephan, Federal Republic of Germany: Verband Deutscher Landwirtschaftlicher Untersuchungs- und Forschungsanstalten

Economic aspects of the use of animal manures

J H M WIJNANDS[a], W F AUSEMS[b] and J M A NIJSSEN[b]
[a] *Agricultural Economics Research Institute, Conradkade 175, 2517 CL The Hague, The Netherlands*
[b] *Research Station for Cattle, Sheep and Horse Husbandry, Runderweg 6, 8219 PK Lelystad, The Netherlands*

Key words: animal manures, plant nutrients, application method, fertilizer costs, storage capacity, manure surpluses

Abstract. This paper deals with some economic aspects of the improvement of the efficiency of the nutrients in animal manures on farms. Animal manures can be an important source of nitrogen (N), phosphorus (P) and potassium (K). An improvement of the efficiency index saves inorganic fertilizers and decreases expenditure on inorganic fertilizers. However, this cost decrease does not always compensate the costs of extra storage capacity and the additional costs of more expensive application systems. The risk of lower roughage production or utilization should not be neglected when manure has been spread in the spring and the summer. It can decrease the gross profit by an amount almost equal to the cost savings on fertilizer. From an economic viewpoint farms with a surplus of manure should use as much manure as (legally and agriculturally) possible. Overdoses of nutrients are less desirable from an environmental viewpoint.

Introduction

An important source of nutrients for the crops can be the nutrients in animal manures. About 50% of the nitrogen (N) used in some Western European countries can be supplied by animal manures. But the N in animal manure is usually less effective than N in inorganic fertilizers. This is indicated by the efficiency index. The efficiency index is the effectiveness of nutrients in animal manures compared with that of inorganic fertilizers. The efficiency index of the nutrients in manures depends on:

- the origin of the manure (e.g. cattle or pig slurry),
- the season of application (e.g. spring or autumn),
- the application technique (e.g. surface application or injection),
- the weather conditions (rainfall and temperature),
- the use of nitrification inhibitors,
- the storage system (whole slurry or separate) and
- storage time.

The efficiency index of N in animal manure, used for dunging grassland, ranges between 10 and 60%. The greater the efficiency index the better substitute manure is for inorganic fertilizer N [1].

H.G. v.d. Meer, et al. (eds.), Animal Manure on Grassland and Fodder Crops.
 ISBN 978-94-010-8137-5.

The objective of this paper is to discuss the economic aspects of assumed improvements of the efficiency indices of nutrients in animal manures used on grassland or fodder maize at farm level. The technical aspects will be discussed in other papers [5, 7, 8]. The availability of nutrients in animal manures and the use of inorganic fertilizers in some Western European countries will be discussed. The problems of manure surpluses in The Netherlands will also be considered. The economic results of the calculations for two different farm types with different intensities of livestock will be discussed next.

Nutrients in animal manures in some Western European countries

Animal manures were traditionally considered valuable fertilizers and crop production was largely dependent on them. Nowadays, a large part of the nutrients is supplied by inorganic fertilizers. The annual production of manure and nutrients per animal is summarized in Table 1. The amounts of nitrogen (N), phosphorus (P) and potassium (K) supplied by animal manures and by inorganic fertilizers in some Western European countries are shown in Table 2.

All figures in Table 2 are annual rates in kg per ha agricultural land (grassland, arable land and permanent crops). Table 2 is only a guide and the figures should not be interpreted as exact values of the quantities of available nutrients. The reasons for this are:

- nutrients in animal manures generally are not as effective as in inorganic fertilizers;
- the manure production of sheep, turkeys, ducks, etc. has been omitted;
- the production and composition of manure per animal depend largely on the feed intake; so there is a difference in the figures mentioned in Table 1 between the countries;
- actual rates of application vary according the need for nutrients of the crops grown and the intensity of stocking and land use.

Table 1. Manure production (in t manure and kg N, P and K) per animal per year or per production cycle. Source: [1]

	Manure	N	P	K
Cattle (livestock unit)	22.0	99.0	16.7	110.7
Fattening pig	1.7	11.6	3.2	9.2
Sow	4.8	19.2	8.2	15.9
Broilers (100 heads)	0.9	23.0	9.4	15.7
Laying hens (100 heads)	5.8	56.0	21.9	33.0

Table 2. Nutrients in animal manures produced in some Western European countries and the consumption of inorganic fertilizers in kg N, P, K per ha agricultural land (permanent crops included). Calculations based on: [1, 2, 3, 4]

	FR Germany	The Netherlands	Belgium/ Luxembourg	United Kingdom	Ireland	Denmark
Nitrogen						
Manure	104	257	185	55	71	100
Fertilizer	111	238	133	86	57	146
Total	215	495	318	141	128	246
Phosphorus						
Manure	21	54	37	10	12	21
Fertilizer	23	18	28	11	11	18
Total	44	72	65	21	23	39
Potassium						
Manure	110	267	194	59	79	101
Fertilizer	56	49	86	24	27	45
Total	166	316	280	83	106	146

Manures from cattle, pigs and poultry are important suppliers of N, P and K. About 48 (in FRG) to 66% (in The Netherlands) of the total P supply originated from animal manures and the share for K varies even between 66 (FRG) and 85% (The Netherlands). In the mentioned European countries inorganic fertilizer is an important source for N: 39 (Belg/Lux) to 58% (UK) is delivered by animal manure.

The share of different animal species in the total amount of N in animal manures is presented in Table 3. Cattle has a share of about 85% in the total amount of N in animal manures. This varies between 69% in Denmark and 97% in Ireland. Pig manure supplies most of the rest with an average share of about 12%, ranging from 3% in Ireland to 30% in Denmark. We have limited our study to cattle and pig manure because of these shares.

There is a relation between the size of the livestock sector and the production of roughage. In the Western European countries about 45% of the agricultural land is grassland and 23% is fodder crops. In Denmark grassland and fodder crops occupy 22% and in Ireland 91% of the agricul-

Table 3. The share (%) of cattle, pigs and poultry in the total amount of N in animal manures

Country	Cattle	Pigs	Poultry
FR Germany	81	16	3
The Netherlands	77	18	5
Belgium/Luxembourg	80	18	3
United Kingdom	88	8	4
Ireland	97	3	0
Denmark	69	30	2

tural land; these figures show some relation with the share of cattle in the total amount of N in animal manures.

Scope of the investigation

The object of this paper is to discuss the economic aspects of an improvement in the efficiency index of nutrients in animal manures. An improvement of the efficiency index can be achieved by various procedures, one possibility is the application of the manure in the spring instead of in the autumn. For example, N from cattle and pig slurry dressed in the autumn and winter may have efficiency indices of only about 10 to 20%. This index may be doubled when the slurry is spread in the spring, at the start of the growing season [1]. A second possibility to improve this index is the application system, e.g. injection instead of surface application.

The economic consequences are calculated here for seven different efficiency indices (Table 4). This range was chosen in order to avoid a technical explanation of the relation between the efficiency index and the application system. Details of these possibilities are discussed in other papers [5, 7, 8].

The application systems in Table 4 are an indication of the possibilities. A different system implies different costs; e.g. injection is more expensive than surface spreading. The changes in the gross profit of a farm are estimated with a model developed by the Research Station for Cattle, Sheep and Horse Husbandry (PR) [11]. The relevant aspects and costs are summarized below:

- N fertilizer: 1.50 guilders per kg N;
- P fertilizer: 3.26 guilders per kg P;
- K fertilizer: 1.03 guilders per kg K;
- surface spreading of slurry: 4.00 guilders per m^3;

Table 4. Efficiency indices (in percentages) of nutrients in animal manures

Efficiency index level	Efficiency index for:						Application system (as an indication)
	N on		P on		K on		
	grass	maize	grass	maize	grass	maize	
1	10	15	100	100	70	90	surface spreading, autumn
2	20	25	100	100	70	90	surface spreading, whole year
3	30	40	100	100	85	95	surface spreading, spring
4	45	40	100	100	85	95	injection, whole year
5	50	40	100	100	85	95	injection, spring
6	55	50	100	100	100	100	"future system"
7	60	50	100	100	100	100	"future system"

- additional storage capacity: 5.00 guilders per m^3;
- transportation of manure surpluses: 5.00 guilders per m^3;
- roughage, bought in case of a shortage: 0.50 guilders per kg dry matter and
- milk production per cow: 5500 kg per year.

Only these aspects and the grazing system can be varied in the calculations, all other factors are fixed. The fertilizer requirement of grass is assumed to be 400 kg N $ha^{-1}yr^{-1}$ from inorganic fertilizers and animal manures. The applied nutrients should be exactly equal to the needs of the grass or fodder maize (Table 5). If the need for P is 34 kg $ha^{-1}yr^{-1}$ and animal manures provide 9 kg, the farmer has to buy 25 kg. However, if animal manures provide 86 kg, an amount of manure which contains 52 kg P has to be used elsewhere. Whether manure can be used as a fertilizer or has to be considered as waste depends on the production of and needs for nutrients. For this reason the investigation takes three livestock densities into account, *viz.* 50, 70 and 90 dairy cows on 25 ha of agricultural land. We assume that 0.27 heifer and 0.30 calf for each cow has to be raised as replacements. It has been indicated that pig manure is also an important source of nutrients and that fodder crops are important for roughage production. For these reasons we consider two farm types, *viz.* 25 ha grassland farms and mixed farms with 20 ha grassland, 5 ha fodder maize and fattening pigs producing 1500 m^3 manure per year. In the last case an improvement of the efficiency index for K can increase the manure surpluses.

Profitability on grassland farms

A consequence of a better efficiency index level of the nutrients in animal manures is a decreased use of inorganic fertilizers. In Table 6 figures are shown for farms with 25 ha grassland and 50, 70 or 90 cows. An increase of the efficiency index means a maximum cost saving of about 3000 guilders per year on the farm with 50 cows and of 6300 guilders on the farm with 90 cows.

Table 5. The (organic and/or inorganic) fertilizer needs of grass and maize in kg N, P, K $ha^{-1}yr^{-1}$. Based on: [11]

	Grassland	Maize
Nitrogen	400	200
Phosphorus	30–47[1]	34
Potassium	215–248[1]	189

[1] The needs depend on the ratio between grazing and cutting for conservation. The manure production of grazing animals is not included.

Table 6. Effect of changes of the efficiency indices (see Table 4) for nutrients in cattle slurry on the costs of inorganic fertilizers (in guilders per year) on 25 ha grassland farms

Efficiency index level	Livestock density		
	50 cows	70 cows	90 cows
1	17,700	16,400	16,000
2	17,300	15,600	15,100
3	16,400	13,900	13,200
4	15,700	12,700	11,900
5	15,500	12,300	11,500
6	14,900	11,100	10,100
7	14,700	10,700	9,700

The reduction of the use of inorganic N takes the largest part of the cost reduction. The savings on P are relatively small, because the animals provide almost the total need for P on grassland. The farms with 50 and 70 cows have no manure surpluses. Only on the farm with 90 cows the need for P is fulfilled before all manure is used. The spreading of manure above the needs of the grassland is forbidden in the calculations. So on this farm about 200 t manure per year are surplus.

The saving on N is a maximum of 118 kg N per ha grassland or 30% of the total needs. In Table 7 the needs of inorganic fertilizer per ha grassland have been summarized for the farm with 90 cows. The cost saving is about 3% of the gross profit. However, the effect on the labour income of the farmer is much higher. Approximately 60 to 80% of the gross profit is necessary to meet for the costs of buildings, equipment, etc. So a better efficiency index level means an improvement of the labour income between 7 and 10%.

A different system of manure application can increase the costs of spreading and/or needs more storage capacity. If manure can only be applied in the spring, storage capacity should be doubled. E.g. if 1000 m^3 manure has to be

Table 7. Effect of changes of the efficiency indices (see Table 4) on the needs of inorganic nutrients ($kg\,ha^{-1}\,yr^{-1}$) on the farm with 90 cows

Efficiency index level	Nitrogen	Phosphorus	Potassium
1	376	0	73
2	352	0	73
3	329	0	36
4	293	0	36
5	281	0	36
6	270	0	0
7	258	0	0

Table 8. Effect of changes of the efficiency indices (see Table 4) on the costs of inorganic fertilizers, slurry application system and additional storage capacity on the farm with 90 cows (in guilders per year)

Efficiency index level	Inorganic fertilizers	Slurry application	Additional storage	Total
1	16,000	5,400	0	21,400
2	15,100	5,400	0	20,500
3	13,200	5,400	3,400	22,000
4	11,900	8,100	0	20,000
5	11,500	8,100	3,400	23,000
6	10,100	8,100	0	18,200
7	9,700	8,100	3,400	21,200

spread, the available storage capacity has been assumed to be 500 m^3. Another 500 m^3 is needed if the manure application is restricted to the spring and/or early summer. This additional storage capacity costs (500 × 5.00 guilders) 2500 guilders per year. An improved application system like injection or the future systems are more expensive than surface spreading. We assume that these systems cost 6.00 instead of 4.00 guilders per m^3 [9]. In reality, the price can also be twice as high as the price of surface spreading. The additional costs for storage capacity and a more expensive spreading system are given in Table 8 for the grassland farm with 90 cows. Table 8 demonstrates that a reduction of the fertilizer costs does not always compensate for the additional costs of the improved application system and storage capacity. The advantage of the improved efficiency index level is mainly environmental.

Spreading or injection of the slurry in spring, mostly necessary for an improvement of the efficiency index level, may also have negative consequences, e.g. scorching of the sward, growth depression and a negative effect on herbage intake. The quantity of roughage available for feeding may decrease and the farmer has to buy (more) roughage. In Table 9 the gross profits are summarized in case the grass production decreases by 5%. The gross profit on the farms decreases by about 2000 to 4000 guilders per year. The losses can abolish the gains of the improvement of the efficiency index

Table 9. Gross profit (in 1000 guilders per farm per year) without and with a decrease of 5% of the grass production

Efficiency index level	50 cows		70 cows		90 cows	
	without	with	without	with	without	with
6	131	127	157	154	175	173
7	130	127	156	153	174	172

level. As a consequence, the risk of a reduction of the grass production is a very important point in the economic evaluation.

Profitability on mixed farms

The supply of nutrients on farms with cattle and pigs is quite different from that on farms with only cattle. The amount of available nutrients will be larger than the needs of the crops. We made the assumption that the farmer also fattens pigs, which produce 1500 m^3 manure per year. The farm land consists of 20 ha grassland and 5 ha fodder maize. We will discuss only the farm with 70 cows.

In Table 10 the effects of an improvement of the efficiency index are summarized. Even with the highest efficiency index level the grass and fodder maize need inorganic N. So an improvement of the efficiency saves inorganic N. The situations for P and K are quite different. The quantity of manure applied to the land is limited, because we made the assumption that it is not permitted to exceed the needs for these nutrients. On grassland the total requirement for P is supplied by animal manure. It is assumed that a part of the P need (17 kg ha^{-1}) of maize has to be inorganic fertilizer (band-application at sowing), the other part is always supplied by manure. The P need limits the application of manure till the efficiency index of K is 95%. In this case K limits the application of manures. It is not permitted to exceed the nutrient requirements of the crops and in that case manure surpluses increase. If the cost of transporting these surpluses to deficit farms increases, the economic balance can decrease even when the efficiency indices improve.

Table 10. Effect of changes of the efficiency indices (see Table 4) on the costs of inorganic fertilizers (guilders per year), the manure surpluses (t per year) and the costs of manure surpluses (guilders per year) for mixed farms with 70 dairy cows and fattening pigs which produce 1500 t slurry per year

Efficiency index level	Fertilizer costs			Manure surplus		Total costs
	grass	maize	total	t	costs	
1	12,300	1,600	13,700	1,577	7,900	21,600
2	11,700	1,500	13,200	1,577	7,900	21,100
3	10,400	1,200	11,600	1,579	7,900	19,500
4	9,400	1,200	10,600	1,579	7,900	18,500
5	9,100	1,200	10,300	1,579	7,900	18,200
6	9.000	800	9,800	1,671	8,400	18,200
7	8,700	800	9,500	1,671	8,400	17,900

Mixed farms and the Dutch environmental policy

In the preceding two sections the discussion on the fertilization of grassland and fodder maize was limited to the requirements of these crops. In The Netherlands a Law on Soil Protection and a Law on Fertilizers will be enacted in 1987 [10]. Only the application of P from animal manure will be restricted: on grassland the limit will be 107 and on maize 150 kg P ha^{-1} yr^{-1}. After 1991 these limits will be lower and after 2000 the limits will show a great similarity with the assumptions in the preceding sections. After 2000 it will not be permitted to exceed the agricultural needs, *i.e.* for fodder maize about 34 kg and for grassland about 30–47 kg P ha^{-1} yr^{-1}. For the near future the legally allowed rates are much higher than the requirements of the crops. As a consequence, on farms where the manure production exceeds the crop requirement, the manure surpluses can be diminished by application of manure above the needs of the crops. In that case the quantity of manure, which can be applied, is restricted by the legally permitted dose or by the maximum dose, which can be applied without damaging the crop. If the quantity of K applied to grassland is too high, the dairy cows can suffer from hypomagnesaemia (grass tetany). The overdose of K on grassland is therefore limited to 41 kg K ha^{-1} yr^{-1}; the total amount of K should be less than the needs of grassland increased by 41 kg. The need of K by grassland is about 215 to 248 kg K ha^{-1} yr^{-1} and depends on the grassland utilization system (see Table 5). The limit of fertilization with K is therefore about 256 to 289 kg K ha^{-1} yr^{-1}.

In the near future the legal limits are very ample and lead to overdoses of P and K. The permitted overdressing with nutrients is shown in Table 11.

Table 11. Dressing with manure (t ha^{-1} yr^{-1}), annual surplus of manure (t per farm) and overdoses of nutrients (kg ha^{-1} yr^{-1}) for the mixed farms with 90 cows and pigs

Efficiency index level	Dressing with manure			Manure surplus	Overdoses				
	grass[1]	maize			grass		maize		
	cattle manure	cattle manure	pig manure		P	K	N	P	K
1	80	1	79	1,104	24	41	0	133	196
2	80	1	79	1,104	24	41	0	133	196
3	66	57	56	1,220	13	41	53	133	344
4	66	57	56	1,220	13	41	53	133	344
5	66	57	56	1,220	13	41	53	133	344
6	56	97	40	1,301	6	41	136	133	464
7	56	97	40	1,301	6	41	136	133	464

[1] Additional to the dung and urine voided by grazing animals.

Table 12. Manure surplus (in t per year) and gross profit (in 1000 guilders per year) on mixed farms with 90 cows and pigs

Efficiency index level	Legally limited doses		Agricultural doses	
	manure surplus	gross profit [1]	manure surplus	gross profit [1]
1	1,104	161	1,911	158
2	1,104	163	1,911	158
3	1,220	164	1,913	161
4	1,220	165	1,913	162
5	1,220	164	1,913	161
6	1,301	165	1,958	163
7	1,301	164	1,958	162

[1] Gross profit only from the cattle and grassland activities, transportation of (pig) manure surpluses included.

The application of manure to grassland is limited by the K dose; the overdose is always the maximum. Therefore, on grassland the legal limit for P will never be limiting. An increase of the efficiency index of K results in a decrease of the application of animal (cattle) manures on grassland. Overdressing with P diminishes at the same time. The legal P doses are only of importance for fodder maize, and on this kind of mixed farms the maximum doses will always be applied. The legal annual dose in the period 1987–1990 is 150 kg P from animal origin per ha maize. The needs of maize are 34 kg P $ha^{-1} yr^{-1}$ of which 17 kg has to be inorganic fertilizer. The overdressing of maize with N increases with the improvement of the efficiency index. The overdosing with K increases not only because of the improvement of the efficiency index but also where there is substitution of pig manure by cattle manure. The ratio between P and K in pig manure is lower than in cattle manure.

The manure surpluses, shown in Table 11 (legally allowed rates of application), are 650 to 800 t lower than when manure application is not allowed to exceed crop requirements. The gross profit in the situation with legal limits is slightly higher (Table 12). But if the transport of the surplus to deficit farms (which is not always possible) costs 10 instead of 5 guilders per m^3, the difference is 5000 guilders per year instead of about 2500. The economic and the environmental approaches are in conflict on these kinds of farms.

Discussion and conclusions

The cost savings on inorganic fertilizers as a consequence of an improvement of the efficiency index of animal manures used on grassland or maize are almost in balance with the extra costs of the storage capacity and spreading systems. Furthermore, if the grass production decreases by 5% the decrease of the gross profit is almost equal to the cost savings on fertilizers. Therefore, the willingness of the farmers to improve the efficiency indices of nutrients in animal manure depends on the increase of storage and spreading costs and also on the risk of a smaller harvest of roughage. The improvement of the efficiency indices should be considered more from the viewpoint of environmental protection than from the viewpoint of the economist. More attention should be paid to the quality of the drinking water. Leaching of nitrate can pollute the groundwater. To purify this polluted groundwater costs 0.20 to 0.40 guilder per m^3 [6]. The run-off of P and other manure components is also an important aspect for the environmental policy.

On farms, where the amount of nutrients in animal manures exceeds the needs of the crops, only an improvement of the efficiency index of N has a positive effect on the gross profit. The supply of N from animal manures is less than the needs of grassland, because the quantity of manure for fertilization purposes is limited by the permitted K dose. In the situation the farm has a surplus of manure, it is economical to use as much as legally or agriculturally is possible. In this case manure is partly waste. A higher efficiency index is always beneficial from the viewpoint of environmental protection.

References

1. Anonymus (1985) Dierlijke mest. Vlugschrift voor de Landbouw nr. 406. Wageningen, The Netherlands: CAD Bodem-, water- en bemestingszaken
2. EUROSTAT (1984) Yearbook of Agricultural Statistics. Luxembourg: BSEG
3. EUROSTAT (1985) Yearbook of regional statistics. Luxembourg: BSEG
4. FAO (1985) Fertilizer Yearbook. Rome: FAO
5. Geurink J H and Verbeek-Overmeer J L R M (1987) The effect of a nitrification inhibitor on the utilization of nitrogen by grass from injected or surface-applied pig slurry (this volume)
6. ICW (1985) Nitraatproblematiek bij grondwaterwinning in Nederland. Wageningen, The Netherlands: ICW
7. Meer H G van der, Thompson R B, Snijders P J M and Geurink J H (1987) Utilization of nitrogen from injected and surface-spread cattle slurry applied to grassland (this volume)
8. Schröder J and Dilz K (1987) Cattle slurry and farmyard manure as fertilizers for forage maize (this volume)
9. Snijders P J M (1986) Via injectie betere benutting runderdrijfmest. Boer en Tuinder 30/10/86
10. Tweede Kamer der Staten Generaal (1986) Regelen inzake het verhandelen van meststoffen en de afvoer van mestoverschotten (meststoffenwet). Den Haag: Tweede Kamer
11. Wieling H (1986) Saldo per hectare of bedrijf. Set voor het analyseren van de saldoberekening rundveehouderij. Lelystad, The Netherlands: PR

The future of animal manures as fertilizer or waste

J K R GASSER
9 Laverstock Park West, Salisbury, Wilts SP1 1QL, UK

Key words: environment, grassland, nutrient utilization, pollution, slurry, waste disposal

Abstract. Intensive animal husbandry can and often does result in the production of more wastes than there is land available to use them effectively. The addition of large amounts of slurry based on a policy of disposal or on nitrogen (N) use by crops will lead to the accumulation of phosphorus (P) in the surface soil and may result in the application of excessive amounts of potassium (K) with deleterious effects on crop quality and animal health. The crops able to use most slurry are grass and forage crops such as maize.

Nitrogen will accumulate in soils with the additions of slurry and will start to decline when no more is added. Much N is lost as ammonia to the atmosphere and as nitrate by leaching and denitrification.

The overall balance of P within a region or country may be calculated from accessions and removals. Steps may need to be taken to limit the excessive accumulation of P in soils by exporting wastes to regions of the same country or other countries which require additional P manuring.

The spreading of slurry can cause several environmental problems such as the release of malodorous compounds and ammonia to the air; the production and entry to the upper atmosphere of nitrous oxide; the loss of slurry by surface run-off; and the leaching of organic matter, N, P and K to drainage and groundwaters.

Background to the problem

The introduction to this Symposium [40] reminded us that the intensive rearing of farm animals, which has become widely practised over the last thirty-five years, has created serious problems of waste disposal. In traditional mixed farming, animals were bedded with cereal straw and the manure resulting from the admixture of urine and faeces with the straw was returned to the land. The principal manures were from cattle and pigs, and, before the advent of the tractor, stable manure from horses. The product was a wet solid which could be handled by the available farm labour and was readily stored and transported. Much or all of the feed for the animals was grown on the farm so that the production of farmyard manure (FYM) both conserved plant nutrients and made possible their return to the land. With the increasing importation of feeding-stuffs and more intensive livestock keeping, some farmers started to produce more manure than they required

H.G. v.d. Meer, et al. (eds.), Animal Manure on Grassland and Fodder Crops.
 ISBN 978-94-010-8137-5.

or were prepared to return to the land, but usually other farmers or horticulturists in the immediate area would take the surplus. These conditions were general until the present expansion in production, when the position has changed dramatically as mechanization and a steadily decreasing labour force have required more intensive methods of animal husbandry with the accompanying use of more imported feeding-stuffs.

Intensification has also been associated with housing more animals for longer periods. Many pigs and poultry are now housed throughout their lives. Dairy cattle are normally housed for part of the time, in Great Britain typically for six months in each year, whereas beef suckler herds may not be housed at all. Most sheep are kept in the fields except for lambing, when they are collected in a sheltered area. There is a recent trend to provide housing immediately before and for a period after lambing. However, because sheep are not kept intensively, no problems have been reported with waste from them. Housing animals leads to the need to dispose of the effluent. From cattle and pigs this is now usually slurry because of the changes in animal husbandry; from battery hens, either slurry or dried manure; and from birds kept in deep-litter houses, a solid poultry manure. Intensive animal enterprises are common throughout the world varying, for example, from the very large cattle lots in North America to more intensive pig production throughout Western Europe and the Far East, so that the problems of waste utilization or disposal are very widespread.

Environmental and health problems

In addition to the problems of disposal or utilization of the wastes, the last decade has seen a steep rise in public awareness of environmental problems, such as odour from animal units and pollution of water supplies by slurry spread under adverse conditions. Many of these problems are discussed in the papers presented to the Symposium. The Commission of the European Communities (CEC) published a directive in 1980 relating to the quality of water intended for human consumption which limited the amounts of many parameters [9]. The directive was fully implemented in 1985. For this discussion, phosphorus (P), potassium (K) and nitrate-N are of particular relevance for animal wastes. The need is re-emphasized for careful and proper application of slurry to prevent pollution of water intended for drinking.

The human and animal health problems associated with the use of slurries are not a matter for discussion here but they have received considerable attention [6].

Economic and social issues

Finally the problems of animal wastes have to be set against the current over-production of cereals, milk, and beef within the European Economic Communities (EEC), which will bear on land use, intensity of production and related questions, and affect the profitability of the enterprises. Therefore the utilization or disposal of animal manures in the future will be much concerned with costs and social consequences as well as with the technical need to improve efficiency of use of the nutrients by crops.

The various questions will be examined in relation to the information put before the Symposium and suggestions will be made concerning the future, particularly the need for further research and development.

Production and distribution of manure

The excess manure produced by intensive livestock farming often poses problems of disposal or use. This theme was developed by the information presented to this Symposium showing the livestock density in different countries and particularly the marked increase in the number of pigs in The Netherlands and Belgium/Luxembourg between 1957–1960 and 1983 [40]. Earlier work had developed a model to maximize the profit from the applications of slurry by assigning values to the nutrients in it [16]. This concept was discussed for dairy farming in The Netherlands [41]. For the future, an econometric model is required which will allow various values of fixed costs, variable costs and labour costs to be assigned and the importance of changes in these on the overall cost or profit of the system employed, to be calculated.

The problem of excess manure was discussed in more detail for an area of The Netherlands, and transport to an adjoining area was used to remove part of the surplus [4]. Where this problem exists, the restriction of intensity of production has been introduced as a solution to prevent an increase of the surplus with proper legislative and administrative controls [4]. However, restricting expansion does not solve the underlying problem of the existing production of an excess of plant nutrients. These nutrients come from imported feeding-stuffs and from the use of artificial fertilizers. If a true balance is required, the final solution to the problem is either to restrict the amount used to the removals and losses or to export the surplus to areas requiring additional plant nutrients.

Handling, storage and spreading of slurry

In between the production of slurry by housed livestock and its use or disposal to land, the material has to be moved from the animals to a store, stored, with treatment if necessary, transported and spread. The major movements are from the animals to the slurry store and from the store to the field. Several methods are used to move the slurry from the animals to the slurry-store, depending on the design of the production unit. Slurries are normally pumped from store but the presence of fibrous material makes this more difficult and the removal of the coarse material before the slurry enters the store makes subsequent pumping easier [15].

Various treatments may be applied to the slurry and these with their advantages and disadvantages are shown in Fig. 1. The separated coarse solids can have some economic value. For example, the fibre from cattle slurry may be composted to produce an organic manure of increased value or it may be cleaned to provide fibre as a raw material for industrial use. The coarse solids of cattle and pig slurries may also be worked by worms to provide a growing medium for plants. Engineering aspects of problems with effluents from livestock have received much attention [15]. The handling and treatment of slurry either by aeration or by anaerobic digestion requires machinery and power and therefore increases production costs.

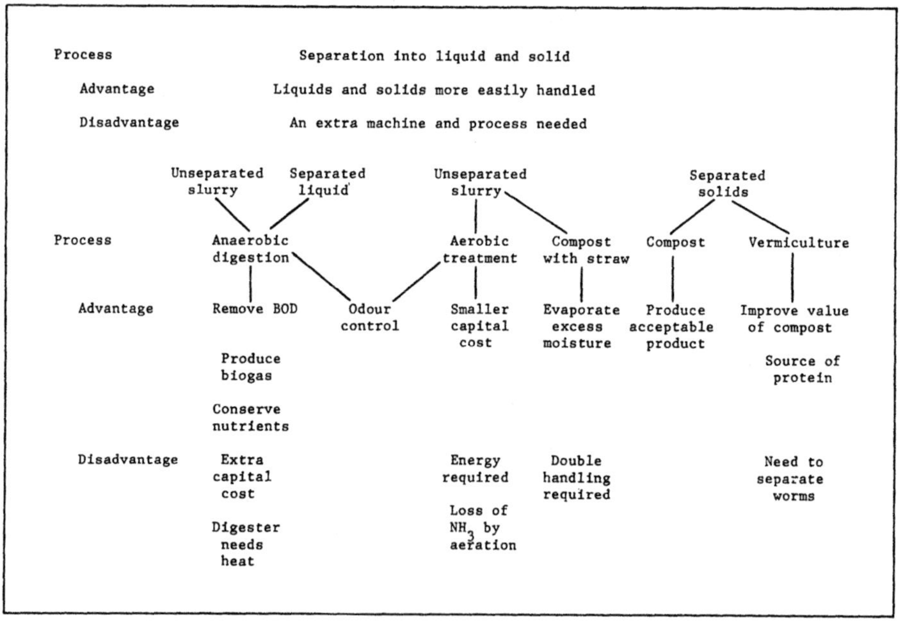

Figure 1. Some major pathways in the handling and treatment of cattle and pig slurries [15].

Economics of waste management

Animal production is not possible without the production of waste. The costs of handling, storage, treatment and spreading are essential parts of the overall costs of animal production and the aim must be to minimize these. One way is to make effective use of the nutrients contained in the slurry. The value of these may be determined by reference to the cost of applying equivalent amounts of available plant nutrients as fertilizers [16]. However, the calculation is not simply that of balancing costs against value, because application for a crop is only one step in the waste disposal system which may well include the need to prevent odour pollution, and environmental pollution after spreading. The application of excessive amounts of some nutrients arising from the unbalanced nature of animal effluents also affects the actual value of the slurry. Some of these aspects have received attention [41]. Work reported to this Symposium showed that maximum or near maximum yields of forage maize were achieved with annual dressings of 200 m^3 slurry per ha [30]. This provides 1000 kg N ha^{-1}, 200 kg P ha^{-1} and 1000 kg K ha^{-1}, all greatly in excess of crop requirements. Similarly, fodder beet utilized up to 240 kg ha^{-1} of inorganic N from slurry [31] leaving a large residue of N mainly in the organic form. The large applications for maize resulted in heavy leaching losses of nitrate-N, which would cause severe water pollution [30]. The ability to apply very large dressings to forage crops is attractive because less area of land is required for disposal, but the consequences for crop composition and the environment require to be fully investigated to allow proper guidelines to be drawn up and, where required, legislation to be enacted.

Future considerations

The system used for storage and handling of slurry cannot be divorced from its value as a fertilizer or cost of disposal as a waste material. For the future, these matters should receive attention during planning and construction of new enterprises. However, with increasing legislation to protect the environment, improved handling, storage, treatment and application may be necessary to enable the farmer to stay in business.

Nutrient supply and ratios of manures

The composition of animal effluents varies with species, age of animals and with the diet [11, 39, 40]; the concentration of the slurry is affected by the amount of extraneous water which is added. However, there are some major differences between cattle, pigs and poultry. Some average UK and Dutch values given in Table 1, show that cattle slurry contains a higher proportion of K relative to N than pig or poultry manure; pig slurry contains slightly more and poultry manure contains a much higher proportion of P than cattle slurry. Similar values for cattle slurry were given at this Symposium, suggesting that these average values will allow valid comparisons to be made [22, 25, 30]. Cattle slurry has a smaller short-term effect than pig or poultry slurry [40]. For fertilizing crops, all manures will require balancing amounts of nutrients, particularly N. In utilizing manures, the decision has to be made whether to use one nutrient, often N, as the limiting one and allow the other nutrients to be in excess or deficiency. If sufficient N is applied to meet crop needs, then normally P will be in excess, and more K than is needed will frequently be added from dressings of cattle slurry [26].

The other approach used is to apply manures so that the nutrient in greatest supply limits the rate of application for the crop concerned [16]. This is usually P, and by limiting the amount added, additional N and K will frequently be needed and excess P will require disposal.

Legislation to be enacted in 1987 in The Netherlands will use the P content of animal manures to limit the rate of application of animal manures [4, 40]. From 1991, this limitation will be for smaller amounts. With smaller amounts of slurry applied, additional N as fertilizer will normally be required and often K as well. The approach of using P as the limiting nutrient has the advantage that the additional N can be applied at times to meet the crop demand, with more efficient use and less risk of loss to the environment. Considering the crop requirements for nutrients, various estimates have been made of the number of animals required to supply the nutrients for one hectare of crop. This is least for cereals and most for grass and fodder crops [26, 40, 41] which increases the problems of disposal for intensive pig and poultry producers in areas of predominantly arable farming.

Table 1. Chemical composition of animal slurries in the UK and The Netherlands [1, 26]

	%N	%P	%K
UK			
Cattle	0.5	0.09	0.50
Pig	0.4	0.09	0.17
Poultry	1.7	0.61	0.58
The Netherlands			
Cattle	0.4	0.08	0.46
Pig	0.7	0.19	0.54
Poultry	1.5	0.70	1.08

The total amount of nutrients entering the farming system can be calculated from the amount of fertilizer used, the imported feeding-stuffs and other sources of plant nutrients. Total removals in products and losses from the system are less easily calculated but reasonable estimates may be made. The net additions or losses are given by the difference between these two amounts. Internal cycling of nutrients does not alter the total, but re-distribution will change the balance between fields and larger areas. The three major elements behave differently and require separate consideration.

Nitrogen is ephemeral and should be applied to be used as efficiently as possible in the year of application. Any inorganic N remaining in the soil at the end of the season, together with any mineralized after an arable crop is removed or grass stops growing, will be nitrified and the nitrate will be at risk of denitrification over the winter [28] or leaching [30]. The continued application of slurry increases the total N in the soil and the amount mineralized each year, so that after several years of large applications considerable amounts of N become available to succeeding crops [30] and correspondingly large amounts of nitrate are formed in arable soils in the autumn. Grass or an autumn sown crop will take up some of this nitrate and lessen the amount at risk over the winter. When applications of organic manures cease, total N content declines rapidly. This decline has been well documented in many countries including the UK [18]. For N, the major problems are in minimizing losses and using it efficiently. Much work is required to improve both these aspects.

Phosphorus is the major crop nutrient with the least mobility in soils and almost the whole of the excess beyond current crop needs will remain in the surface soil. Much of the P is in the inorganic form when slurry is applied to the soil [11] and the balance is mineralized quite quickly. Some accumulation is beneficial in increasing soil fertility [27]. With large annual additions of P, the adsorptive capacity of the surface layer will become saturated and the idea is developed that the saturation front will move downwards [27], with the rate of movement depending particularly on the sorption capacity of the soil. This model should be tested using the measured distribution of P in soils which have a known history of heavy slurry application because earlier results with FYM and fertilizer P applied to a sandy loam soil showed that P from FYM had moved to 450 mm, whereas P from fertilizer had not moved to below the cultivated layer, 225 mm [18].

Potassium in animal manures is freely available to crops and excessive applications to arable crops can result in unnecessary uptake and possibly nutrient imbalance. When applied to pastures, excess K can lead to hypomagnesaemia in cattle [25], so that excess should be avoided. However, K is retained well by heavier soils and to a lesser extent by sandy soils so that residual K will be available to succeeding crops. Leaching of excess K from sandy soils will occur and further information is necessary to be able to predict when the amounts will be large enough to cause pollution of water supplies [9].

When copper (Cu) and zinc (Zn) are added to pig feed their concentrations in slurry will be increased which, when the slurry is applied to soils, will raise their Cu and Zn contents. Table 2 gives results from Northern Ireland which show how the extractable Cu was increased by the addition of about 1700 t pig slurry per ha over a period of 8 years [21]. The specific problems of Cu in animal wastes and sewage sludge were discussed at a CEC Workshop, when one conclusion was, "if the present policy of disposing of animal wastes on limited areas is continued, the application of slurry from pigs receiving supplementary copper could increase the concentrations in the soil to unacceptably high levels within a finite time. The use of lower levels of supplementation will increase the time before toxic amounts are present but the major factor in determining the amount of copper applied to the land is the intensity of pig production in the area." Therefore, in those areas where extra additions of Cu or Zn to the soil are not desirable, these metals should not be added to the pig feed. However, there are large areas of Cu deficient soils which would be expected to benefit from additions of Cu. These differences in requirements make the framing of regulations very difficult if they are to have general implementation.

Effects on soil properties

The effects of continued dressings of pig slurry on the chemical properties of a grassland soil are illustrated by results in Table 2 [21]. Increases in

Table 2. Chemical properties of a grassland soil, A – untreated and B – treated with 1700 t pig slurry per ha over a period of 8 years [21]

Soil depth (mm)	Total				Extractable*					
	C (% of air dry soil)		N		P (mg per kg air dry soil)		K		Cu	
	A	B	A	B	A	B	A	B	A	B
0–150	4.47	5.08	0.58	0.71	15	154	135	570	16	38
150–300	2.03	2.16	0.21	0.25	2	15	50	115	11	15
300–460	0.77	0.87	0.07	0.11	1	2	37	60	5	8
460–610	0.38	0.46	0.03	0.04	1	1	39	62	4	5
610–760	0.27	0.37	0.02	0.03	2	1	41	79	3	4

* Nutrients extracted by 1 *N* ammonium acetate/acetic acid solution at pH 4.2

total C and total N and extractable P and Cu were almost entirely in the top 300 mm, with very small amounts below this, whereas K had moved to the full depth sampled, 760 mm. Similarly, cattle slurry applied for forage maize had increased the total N content of the surface layer and of the 200–600 mm layer of soil and P had accumulated in the surface soil and moved to a depth greater than 400 mm [30]. Organic manures generally increase the extractable P in soils [18]. The continued application of slurry will increase production of the crop to which it is applied and the residual value will improve the yield of following crops. Results presented to the Symposium described how from 6 to 10 dressings of slurry to maize both increased its yields and the residual values improved the DM yield of following grass at all rates of fertilizer N applied to the grass. The benefits increased with increasing amounts of slurry applied [30].

The chemistry of P in soils is highly complex and some aspects in relation to animal manures have been touched on at this Symposium [11, 27]. However, the understanding of the processes and their application to problems in the field require longer-term investigations. The major questions to be answered in the future are:

i) What is the greatest P content that soils should have and should this be related to texture and other soil characteristics?

ii) What is the largest single addition of P that should be made?

In the experiments on forage maize, the slurry applied contained up to ten times the amount of P required by the crop but smaller excesses of N and K, suggesting that N and K are likely to limit production before P does [30, 40]. For crop production the question of the maximum desirable P concentration in soils cannot be answered on present evidence. Fresh applications of slurry increased the P content of crops [32]; within two years the effect had gone although large amounts of the added P still remained. These often increase the amount of soluble P [18]. Even with the largest amount of slurry applied the P content in the herbage did not exceed the optimum concentration [32]. Attention needs to be paid to the differing effects of animal manures on grassland, with surface accumulation, and arable soils, with mixing. The role of injection of slurry on the availability of P also requires further elucidation [22, 25], particularly why grass does not take up P in the first cut whereas K is available [32].

Damage to grassland by grazing cattle and applying slurry

Grassland can suffer damage in three ways:

i) with grazing animals – from urine scorch and smothering by dung pats;

ii) with slurry spread – scorch and smothering may occur especially with cattle slurry and the soil surface may become sealed;

iii) with injection of slurry – mechanical damage to the sward.

Urine patches tend to scorch the herbage, but the regrowth of the herbage is attractive to the grazing animal. In contrast, cattle avoid herbage growth round the edge of dung pats. Increasing the intensity of grazing increases the proportion of the area adversely affected by dung although the effect on production may be slight [19]. There is the related question whether dung pats should be spread at the end of a grazing period for rotationally grazed grassland to make the distribution more uniform. For continuous grazing, there is inevitably a compromise between stocking rate, herbage utilization and performance of individual animals. Within these constraints the effects of dung and urine patches have to be considered, because the return of nutrients is very uneven [19]. The production of herbage varies from very good in areas receiving urine, through areas of lesser output to severe temporary damage under dunged and scorched areas, with considerable variations in the quality of the herbage. These differences also affect the behaviour and performance of the grazing animal, so that often only the total animal production of the system can be measured.

The effects of smothering by slurry have been considered to be one cause of the poorer performance of cattle slurry compared with pig slurry [25, 37].

Injection of slurry into grassland has received attention [22, 25, 32] and the general conclusion reached is that injection improves the effectiveness of slurry N. The spacing should not exceed 300 mm for optimum utilization and maximum yield. However, injection at spacings closer than 500 mm is not practical at present because of damage to the sward [22] and power requirements. Uptakes of P and N [22] are much affected by increasing distance from the point of injection [25]. The mechanical damage due to the passage of the injectors considerably depresses yield at the first cut and must be considered when measuring the overall effects of injection [25]. More dry matter is produced and greater N uptake is obtained by injecting slurry compared with surface spreading, with the loss of yield from mechanical damage apparently compensated for by improved nutrient utilization. The distribution of yield between cuts is also altered. The injection of slurry is attractive for minimizing odour problems and virtually eliminates volatilization of ammonia and the disadvantages associated with surface spreading of slurry, like smothering and fouling of grass. A number of points appear to require further investigation:

i) Can the technique of injection be improved to allow closer and shallower spacing with less power requirement without damage to the sward, thus making it more cost effective?

ii) How is sward productivity affected by root disturbance and the very unequal distribution of organic matter and nutrients below the surface layer? Slurry is normally injected below the main root zone of grass and this may be expected to affect the availability of nutrients, the growth of roots into the slurry and re-growth of the severed deeper roots. These physiological effects may be partly or wholly the cause of the poorer growth of the first cut from injected slurry.

iii) How does injecting slurry affect the drainage of the surface soil and the movement of organic matter and nutrients into the drainage and groundwaters? In the absence of additional fertilizer N, both injected and surface-spread slurry increased the nitrate concentration at 1 m depth in the autumn [17], but all values were below the new maximum EEC limits [9]. With about 375 kg N $ha^{-1} yr^{-1}$ as fertilizer, all concentrations exceeded the limits whether slurry was applied or not, although the effects of the two smaller amounts of injected slurry were irregular [17].

iv) More information is required on the effects of injected slurry and its interaction with fertilizer N on a) yields of herbage, particularly the first cut; b) composition of the grass; c) losses of nitrate to the drainage water.

Nitrogen cycles and modelling

Many models for N cycles have been suggested and the modelling of N from farm wastes was itself the subject of a CEC Workshop [7]. The papers included a very useful discussion on mathematical modelling of individual processes as well as larger models of N transformations in soil. Fig. 2 illustrates how these may be expressed in a simplified scheme for N fluxes in soil [20] and shows that:

i) effluent enters a number of pools of N of the system,
ii) a number of processes allow interchange between pools,
iii) only ammonium and nitrate allow exit from the system.

The last conclusion assumes total retention by the soil so that there is no mass flow of effluent in drainage or surface run-off or accumulation of nitrite, which if present will behave similarly to nitrate, except that nitrite is phytotoxic. The processes were discussed and values calculated for some of the parameters. Additional material has been provided here which enhances the value of such models [11, 17]. The use of models is very valuable in dealing with systems such as the production and use of slurries by providing a conceptual framework for the results, indicating gaps in the information and providing an assessment of the importance of the processes. Models may be used for both scientific purposes and to aid the farmer in devising

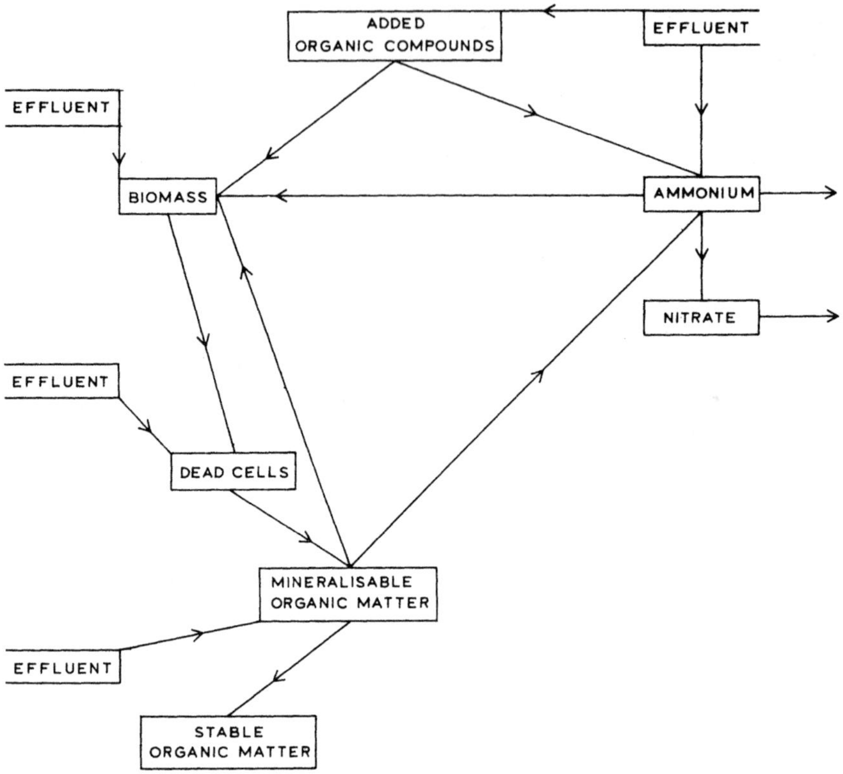

Figure 2. Simplified scheme for nitrogen fluxes in the soil [20].

safe systems for disposal [7, 8, 23], and further development is required of both aspects.

Losses of nitrogen in the field

With cattle and sheep grazing in the field, dung and urine normally fall on separate patches of herbage. The forms of N in them and their physical natures differ. The main form of N in urine is urea, whereas in dung, the N is mainly in more complex organic compounds. The losses of ammonia from urea following decomposition have been well documented [17]. Some ammonia is lost from dung pats, but considerable nitrification of ammonia formed on decomposition of the organic N occurs. The nitrate formed will leach into the underlying soil, where the zone under the dung pat is intensely reducing and much will be reduced. Of the balance, some will be taken up by the crop, some will be leached to the drainage water and some more will be denitrified [5]. Losses of N as ammonia from urine and dung and of nitrate by leaching and denitrification from grazed swards are inevitable [15, 17, 28, 29]. They can be controlled to a limited extent by using zero-grazing and returning the slurry to the sward under optimum conditions, but the efficient use of N under grazing conditions is still a major problem to be solved.

Losses as ammonia

Ammonia in the atmosphere comes principally from animal wastes, industry and domestic coal burning. It has a relatively short residence time as evidenced by the increased concentration round industrial plants and urban areas, compared with the general background value. Ammonia losses are particularly large from heavily grazed grasslands [35] and surface-spread animal slurries. Some will be re-absorbed in the leaf canopy, some will be deposited by dry deposition and the balance will enter the upper atmosphere and be re-precipitated with rainfall [35]. The proportions from animal wastes and from urban and industrial sources are difficult to allocate but concentrations in rainfall in Glasgow are 0.60 mg NH_4^+-N l^{-1}compared with the average of 0.27 mg NH_4^+-N l^{-1} in Northern England [35]. At Rothamsted Experimental Station, Harpenden, in an area of mixed farming, the average rainfall 1969–73 was 619 mm and the average NH_4^+-N concentration was 1.71 mg l^{-1} with a range from < 0.05 to 8.5 mg l^{-1}. The mean annual deposition was 10.7 kg NH_4^+-N ha^{-1} [42]. In The Netherlands, the total annual deposition was 28.4 kg NH_4^+-N ha^{-1} (12.2 kg NH_4^+-N ha^{-1} as wet deposition and 16.2 kg NH_4^+-N ha^{-1} as dry deposition). Eighty five percent of the total is ascribed to animal manures [40].

Allowing for uniform distribution of ammonia from animals, these amounts represent an important addition of N in natural areas. However, in the UK it is less than the oxides of N produced by burning fossil fuel [35]. There is still an urgent need to devise methods for minimizing losses of ammonia both to conserve N and to lessen pollution, particularly the acidifying effect of the ammonia and ammonium sulphate falling on natural areas with poorly buffered soils.

Losses by denitrification

Denitrification is a common process in agricultural soils carried out by obligate anaerobes under reducing conditions to give N_2 and nitrous oxide (N_2O). A similar process leading to the formation of N_2O during the nitrification of ammonium N was reported over thirty years ago [2] and the conditions of its formation have recently been investigated in great detail [3]. The amounts produced during nitrification of added ammonium can exceed those produced by denitrification of nitrate [3].

The presence of dung and slurries can create conditions that locally favour denitrification and some values have been presented to show the magnitude of these effects [5, 24, 34]. The effects of organic matter on the production of N_2 and N_2O during nitrification are not known. There is a need to determine whether it is better to apply slurry N in the ammonium form with subsequent oxidation or to allow nitrification to proceed before application. This question is also relevant to the possible use of nitrification inhibitors, which is discussed later.

Denitrification represents a loss to the farmer of a valuable plant nutrient. However, where excess nitrate is at risk of causing pollution of water the process is beneficial in removing some or all of the nitrate. Recent evidence suggests that draining grassland on a heavy soil increased the amount of nitrate in the drainage water [12]. The processes leading to the formation of nitrate and its loss by denitrification under dung pats or heavy dressings of slurry have been described [5, 29]. The losses of N by denitrification can decrease the manurial value of slurry, therefore, in the less usual case where much of the N in slurry is in the nitrate form, care should be taken to ensure that adequate aeration is maintained.

Losses by leaching of nitrate

The increasing intensification of agriculture has resulted in larger amounts of N circulating with increases in the size of the nitrate-N pool (Fig. 2) so that when uptake by the crop is low there is increased risk of loss by leaching. Excess nitrate is present when excess slurry or fertilizer N is applied to grassland, when slurry is applied in autumn and winter and with increasing intensity of stocking of grazed grassland. Losses from grazed grassland are greater than from cut grass [17]. Much further work is required in order to minimize risk to the environment from excess nitrate, to improve its efficiency of use by crops and to prevent denitrification, where this results in loss to the farmer.

Prevention of pollution of the environment

The use of slurry for crop production or its disposal by land-spreading can lead to severe problems of contamination of the environment, particularly if the spreading is not done in accordance with good practice. The major causes of pollution are surface run-off, losses of ammonia and nitrous oxide to the atmosphere and movement of nitrate to drainage and groundwaters. The N aspects of surface run-off received attention in a Workshop of the EEC Programme of Co-ordination of Research on Effluents from Livestock [10]. Other related aspects were discussed in an earlier Seminar on Modelling Nitrogen from Farm Wastes [7]. There will be corresponding losses of other constituents of slurry including P, K and organic matter. The papers highlight two aspects of behaviour that affect pollution. The first is the need to apply slurry only when conditions minimize the risk of surface run-off. The second is not to apply more slurry than the soil can accept. Both these points have received attention in the development of an algorithm for applying pig slurry to grassland [23]. The amount that could be applied was limited either by the soil moisture deficit, which had to exceed 15 mm before application was allowed, or by the amount of N in the slurry. The use of such algorithms incorporating the latest findings on the use of slurry should be developed and extended as an aid to management, and could include P and Cu and other heavy metals as limiting factors [27].

Minimizing losses of ammonia from slurry

Nitrogen is lost from slurry as ammonia after the decomposition of urea and other organic compounds to yield ammonia. One way to minimize loss at this stage would be to prevent breakdown to ammonia by the use of a urease inhibitor. At present, no effective inhibitors are available so that this route to conserving N is not available [24].

Nitrogen is lost from ammoniacal solutions as ammonia when the pH becomes alkaline. Attempts have been made to decrease losses by adding acidic materials and adsorbents, such as clay, to the slurry. The use of phosphoric acid or superphosphate decreased losses but was not considered to be economic; sulphuric acid and various clay minerals have also been used [24]. A complete aerobic treatment of piggery wastes by acidic sludge separation has been developed [38]. The addition of sulphuric acid precipitates the proteinaceous material to leave a clear supernatant liquid. The excess acid is then neutralized. Adding acid to unseparated slurry suffers from the disadvantage that the large buffering capacity of slurry will prevent large changes in pH for reasonable additions of acid. Also the acid may flocculate the slurry thus changing its physical nature. Sulphuric acid will also increase the acidity when applied of the soil. To add extra phosphoric acid will further increase the nutrient imbalance as well as being more expensive. The best solutions appear to be either to incorporate the slurry into the soil immediately after the application or to inject it into the soil. The latter technique is receiving considerable attention, and should be further investigated.

Minimizing losses of nitrate

In aerated slurry or in the soil, ammonium is oxidized to nitrite and then to nitrate. Nitrite only accumulates occasionally, but both it and nitrate are susceptible to loss by leaching or denitrification. The use of nitrification inhibitors is attractive to prevent the transformation of ammonium to nitrite. With slurry, limited success has been achieved in inhibiting nitrification by the use of dicyandiamide and N-Serve, also called Nitrapyrin. Dicyandiamide has the advantage that it provides additional N on breakdown. However, a more specific and more active inhibitor appears to be required before its use becomes viable for preventing losses through denitrification of the nitrate produced [14].

Nitrous oxide, formed both during the nitrification of ammonium and denitrification of nitrate, is important in the chemistry of the upper atmosphere because it produces nitric oxide via a reaction chain. Nitric oxide is in equilibrium with nitrogen dioxide and both largely control the distribution and abundance of atmospheric ozone [35]. Therefore, increased N_2O emission is one factor influencing ozone concentrations, but the amount calculated to come from fertilizer applications and agricultural soils is no greater than from natural soils or the oceans [35]. Not all N_2O produced enters the upper atmosphere, for much is retained by the soil and more is removed by rainfall [28], nevertheless the role of slurry in increasing the N_2O in the upper atmosphere requires quantification.

Losses to drainage and groundwaters

Pollution of drainage and groundwater occurs by the leaching of soluble compounds not retained by the soil, particularly nitrate and by flow through cracks, fissures and other structural faults. For example, when pig slurry was applied to grassland in summer with large moisture deficits, the biological oxygen demand (BOD), and N and P contents of the drainage water increased sharply within one day of applying the slurry. They returned to their base values after several days [23]. The rapid entry was considered to be due to the flow of slurry down soil fissures to the drains, a consequence of the normal characteristics of the heavy soil of the site. In another investigation with pig slurry [21], the depth of drains and nature of the backfill were also important. Two systems had drains either 1.25 m deep backfilled with soil (A) or 0.5 m deep backfilled with stones to a level close to the surface (B).

After the application of pig slurry, the composition of the water from the two drainage systems differed markedly. Water from the deep drains (A) contained less organic matter, N and P in the period shortly after applying pig slurry than did water flowing from shallow drains (B). The flows were similar. The movement of P has been assumed to be piston like in character [27], although other workers have demonstrated preferential flow through undisturbed columns of soil [33]. The relative importance of the two processes needs to be more clearly understood. All these results point to the need for more information on the behaviour of slurry applied to soils of varying texture and other physical attributes in order that the algorithms used to determine the suitability of spreading may be further improved.

Application of slurry in the autumn and winter leads to increased pollution of drainage and groundwaters, particularly by nitrate, and of surface waters by run-off of whole slurry. However, with grazed swards and the application of organic manures to arable soils, some different effects have been recorded. With grazing cattle, maximum concentrations of nitrate-N in drainage water occurred in late winter and early spring showing that decomposition of organic residues occurs over the winter [36]. The fate of this N needs to be studied. Similar results were found in arable soils, because the early results at Rothamsted Experimental Station on Broadbalk field showed that plots treated annually with FYM since 1843 contained much more nitrate in the autumn and more mineralizable N than the plots with fertilizer [13]. The drainage water from the plot receiving FYM also contained more nitrate than that from plots with fertilizer N. The magnitude of these effects needs to be measured on other soils and with different climates. The application of organic manures to grazed pastures or arable soils will increase the amount of nitrate lost into the drainage and groundwaters, although grass and autumn sown crops will remove some until growth ceases. We need to be able to predict the effects so that applications of slurry do not increase the concentration of nitrate in the drainage water above the acceptable limit [9].

Conclusions

- The intensification of livestock production has changed the form of the wastes from solid manure to slurries with the need for different systems of collection, storage and spreading. Intensification has also allowed the production of more wastes than there is land available for optimum use, so that either excessive amounts have to be applied or the manure has to be transported to areas with a demand for plant nutrients.

- The rate of application of slurry is usually determined by the N requirement of the crop, so that excess P is often applied, and sometimes excess K.
- The increased use of fertilizers and imported feeding-stuffs has increased the total N, P and K entering and in use in agriculture.
- Total N in soils receiving heavy dressings of slurry is increased by much less than the residual amount and therefore, the excess N not used in production is mainly lost as ammonia to the atmosphere or as nitrate by leaching or denitrification.
- Phosphorus is almost immobile and excess P accumulates in the surface layers of soil and in some areas the increased P content is causing concern.
- Potassium may be applied in sufficient excess to cause animal health problems (hypomagnesaemia) or imbalance in crop nutrition.
- Losses of N, P and K by various routes to water used for drinking purposes can be considerable and may cause pollution.
- The development of models predicting the behaviour of nutrients in slurry in soils will aid in its better use with less adverse effects on crops, animals and environment.
- A better understanding of the processes causing pollution of the environment is required in order to prevent them. More research is required to provide the basic information for both simple farm system models and more complex scientific models.

References

1. Anon (1985) Dierlijke mest. Vlugschrift voor de Landbouw nr. 406. The Hague: Ministerie van Landbouw en Visserij
2. Arnold P W (1954) Loss of nitrous oxide from soil. Journal of Soil Science 5, 116–128
3. Blackmer A M, Bremner J M and Schmidt E L (1980) Production of nitrous oxide by ammonia-oxidising chemoautotrophic microorganisms in soil. Applied and Environmental Microbiology 40, 1060–1066
4. Boheemen P J M van (1987) Extent, effects and tackling of a regional manure surplus: a case study for a Dutch region (this volume)
5. Burford J R, Greenland D J and Pain B F (1976) Effects of heavy dressings of slurry and inorganic fertilizers applied to grassland on the composition of drainage waters and the soil atmosphere. In Agriculture and Water Quality, Technical Bulletin No. 32, pp 432–443, Ministry of Agriculture, Fisheries and Food. London: HMSO
6. Commission of the European Communities (1978) Kelly W R, ed. Animal and human health hazards associated with the utilization of animal effluents. EUR 6009 EN. Luxembourg: Commission of the European Communities
7. Commission of the European Communities (1979) Gasser J K R, ed. Modelling nitrogen from farm wastes. London: Applied Science Publishers

8. Commission of the European Communities (1980) Gasser J K R, ed. Effluents from livestock. London: Applied Science Publishers
9. Commission of the European Communities (1980) Council Directive of 15 July 1980 relating to the quality of water intended for human consumption. Official Journal of the European Communities, L229 Volume 23, 30 August 1980, pp 11–29. Luxembourg: Commission of the European Communities.
10. Commission of the European Communities (1981) Brogan J C, ed. Nitrogen losses and surface run-off from landspreading of manures. Developments in Plant and Soil Sciences, Vol. 2. The Hague: Martinus Nijhoff/Dr Junk
11. Faassen H G van and Dijk H van (1987) Manure as a source of nitrogen and phosphorus in soils (this volume)
12. Garwood E A, Ryden J C and Tyson K C (1986) Nitrogen losses from drained grassland. Occasional Symposium No. 20 of the British Grassland Society, pp 70–74. Hurley, UK: The British Grassland Society
13. Gasser J K R (1962) Effects of long continued treatment on the mineral nitrogen content and mineralisable nitrogen of soil from selected plots of the Broadbalk experiment on continuous wheat, Rothamsted. Plant and Soil 17, 209–220
14. Gasser J K R (1970) Nitrification inhibitors – their occurrence, production and effects of their use on crop yields and composition. Soils and Fertilizers 33, 547–554
15. Gasser J K R (1984) Disposal of effluents from intensively housed livestock. Outlook on Agriculture 13, No. 2, 80–86
16. Herlihy P D (1980) The use of mathematical models in studying the land spreading of animal manures. In Gasser J K R, ed. Effluents from livestock, pp 446–458. London: Applied Science Publishers
17. Jarvis S C, Sherwood M and Steenvoorden J H A M (1987) Nitrogen losses from animal manures: from grazed pastures and from applied slurry (this volume)
18. Johnston A E (1975) The Woburn Market Garden experiment, 1942–69. II. The effects of the treatments on soil pH, soil carbon, nitrogen, phosphorus and potassium. Report of Rothamsted Experimental Station for 1974, Part 2, pp 102–130. Harpenden, UK: Rothamsted Experimental Station
19. Lantinga E A, Keuning J A, Groenwold J and Deenen P J A G (1987) Distribution of excreted nitrogen by grazing cattle and its effects on sward quality, herbage production and utilization (this volume)
20. Laudelout H and Lambert R (1980) Simulation of environmental pollution by land spreading of manure. In Gasser J K R, ed. Effluents from livestock, pp 433–445. London: Applied Science Publishers
21. McAllister J S V (1976) Studies in Northern Ireland on problems related to the disposal of slurry. In Agriculture and Water Quality, Technical Bulletin No. 32, pp 418–431, Ministry of Agriculture, Fisheries and Food. London: HMSO
22. Meer H G van der, Thompson R B, Snijders P J M and Geurink J H (1987) Utilization of nitrogen from injected and surface-spread cattle slurry applied to grassland (this volume)
23. O'Callaghan J R and Pollock K A (1976) A land spreading trial with pig slurry. In Agriculture and Water Quality, Technical Bulletin No. 32, pp 371–399, Ministry of Agriculture, Fisheries and Food. London: HMSO
24. Pain B F, Thompson R B, Lande Cremer L C N de la and Holte L ten (1987) The use of additives in livestock slurries to improve their flow properties, conserve nitrogen and reduce odours (this volume)
25. Prins W H and Snijders P J M (1987) Negative effects of animal manure on grassland due to surface spreading and injection (this volume)

26. Richardson S J (1976) Animal manures as potential pollutants. In Agriculture and Water Quality, Technical Bulletin No. 32, pp 405–417, Ministry of Agriculture, Fisheries and Food. London: HMSO
27. Riemsdijk W H van, Lexmond Th M, Enfield C G and Zee S E A T M van der (1987) Phosphorus and heavy metals: accumulation and consequences (this volume)
28. Ryden J C (1984) The flow of nitrogen in grassland. Proceedings No. 229. London, UK: The Fertiliser Society
29. Ryden J C, Ball P R and Garwood E A (1984) Nitrate leaching from grassland. Nature (London) 311, 50–53
30. Schröder J and Dilz K (1987) Cattle slurry and farmyard manure as fertilizers for forage maize (this volume)
31. Simán G, Dam Kofoed A, Larsen K E, Kemppainen E and Steineck S (1987) Utilization of nitrogen from slurry applied to fodder crops (this volume)
32. Smith K A and Dijk T A van (1987) Utilization of phosphorus and potassium from animal manures on grassland and forage crops (this volume)
33. Starr J L, Parlange J-Y and Nielsen D R (1979) The use of open soil systems and mathematical models to study N reactions and movement in soils. In Gasser J K R, ed. Modelling nitrogen from farm wastes, pp 1–34. London: Applied Science Publishers
34. Stevens R J and Cornforth I S (1974) The effect of pig slurry applied to a soil surface on the composition of the soil atmosphere. Journal of the Science of Food and Agriculture 25, 1263–1272
35. The Royal Society (1983) The nitrogen cycle of the United Kingdom. Report of a Royal Society Study Group. London: The Royal Society
36. Troake R P, Troake L E and Walling D E (1976) Nitrate loads of South Devon streams. In Agriculture and Water Quality, Technical Bulletin No. 32, pp 340–351, Ministry of Agriculture, Fisheries and Food. London: HMSO
37. Tunney H, Molloy S and Codd F (1980) Effects of cattle slurry, pig slurry and fertilizer on yield and quality of grass silage. In Gasser J K R, ed. Effluents from livestock, pp 327–343. London: Applied Science Publishers
38. Vanstaen H, Neukermans G, Debruyckere M and Verstraete W (1979) Complete aerobic treatment of piggery wastewaters by the acidic sludge separation process. In Hawkins J C, ed. Engineering problems with effluents from livestock, pp 273–282. EUR 6249 EN. Luxembourg, Commission of the European Communities
39. Vuuren A M van and Meijs J A C (1987) Effects of herbage composition and supplement feeding on the excretion of nitrogen in dung and urine by grazing dairy cows (this volume)
40. Wadman W P, Sluijsmans C M J and Lande Cremer L C N de la (1987) Value of animal manures: changes in perception (this volume)
41. Wijnands J H M, Ausems W F and Nijssen J M A (1987) Economic aspects of the use of animal manures (this volume)
42. Williams R J B (1976) The chemical composition of rain, land drainage and borehole water from Rothamsted, Broom's Barn, Saxmundham and Woburn Experimental Stations. In Agriculture, and Water Quality, Technical Bulletin No. 32, pp 174–200, Ministry of Agriculture Fisheries and Food. London: HMSO

Utilization of stored, aerated or anaerobically digested dairy cattle and pig slurries on sown grassland

J M BESSON[a], V LEHMANN[a], M SODER[a], J DEGALLIER[b] and L RAVENEL[b]

[a] *Swiss Federal Research Station for Agricultural Chemistry and Hygiene of Environment, CH-3097 Liebefeld-Bern, Switzerland*

[b] *Cantonal Agricultural School, Grangeverney, CH-1510 Moudon, Switzerland*

Key words: slurry, storage, aeration, anaerobic digestion, sown grassland

Abstract. The effect of slurry treatments (storage, aeration, anaerobic digestion) on the N content and N forms in slurry, the N efficiency, the yield and the botanical composition of a sown meadow, is clearly lower than the effect of the kind (cattle or pig) and rate (50 or $100\,m^3\,ha^{-1}\,yr^{-1}$) of the utilized slurries on the above mentioned parameters.

Introduction

The influence of storage, aeration and anaerobic digestion of dairy cattle and pig slurries on their composition, has already been described in earlier publications [3, 4]. The differences before treatment between both kinds of slurries remain after treatment, and the influence of treatment on N content and forms is proportionally slight. The N efficiency in treated slurries as measured in pot trials with *Lolium multiflorum*, was given by the ammonium-N content of the slurries: the kind and the rate of applied slurries are more important effects than slurry treatments [1, 2].

Materials and methods

The treated slurries were spread on a meadow, sown with a standard mixture (4 grass and 2 clover species). The field trial was carried out from August 1978 (sowing) to October 1984 (last cut). Slurry at 50 or $100\,m^3\,ha^{-1}\,yr^{-1}$ was spread in two batches, the first one at the beginning of growth, the other after the second cut. No manure was applied in 1984 in order to test the residual effect.

H.G. v.d. Meer, et al. (eds.), Animal Manure on Grassland and Fodder Crops.
 ISBN 978-94-010-8137-5.

Results

The effects of kind and rate of slurry on the dry matter yield of the sown meadow were slightly greater than those of the slurry treatment, which were especially low. However, the decrease of the yields with time from 1979 to 1984 evolved as expected. There are some changes in the botanical composition, depending on the different treatments, but the influence of the slurry treatments is smaller than that of kind or rate of slurry. From 1980 to 1984 *Dactylis glomerata* dominated; it represented more than 60% of the sown grasses. *Trifolium pratense* decreased rapidly (less than 2% in the third cut of 1980) and disappeared almost completely after two years so that practically only *Trifolium repens* was left from the seeded Leguminosae. In the group of colonizing species, *Taraxacum officinalis* dominated: 85–95% of this group in the first cut of 1984, together with 5–10% of *Poa trivialis*; in the third cut of the same year *Taraxacum officinalis* contributed to more than 95% of the yield of the unsown species. The sward was denser after spreading cattle rather than pig slurry. At higher manure rates, there was an irregular ground cover of compact and rough wisps of *Dactylis glomerata* between which the soil was often bare. The effect of the slurry processing was also perceptible but appreciably less important.

Conclusions

Our results [as confirmed by 5, 6, 7] show that the influence of slurry treatment is clearly less important than the kind or the amount of slurry applied. For an optimal valorization of animal manures for crop and grassland production as well as for environmental protection, one should establish a manure and fertilization plan. The aim of such a plan is to account for the quantity of animal manure produced at the farm, and to decide how it should be applied: kind of manure, rate, time and physical distribution, application methods, etc. These problems are much more important and should be dealt with before those of slurry treatment.

References

1. Besson J M, Daniel R and Lischer P (1984) Recherche Agronomique Suisse 23, 249–268
2. Besson J M, Daniel R and Lischer P (1985) In Williams J H, Guidi G and Hermite P l', eds. Long-term effects of sewage sludge and farm slurries applications, pp 22–32. London and New York: Elsevier
3. Besson J M, Lehmann V and Roulet M (1985) Revue Suisse d'Agriculture 17, 231–238

4. Besson J M, Roulet M and Edelmann W (1982) Revue Suisse d'Agriculture 14, 327–355
5. Buchgraber K (1983) Institut für Pflanzenbau und Pflanzenzüchtung der Universität für Bodenkultur, Wien, Dissertation, 1–213
6. Schechtner G (1980) Der Forderungsdienst 28/5, 25–31
7. Thalmann H (1985) Lehrstuhl für Grünland und Futterbau der technischen Universität München Freising-Weihenstephan, Dissertation, 1–346

Utilization of nitrogen from slurry applied to permanent grassland

P ERNST

Institute of Ecology, Landscape Development and Forest Planning Northrhine-Westphalia, Section Pasture and Fodder Research, Kleve-Kellen, FRG

Key words: permanent grassland, slurry, nitrogen fertilization, nitrogen utilization

Abstract. A longterm slurry fertilization experiment on permanent grassland gave the following results: 1. Equal nitrogen (N) efficiency of slurry applied in spring and autumn. 2. Improved slurry efficiency caused by additional inorganic N fertilization for the subsequent growth. 3. Equal utilization of ammonium-N (NH_4^+-N) from slurry and inorganic N fertilizers.

Introduction

When applying slurry to agricultural land the effect of nitrogen (N) can vary considerably [2]. Losses of N pollute the environment [3]. In order to minimize losses of N, the slurry should be applied according to crop requirements. Guidelines for this can only be obtained by evaluating experimental results.

Materials and methods

The N effects of slurry are examined by means of an experiment which has been running since 1978. The time of application, the quantity applied and additional N fertilization are considered. The test was carried out on good quality grassland on a loamy sand. The yield was determined by mowing at five cuts.

Results and discussion

Without supplementary inorganic N fertilization, the longterm slurry N effect was from 5 to 8 kg DM per kg slurry N (Table 1). No effect of the quantity of slurry N applied could be found. With 40 m^3 cattle slurry per ha,

H.G. v.d. Meer, et al. (eds.), Animal Manure on Grassland and Fodder Crops.
 ISBN 978-94-010-8137-5.

Table 1. Effect of slurry N on grass DM response (kg DM per kg slurry N) according to application time, quantity applied and additional inorganic N fertilization; 1979–1986 (a = autumn; s = spring)

	No slurry	Cattle slurry, $m^3 ha^{-1} yr^{-1}$				Pig slurry $m^3 ha^{-1} yr^{-1}$
		40 a	40 s	80 (40a + 40s)	120 (60a + 60s)	120 (60a + 60s)
total slurry N, $kg ha^{-1} yr^{-1}$	0	98	114	213	319	654
without fertilizer N	–	5.2	5.0	5.5	7.9	6.6
200 $kg ha^{-1} yr^{-1}$ fertilizer N	12.3	11.7	11.8	5.4	7.3	5.3
400 $kg ha^{-1} yr^{-1}$ fertilizer N	10.5	9.0	9.8	4.6	5.9	3.9

the supplementary inorganic N fertilization caused a duplication of the effect of slurry N on DM yield [1]. With larger quantities of slurry, as well as with the highest rate of supplementary N fertilization, the N effect of the slurry decreased. With 40 m^3 slurry per ha, no difference in the N efficiency between spring and autumn application was recorded [1].

A longterm N evaluation shows (Fig. 1), that with more than 300 kg N $ha^{-1} yr^{-1}$ (fertilizer N and slurry NH_4^+-N) N "losses" have been incurred. From the linear regression between quantity of N fertilization and quantity of N retained in plant and soil, an equal utilization of fertilizer N and NH_4^+-N in slurry can be deduced. It is concluded from this, that the NH_4^+-N

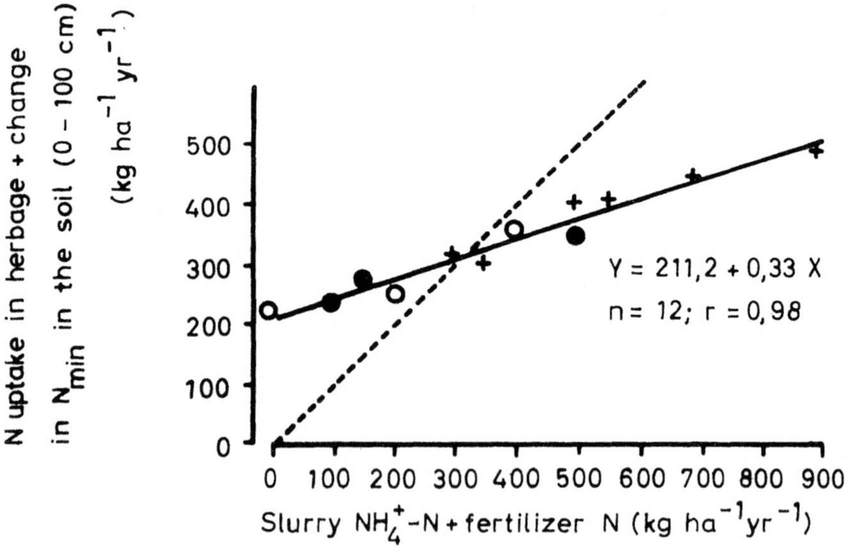

Figure 1. N balance in relation to N fertilization; average of 7 years (1979–1985). ○ fertilizer N; ● slurry NH_4^+-N; + fertilizer N and slurry NH_4^+-N.

of slurry, independent of the time of application, should be fully considered in the assessment of N fertilization.

References

1. Schechtner G (1981) Bericht 7. Gülletagung "Fragen der Güllerei" Gumpenstein, pp 135–196. Gumpenstein, Österreich: Bundesanstalt für alpenländische Landwirtschaft
2. Tunney H (1981) In Brogan J C, ed. Nitrogen losses and surface runoff from landspreading of manures, pp 161–167. The Hague: Martinus Nijhoff/Dr W Junk
3. Vetter H and Steffens G (1983) In Arbeiten der DLG "Nitrat – ein Problem für unsere Trinkwasserversorgung, pp 121–133. Frankfurt, FRG: DLG-Verlag

The effectiveness of high doses of cattle slurry as pasture fertilizer in a submontane region

G MICHNA
Institute of Zootechnics, Experimental Station in Grodziec Slaski, Poland

Key words: slurry doses, grass yield, chemical composition

Abstract. Intensive fertilization with slurry has been shown to produce high yields of dry matter and a favourable nutritive content of forage.

Introduction

The work studied the effectiveness of high doses of cattle slurry applied to permanent pasture in a submontane region. The dry matter (DM) yields, botanical composition and the more important nutritive components in the forage were determined.

Materials and methods

Trials were carried out at 360 m above sea level on a loamy soil of pH 6.0. The permanent pasture was predominantly *Poa pratensis* and *Lolium perenne*. Rainfall and temperature varied from year to year. Annual average rainfall is 1068 mm with 789 mm during the growing season (April–October). The corresponding average temperatures are 7.6 °C and 13.0 °C. Experiments were carried out for four years on plots of 500 m^2. The fertilizer rates tested were 200, 300 and 400 kg N $ha^{-1} yr^{-1}$ as slurry or fertilizer complemented with phosphate and potash. All treatments were replicated four times. The slurry was applied each year in three equal doses in winter, in spring and after the second cut. Nitrogen fertilizer was applied in spring (50%) and after the second and third cut (25% each).

Results and discussion

The effectiveness of slurry N expressed as kg DM produced per kg N applied, was lower than that of fertilizer at equal application rates (Table 1).

H.G. v.d. Meer, et al. (eds.), Animal Manure on Grassland and Fodder Crops.
 ISBN 978-94-010-8137-5.

Table 1. Dry matter yields ($t\,ha^{-1}\,yr^{-1}$) and response to N (kg DM per kg N applied)

N application ($kg\,ha^{-1}\,yr^{-1}$)	DM yield ($t\,ha^{-1}\,yr^{-1}$)		DM response (kg DM per kg N)		Apparent equivalence of slurry
	slurry	fertilizer	slurry	fertilizer	
0	4.9	4.9			
200	7.2	9.6	11.5	23.5	48.9
300	8.5	9.7	12.0	16.0	75.0
400	9.2	10.2	10.8	13.2	81.8

The sward did not respond to N above 200 kg N $ha^{-1}\,yr^{-1}$ as fertilizer. The apparent equivalence of slurry N calculated as the relative rates of DM response, increased with the rate of slurry applied. However, the highest rate of slurry produced less grass than the lowest rate of fertilizer indicating an overall efficiency of slurry N of less than 50%. In dry, and especially in warmer years the effectiveness of slurry was considerably lower. Ammonia volatilization, which in summer time can reach 35% [2], was considered responsible. Slurry has also been shown to have a low fertilizer value [1] under conditions of moisture stress. Cool weather had a positive influence, reducing N losses and promoting a better utilization of summer doses of slurry N.

At high doses of slurry the reduction caused in the Papilionaceae, to about 8% in the sward, was less than that from fertilizer where only a few plants of this group were found.

The slurry manured sward had a lower crude protein and a higher carbohydrate content than the fertilized sward. Sometimes the P, Mg and Mn contents were also less. Potassium content was higher in slurry dressed plants, particularly in early season.

References

1. Kalembasa S (1974) Szczec. Tow. Nauk., Wydz. Nauk. Przyr. Rol., vol. XXXIX, 2. Warszawa-Poznan: PWRiL
2. Schechtner G, Tunney H, Arnold G H and Keuning J A (1980) Proceedings international Symposium of the European Grassland Federation (Wageningen), pp 77–93

Yield of hay from grassland treated with mineral fertilizers and slurry

J GAJDA and J SAWICKI
Grassland Farming Department, Agricultural Academy, Poland

Key words: grassland, yield, slurry, mineral fertilizers

Abstract. In the years from 1976–1981 four experiments were carried out treating grassland with cattle slurry and mineral fertilizers. Every year a higher yield of hay was obtained on plots receiving mineral fertilizers.

Introduction

The best way to use animal slurry is by agricultural utilization to fertilize grassland and forage crops.

Materials and methods

In the region of Lublin Polesie two types of studies were carried out. Each consisted of two randomized block experiments. In the first type of study (I) the same fertilizer treatments were applied during the first three years and in the fourth year the residual effects were determined. In the second type of study (II) two cycles were established consisting of one year of fertilization and one year of residual effect. Cattle slurry contained on average 7.9% dry matter, 0.277% N, 0.045% P, 0.217% K, 0.085% Ca and 0.044% Mg. Annual doses of slurry and mineral fertilizers were applied in three equal parts. Results were Subject to Student "t" test.

Results and discussion

Higher yields of hay were obtained from plots where mineral fertilizers were applied and from the meadow muck-peat soil (Fig. 1). The yield of hay fertilized with the slurry was dependent on the level of precipitation during the growth period, which has been confirmed by other authors [1, 2, 3]. A

H.G. v.d. Meer, et al. (eds.), Animal Manure on Grassland and Fodder Crops.
 ISBN 978-94-010-8137-5.

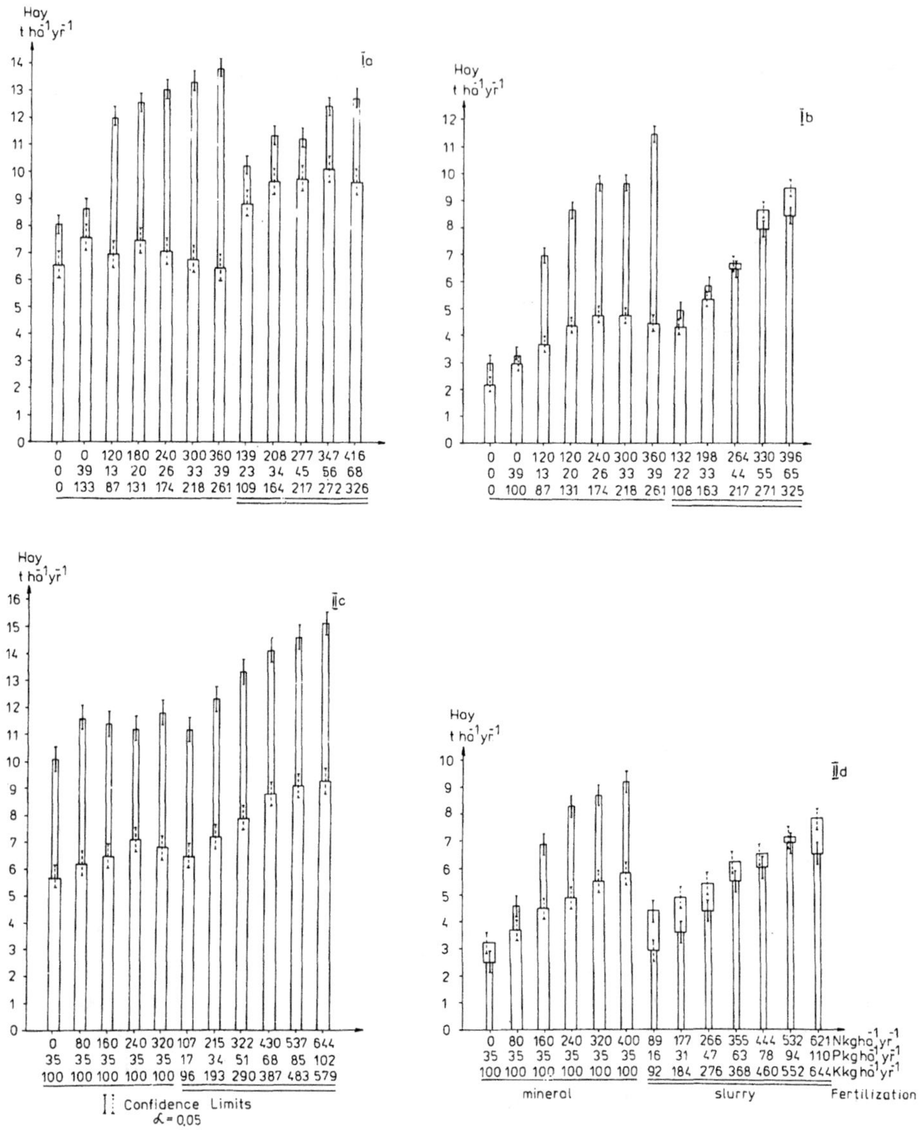

Figure 1. Average yield of hay from the muck-peat soil (Ia, IIc) and from the black-sand soil (Ib, IId), in the years of fertilization (narrow columns) and due to a residual effect (wide columns). Type of studies: a, b = I; c, d = II (see Materials and methods).

higher yield from the muck-peat soil may be explained by its higher moisture reserves. The residual effects of slurry on yield were, on average, twice as high as those of mineral fertilizers. There was a markedly higher residual effect on the black-sand soil, which is in agreement with other studies [4, 5].

High doses of slurry (over $125\,m^3\,ha^{-1}\,yr^{-1}$) caused the disappearance of valuable species of plants. It is recommended to apply lower doses of slurry (about $50–80\,m^3\,ha^{-1}\,yr^{-1}$ in three doses) and to supplement the nutrients applied with mineral fertilizers.

Conclusions

1. Slurry is a good organic fertilizer for use on grassland.
2. A more favourable response to slurry was noticed in years with higher precipitation.
3. The residual effect of slurry is twice as high as that from plots treated with mineral fertilizers.

References

1. Flacht B (1976) Producter agric. 183, 21–22
2. Gajda J and Sawicki J (1986) Zesz. Probl. Nauk Rol. 284, 759–770
3. Herriott J B D, Wells D A and Crooks P (1966) J. Brit. Grassl. Soc. 21, 85–92
4. Höde M (1967) Arb. Inst. Meliorationswes. u. Grünland Friedrich-Schiller-Univ., Jena T 2, 76–83
5. Sawicki J (1986) Rocz. Nauk Rol. S.A 106 (2), 103–115

Yields from swards of varying botanical composition fertilized with cattle slurry

M GRYNIA and A KRYSZAK
Academy of Agriculture, Poznan, Poland

Key words: meadow mixtures, yields, cattle slurry

Abstract. In the years 1977–1984 studies were carried out in the AES-Brody near Poznan (West Poland) on muck soils formed from shallow black peat. The botanical composition and the yields of green and dry matter of three sown meadow mixtures (M1, M2 and M3), fertilized with varying doses, 50–200 $m^3 ha^{-1} yr^{-1}$, of slurry were investigated. The highest yields of green and dry matter were obtained from mixtures M2 and M1. The most favourable level of slurry for grass production and fodder quality proved to be 50 $m^3 ha^{-1} yr^{-1}$.

Introduction

The literature referring to the chemical composition of slurry and its use on perennial grassland is comprehensive and refers to many aspects [1, 2, 3]. The problems of producing adequate grass from swards fertilized by slurry, and the resulting fodder quality have not been satisfactorily investigated yet [4].

Methods

A randomized block experiment with four replicates started in 1977. Three meadow mixtures: M1, M2 and M3 (Table 1) were sown.

The meadow was fertilized with cattle slurry in doses of 50, 100, 150 and 200 $m^3 ha^{-1} yr^{-1}$. The chemical composition of the slurry was: DM: 3.55%, N: 0.23%, P: 0.048%, K: 0.25%, Ca: 0.09%, Mg: 0.03%. Supplementary mineral fertilizers were applied at 125 kg N $ha^{-1} yr^{-1}$ and 27 kg P $ha^{-1} yr^{-1}$. In the years 1977–1984 the botanical composition of the sown mixtures was investigated and the yields of the green and the dry matter were determined.

Results and discussion

The results of the seven year study are shown in Fig. 1. The best yielding meadow mixtures were M2 and M1 (except in 1984 when high slurry

H.G. v.d. Meer, et al. (eds.), Animal Manure on Grassland and Fodder Crops.
 ISBN 978-94-010-8137-5.

Table 1. Species composition of meadow mixtures (%)

Species	M1	M2	M3
Dactylis glomerata	15.0	15.0	–
Alopecurus pratensis	20.0	30.0	20.0
Poa pratensis	15.0	20.0	20.0
Agrostis gigantea	10.0	–	10.0
Poa palustris	15.0	–	–
Phalaris arundinacea	25.0	–	–
Festuca arundinacea	–	15.0	–
Phleum pratense	–	20.0	20.0
Festuca pratensis	–	–	20.0
Bromus inermis	–	–	10.0

applications gave high M3 yields) which is connected with the nitrophilous character of grasses such as *Dactylis glomerata, Alopecurus pratensis, Festuca pratensis, Phalaris arundinacea, Poa pratensis*, which were predominant. These meadow mixtures indicate the greatest suitability for the soil and climatic conditions of Poznan region. They can be recommended for the land fertilized by slurry in this area.

The most effective rate of slurry in the conditions of the Poznan region is $50 \, m^3 \, ha^{-1} \, yr^{-1}$, which provides at the same time the most suitable chemical

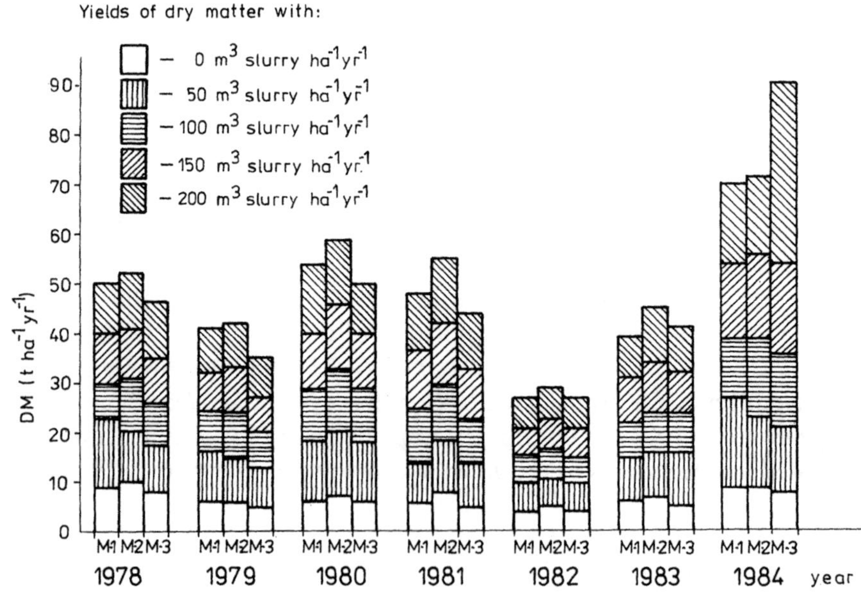

Figure 1. Yields of dry matter of three meadow mixtures (M1, M2, M3) fertilized with different rates of cattle slurry in the period 1978–1984.

composition of the sward. In 1984, when high rainfalls were recorded, the highest yields from all mixtures were obtained from 200 m^3 slurry ha^{-1} yr^{-1}. However, 50 m^3 ha^{-1} yr^{-1} is regarded as safe taking into account the need for environmental protection.

References

1. Biniaś J, Grynia M and Szoszkiewicz J (1983) Materials of scientific Conference Polish Academy of Science – the Committee of Plant Cultivation – Conference of Kalsk, near Zielona Gora
2. Gajda J and Sawicki (1979) Melioration and Meadow News 5, Warsaw
3. Kurzweg B (1976) Agriculture and Silviculture State Edition, Warsaw
4. Vedekind P and Sussenbach D (1972) Wasserwirtschaft und Wassertechnik 6

Mid-season application of organic manures — a waste of resources?

H I GRACEY

Greenmount College of Agriculture and Horticulture, Antrim BT41 4PU, Northern Ireland

Key words: organic manures, grassland, silage, rainfall

Abstract. Results are reported from a series of field trials which measured the response of grassland swards to mid-season application of organic manures. A cattle slurry treatment, supplying the equivalent of 50 kg total N ha^{-1} was superimposed on grassland swards fertilized with 100 kg inorganic N ha^{-1} and yields of second cut silage measured some 40–60 days later. Organic manure had a positive effect in 4 years, *i.e.* herbage dry matter yields were increased, whereas in the other 4 years the effect was negative. There was a significant correlation between the amount of rainfall within a 72 h period of the slurry being applied in each year and the DM response.

Introduction

In Northern Ireland one third of the total nitrogen (N) and potassium (K) and one half of the phosphorus (P) supplied to farmland is in the form of organic manures [3] and provided they are properly stored, they are a valuable source of plant nutrients for both legumes and grasses [1, 2, 4]. In trials at Greenmount with first cut silage swards receiving 100 kg inorganic N ha^{-1}, herbage DM yields were of the order November > January > March slurry applications, and in further trials yields were of the order February < March < April slurry applications [4, 6]. However, the question is what to do with slurry collected during the remainder of the housing period, *i.e.* until stock are out at grass. A series of trials was initiated to assess the response of cattle slurry applied after first cut silage on grassland swards receiving inorganic N.

Materials and methods

A cattle slurry treatment, supplying the equivalent of 50 kg total N ha^{-1} was superimposed on a series of trials to investigate responses to inorganic P and K, and the effect of inorganic N source [5, 7]. The dry matter (DM) of the

H.G. v.d. Meer, et al. (eds.), Animal Manure on Grassland and Fodder Crops.
 ISBN 978-94-010-8137-5.

slurry ranged from 50–80 g kg^{-1}. Slurry was applied immediately after field plots on established grassland swards had been harvested to determine first cut silage yields. The slurry was well mixed before spreading, and in addition to the cattle slurry, all treatments received 100 kg inorganic N ha^{-1}. Herbage DM yields were determined after 40–60 days regrowth.

Results and discussion

The additional effect of the slurry treatment on yield of second cut silage is illustrated in Fig. 1. Cattle slurry had a positive effect in 4 years, *i.e.* yields were increased, whereas in the other 4 years, this relatively light application of cattle slurry had a negative effect on yield (the rate of slurry application was approximately 15 m^3 ha^{-1}, although this varied to take account of the N content). Rainfall within a 72 h period of the slurry being applied each year was as follows:

Year	1979	1980	1981	1982	1983	1984	1985	1986
Rainfall, mm	1.3	23.6	23.1	4.5	0.0	1.1	14.3	2.8

There was a significant correlation (r = 0.86**) between the amount of rainfall and the response to cattle slurry, with the best responses obtained when slurry nutrients were well washed into the soil.

These trials clearly demonstrate the limitations in terms of DM response to mid-season slurry applications, particularly when no rain is expected.

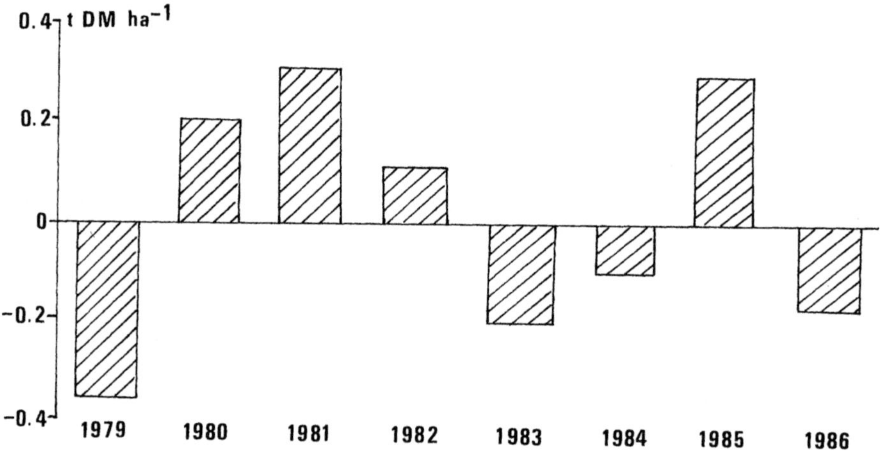

Figure 1. Effect of cattle slurry on DM yields of second cut silage.

Stevens (unpublished data) has shown that this inefficiency can largely be accounted for by losses of gaseous N from slurry to the atmosphere.

References

1. Gracey H I (1979) Experimental Husbandry 35, 47–51
2. Gracey H I (1981) Grass and Forage Science 36, 291–295
3. Gracey H I (1981) PhD Thesis. The Queen's University of Belfast
4. Gracey H I (1982) Proceedings of Efficient Grassland Farming, Occasional Symposium No. 14 British Grassland Society
5. Gracey H I (1984) Annual Report of Research and Technical Work. Greenmount College, 3–4
6. Gracey H I (1986) British Grassland Society Winter Meeting, 5 December 1985 (in press)
7. Gracey H I and O'Neill D G (1983) Agriculture in N. Ireland 57, 370–373

Long term effects of slurry on grassland

P CHRISTIE
Agricultural and Food Chemistry Department, The Queen's University of Belfast and Department of Agriculture for Northern Ireland, Newforge Lane, Belfast BT9 5PX, UK

Key words: pig slurry, cow slurry, *Lolium perenne*, herbage yield, mineral composition

Abstract. The herbage yield of a ryegrass (*Lolium perenne*) sward receiving fertilizer, pig slurry or cow slurry applications for 16 years showed a typical response curve to the amount of soluble nitrogen (N) applied. Pig slurry supplied more phosphorus (P) than did cow slurry while the reverse was found for potassium (K). Herbage P and K contents were correlated with the amounts applied. Herbage sodium (Na) content declined under the largest pig slurry dressing and all dressings of cow slurry, probably due to antagonism between K and Na and also to changes in sward botanical composition. Large slurry dressings reduced the proportion of *L. perenne* in the sward. Pig slurry had an acidifying effect on the soil but only the largest dressing lowered soil pH more than did the fertilizer treatment. The largest dressing of cow slurry maintained soil pH at the original value of 6.3. Substantial accumulations of soil P from pig slurry and K from cow slurry occurred in the top 15 cm of soil, especially at the highest slurry application rate (200 $m^3\,ha^{-1}\,yr^{-1}$).

Introduction

Concern has been expressed that repeated slurry applications could damage the land in various ways [2]. The aim of this field experiment was to study the long term effects of regular and frequent slurry applications to grassland on herbage yield and mineral composition and on soil physical and chemical properties. Some early results have been published [3, 4] and some results for the first 16 years have been reported [1].

Materials and methods

Eight treatments were applied to an established perennial ryegrass sward: (1) fertilizer supplying 200 kg N (as ammonium nitrate for 4 years, and urea for the subsequent 12 years), 32 kg P as superphosphate and 160 kg K $ha^{-1}\,yr^{-1}$ as muriate of potash; (2) fertilizer applied as above from 1970 to 1972 and then no treatment (control); (3), (4), (5) pig slurry applied at 50, 100 or 200 $m^3\,ha^{-1}\,yr^{-1}$; (6), (7), (8) cow slurry at the same three rates. There were

H.G. v.d. Meer, et al. (eds.), Animal Manure on Grassland and Fodder Crops.
 ISBN 978-94-010-8137-5.

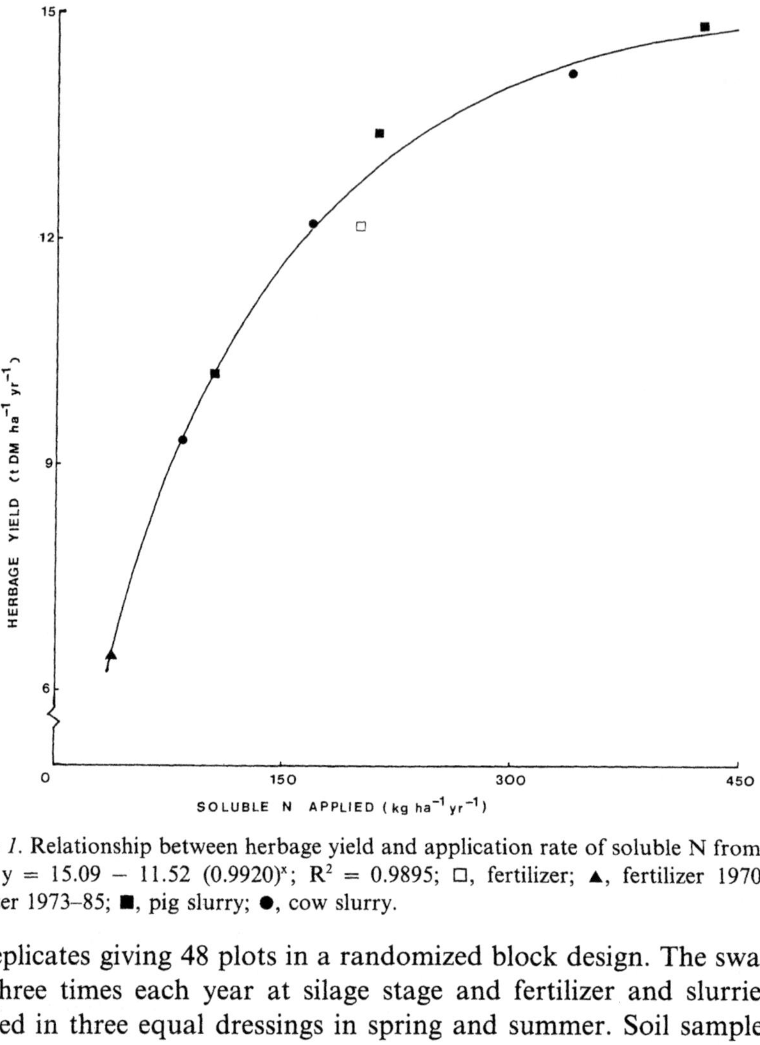

Figure 1. Relationship between herbage yield and application rate of soluble N from 1970 to 1985; $y = 15.09 - 11.52\ (0.9920)^x$; $R^2 = 0.9895$; □, fertilizer; ▲, fertilizer 1970–72, no fertilizer 1973–85; ■, pig slurry; ●, cow slurry.

six replicates giving 48 plots in a randomized block design. The sward was cut three times each year at silage stage and fertilizer and slurries were applied in three equal dressings in spring and summer. Soil samples were taken every February. At all three cuts in 1981 and the first cut of 1982 extra sub-samples of herbage were collected to determine the proportions of the principal plant species. All methods have been described in more detail [1].

Results and discussion

Fig. 1 shows the yield response curve fitted by asymptotic regression between mean annual herbage yield and mean annual dressing of soluble N (*i.e.* inorganic N in fertilizer and NH_4^+-N in slurries). A good regression ($R^2 = 0.9895$) was obtained when results for all fertilizer and slurry treat-

Table 1. Mean annual rates of application of P and K and mean herbage contents from 1970 to 1985, herbage mineral composition in 1985 (weighted averages), soil analysis in February 1985, and proportions of the two dominant grass species at the first cut of 1982. All herbage results expressed on dry matter basis. Soil P was extracted with 0.5 *M* $NaHCO_3$ and K and Mg with 1 *M* ammonium acetate

Treatment		P		K		Herbage mineral content in 1985						Soil analysis, February 1985, top 15 cm:				Proportions of grass species at the first cut of 1982:	
		Applied ($kg\,ha^{-1}\,yr^{-1}$)	Mean herbage content (%)	Applied ($kg\,ha^{-1}\,yr^{-1}$)	Mean herbage content (%)	N	P	K	Na	Ca	Mg	pH (in water)	extractable			*L. perenne*	*A. stolonifera*
						(%)							P	K	Mg	(percent of herbage DM)	
													($mg\,kg^{-1}$)				
Fertilizer		32	0.28	160	1.67	1.80	0.31	1.67	0.32	0.50	0.11	5.5	30	81	55	56	16
Control		6	0.27	30	1.47	1.62	0.23	1.11	0.49	0.52	0.13	5.8	15	71	71	46	7
Pig slurry rate	low	38	0.29	98	1.41	1.83	0.33	1.39	0.51	0.53	0.15	5.9	35	81	145	62	8
	medium	76	0.30	196	1.71	2.31	0.37	2.19	0.38	0.47	0.15	5.8	82	129	193	35	32
	high	152	0.33	392	2.25	3.01	0.43	3.34	0.16	0.42	0.15	5.4	190	300	222	11	51
Cow slurry rate	low	21	0.28	136	1.84	1.61	0.27	2.13	0.28	0.43	0.13	5.9	19	129	136	57	9
	medium	42	0.28	272	2.27	1.91	0.32	2.72	0.19	0.38	0.13	6.1	29	221	213	56	11
	high	84	0.30	544	2.71	2.24	0.37	3.31	0.15	0.34	0.14	6.3	62	665	317	13	48
												(6.3	23	63	ND)*		

* Soil analysis in November 1969 (before treatments were applied); ND not determined

ments were combined. This agrees with the close relationship between effectiveness of slurry N and the 'soluble' N (later, NH_4^+-N) content which has been shown consistently in Northern Ireland [4]. The P and K contents of herbage can be used to compare the efficiency of slurries with that of fertilizer as sources of P and K (Table 1). Herbage responses to applied P were small, with little difference whether it was applied as fertilizer or slurry and mean annual dressing of P was linearly correlated with mean P content of herbage from 1970 to 1985 ($r = 0.983$, $P < 0.001$). Larger differences between treatments in herbage K content were found, with mean annual dressing of K and mean herbage K content also highly correlated ($r = 0.948$, $P < 0.001$). The cow slurry K appeared to be the most effective. For example, the smallest dressing supplied 136 kg K $ha^{-1} yr^{-1}$ but produced a higher K concentration in the herbage than did 160 kg K $ha^{-1} yr^{-1}$ as fertilizer. Herbage Na content declined under the largest pig slurry dressing and all dressings of cow slurry. This may have been due in part to antagonism in the uptake of K and Na, and partly to changes in sward botanical composition. Increasing slurry application rates reduced the proportion of *L. perenne* and increased the abundance of *Agrostis stolonifera*. No lime was applied to any plot, consequently soil pH fell in all treatments except the largest dressing of cow slurry. Large dressings of pig and cow slurries built up soil reserves of P and K, respectively, but analysis of soil samples from different depths within the top 15 cm (not presented in Table 1) showed that these nutrients penetrated markedly below 10 cm depth only under the heavy slurry dressings. In contrast, soil Mg accumulated greatly at all depths examined. In conclusion, soil P and K accumulation and changes in sward botanical composition were very marked only under the largest dressing of both slurries which was included to test extreme conditions. Even under these treatments herbage dry matter yield continued to increase in response to the NH_4^+-N content of the slurries.

References

1. Christie P (1987) The Journal of Agricultural Science (Cambridge) 108 (in press)
2. McAllister J S V (1976) In Agriculture and water quality, pp 418–431. London: HMSO
3. McAllister J S V (1977) In Voorburg J H, ed. Utilization of manure by land spreading, pp 87–103. Proceedings of an EEC Seminar. Luxembourg: CEC
4. McAllister J S V (1981) In Brogan J C, ed. Nitrogen losses and surface run-off from land-spreading of manures, pp 389–393. The Hague: Martinus Nijhoff/Dr Junk

The utilization of slurry on light land

C P BECKWITH and P F HEMINGWAY
Harper Adams Agricultural College, Newport, Shropshire TF10 8NB, UK

Key words: slurry, nitrogen utilization, copper, nitrogen, potassium, leaching

Abstract. Samples were taken at two depths from a sandy soil receiving slurry or farmyard manure (FYM). Analysis indicated that inorganic nitrogen (N) and to a lesser extent potassium (K), moved rapidly through the soil, particularly in the winter months. Slurry was separated into liquid and semi-solid fractions. The liquid could be applied to a growing crop through an irrigation system. It was applied to Italian ryegrass (cv. Augusta) at a rate of $50\,m^3\,ha^{-1}\,yr^{-1}$ in three applications and gave a yield response equivalent to about $180\,kg\,N\,ha^{-1}$ supplied as ammonium nitrate. However, the liquid fraction applied late in the season significantly increased the copper (Cu) content of the grass.

Introduction

Livestock wastes are often considered to be a nuisance and disposed of onto the land at any convenient time. Alternatively, slurry can be stored and applied to the land at a time suitable for crop growth, in spring or summer. Slurry which has been separated into a liquid fraction of low dry matter (DM), has the advantage of being more easily applied to the crop through an irrigation system. On sandy soils the disposal of wastes may lead to an increased leaching of nutrients and the risks of pollution of ground or surface water [3]. Pig slurry applied to growing grass may increase its copper (Cu) content [1]. This paper looks at some of the consequences of applying manures to very light land. Data are presented illustrating the use of separated pig/cow slurry as a fertilizer on grassland.

Materials and methods

Soil samples were taken on a monthly basis at two depths, 0–15 cm and 45–60 cm from a field of loamy sand texture. This field received applications of FYM and pig/cow slurry which reflected the need to dispose of farm wastes on a regular basis. Inorganic N and extractable K were determined on the soil samples [4].

H.G. v.d. Meer, et al. (eds.), Animal Manure on Grassland and Fodder Crops.

The waste handling system used included a mechanical separator which converted raw slurry of average DM of 4% into a 2% DM liquid fraction (liquor) and a 10% DM semi-solid fraction. Analysis figures for the liquor varied as a result of seasonal variations in the proportions of cow and pig slurry and also in the degree of dilution of the raw slurry by extraneous water. On average, the slurry liquor had a total N content of 0.19%, an inorganic N content of 0.11%, a K content of 0.16% and a Cu and Zn content of 7.5 and 10.5 mg l^{-1}, respectively.

Liquor was applied to Italian ryegrass (cv. Augusta) in five applications of 50 m^3 ha^{-1} over two years and compared with ammonium nitrate fertilizer supplying either 30, 60 or 120 kg N ha^{-1} (Table 1). Analysis of grass and liquor samples for minerals was performed following wet digestion with perchloric acid [4]. Total N in the grass and slurry was determined by the Kjeldahl procedure. The trial was performed on a sandy loam soil.

Results and discussion

Inorganic N is very susceptible to leaching in light soils during the winter months [2, 3]. Following winter applications of slurry and 100 t ha^{-1} of FYM to an uncropped soil, 100 kg N ha^{-1} was found at a depth of 0–15 cm and 56 kg N ha^{-1} was found at a depth of 45–60 cm in January. A month later the inorganic N content at the two depths was 4 kg N ha^{-1} and 18 kg N ha^{-1}, respectively. A similar pattern of behaviour was noted with K. Water samples taken from a borehole within this area in April indicated a total oxidizable N content of 58 mg l^{-1} and a K content of 37 mg l^{-1}.

Slurry liquor applied to Italian ryegrass at a rate per application of

Table 1. Yields of DM and N in response to applications of ammonium nitrate and slurry liquor to Italian ryegrass (cv. Augusta)

Date of fert. or slurry application	9.4	4.6	14.7				
Date of cut	5.4	3.6	13.7	6.9	28.10	Total annual yield of	
Treatment at each application	DM yield at each cut (t ha^{-1})					DM (t ha^{-1})	N (kg ha^{-1})
Slurry liquor 50 m^3 ha^{-1}	1.57	6.45	3.59	2.32	0.56	14.48	226
30 kg N ha^{-1}	1.36	5.59	2.71	1.33	0.19	11.15	146
60 kg N ha^{-1}	1.42	7.17	3.54	1.84	0.31	14.25	218
120 kg N ha^{-1}	1.53	7.63	4.06	2.09	0.56	15.89	284
L.S.D. at $p = 0.05$	n.s.	0.622	0.253	0.228	0.154	0.882	22.9

Table 2. Analysis of Italian ryegrass (cv. Augusta) in late summer following applications of either slurry liquor or inorganic fertilizer

	Italian ryegrass treated with:		
	slurry liquor	inorganic fertilizer	S.E.D.
K (%)	1.30	1.30	±0.041
P (%)	0.37	0.30	±0.025
Ca ($g\,kg^{-1}$)	0.48	0.44	±0.022
Mg ($g\,kg^{-1}$)	0.16	0.14	±0.008
Cu ($mg\,kg^{-1}$)	30.65	21.57	±2.80
Zn ($mg\,kg^{-1}$)	30.00	22.50	±0.025

$50\,m^3\,ha^{-1}$ produced a yield response equivalent to about $60\,kg\,N\,ha^{-1}$ in each of three applications, applied as ammonium nitrate (Table 1).

Grass analysed at the end of the first year of the trial (Table 2), when the overall mineral content of the grass was declining, showed that the application of slurry liquor significantly raised the Cu and Zn content of the grass. The Cu content of the grass would cause some concern if the grass was grazed by sheep [1].

Conclusions

The results of this study emphasize the importance of avoiding winter application of manures to light soils if pollution of groundwater is to be avoided. Applied to grass in spring or early summer, separated cow/pig slurry liquor can provide useful quantities of N, but the Cu content of the grass needs to be monitored if it is to be grazed by sheep.

Acknowledgement

The authors wish to thank R C Anslow for his help and advice.

References

1. Batey T, Berryman C and Line C (1972) Journal of the British Grassland Society 27, 139–143
2. Burns I G (1977) Outlook on Agriculture 9 No. 3, 144–148
3. Kolenbrander G J (1981) Developments in Plant and Soil Sciences 2, 199–216
4. Ministry of Agriculture, Fisheries and Food (1981) London: HMSO RB427

Effects of organic fertilizer application on hay meadow quality in the French Northern Alps

P FLEURY[a], B JEANNIN[a] and J M DORIOZ[b]
[a] *INRA — SAD, Route de St Cyr, 78000 Versailles, France*
[b] *INRA — Science du Sol, 74200 Thonon, France*

Key words: permanent pasture, organic fertilizing, botanical composition, morphology, nutritional value, yield

Abstract. In the first growth cycle of mountain hay meadows, increasing rates of organic fertilization produce an increase in dry matter (DM) yield, a slight fall in total protein content (TPC) and in particular an increasingly rapid fall in organic matter digestibility (OMD) with growth stage.

Introduction

In order to demonstrate the effects of organic fertilization on the quantitative and qualitative yield of mountain hay meadows, results are presented corresponding to 3 major meadow types which reflect significantly different levels of organic fertilizer dressings (Table 1). Vegetation in these

Table 1. Plant species and ecological determinants of three mountain hay meadow types

Meadow type	Dominant species	Usual treatment		Pedoclimatic situation (gradient of available water in summer)
		Organic fertilizer dressings $t\,ha^{-1}\,yr^{-1}$	Cutting date (*D. glomerata* stage)	
A	*D. glomerata* *Poa trivialis*	high 20 to 40	early (SFHE to SFl)	cool to medium
B	*D. glomerata* *Trisetum flavescens* *Festuca pratensis*	medium 10 to 30	early to slightly late (SFHE to SFr)	medium to dry
C	*Festuca rubra* *Agrostis tenuis* *Geranium silvaticum*	low 0 to 10	late (SFr to SSh)	cool

H.G. v.d. Meer, et al. (eds.), Animal Manure on Grassland and Fodder Crops.
 ISBN 978-94-010-8137-5.

meadows is also influenced by other ecological factors (see Table 1), but their influence on quantitative and qualitative production is secondary and they are not discussed here.

Materials and methods

The production of some 30 meadows submitted to similar agricultural treatment for more than 5 years has been investigated over a period of 3 years. On these sites, during the first growth cycle, samples were taken every 6 to 8 days from the onset of the vegetation period to shedding of *Dactylis glomerata* seeds. Mean dry matter (DM) yields and total protein content (TPC) were systematically assessed. Pepsin cellulase digestibility [1] was also assayed in order to determine organic matter digestibility (OMD) at the following stages: "10 cm stage" (MM), start of flower head emergence (SFHE), full development of flower head (FDFH), start of flowering (SFl), start of fructification (SFr) and start of seed shedding (SSh) in *Dactylis glomerata.*

The results obtained were then adjusted for each meadow type (same vegetation, same agricultural treatment, same physical environment), using a linear regression in relation with the number of days elapsed since stage SFHE of *Dactylis glomerata.* The models assume a linear or parabolic function (Fig. 1). The number of observations (n), correlation coefficient (R) and residual standard deviation (SR) are indicated for each model.

Results and discussion

Fertilization has a strong positive effect on DM yield (Fig. 1a): the difference between extreme types (A and C) varies from 1 t DM ha^{-1} at stage SFHE in *Dactylis glomerata* to 2 t DM ha^{-1} at stage SSh.

Changes in the botanical composition (Table 1) as well as morphological variations induced by fertilizer application have a significant effect on nutritional value (Fig. 1b and 1c). Thus on heavily fertilized meadows (Type A), *Dactylis glomerata* grows in thick tufts with numerous stems in which there is much dead tissue as early as stage SFl. In mesotrophic meadows (Type B), *Dactylis glomerata* and other grasses grow in smaller tufts which are richer in herbaceous tillers and less susceptible to competition. In poor swards (Type C), the species which are abundant produce a lot of leaves and due to low vegetation density there is no significant dieback [3]. It has been shown that for a given species, at the same stage, nitrogen content and

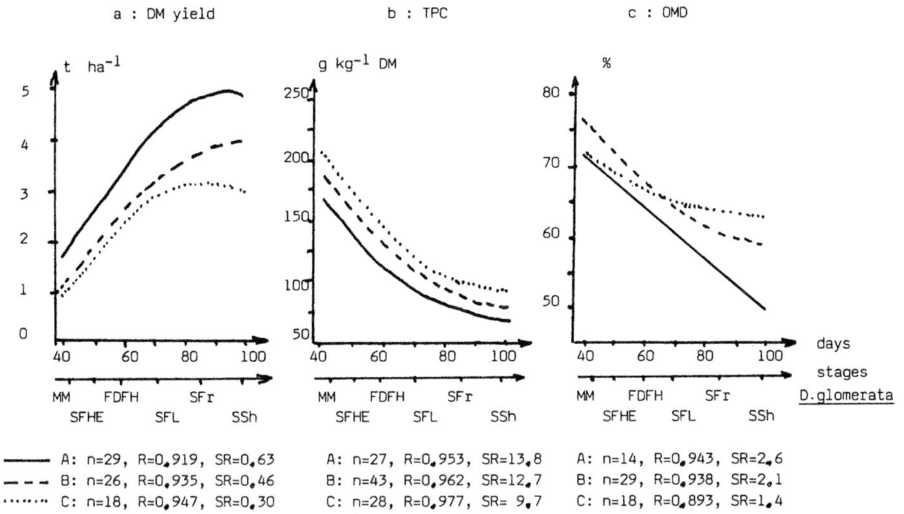

Figure 1. Evolution of dry matter (DM) yield, total protein control (TPC) and organic matter digestibility (OMD) during the first growth cycle in three types of permanent mountain meadows (adjusted values).

digestibility are related positively with the leaf/stem ratio [2, 4]. This trend seems to hold in the case of these complex plant communities. It has thus been noted that TP contents (Fig. 1b) are slightly lower, and in particular that the decrease in OMD is much faster in eutrophic swards (Type A) than it is in mesophilic swards (Type B) and especially in poor swards (Type C).

Conclusion

These results have a bearing on meadow exploitation. Yield increases resulting from high dressings of organic manure have as a corollary a decrease in flexibility of cutting opportunities and in particular require earlier mowing in order to obtain herbage with the same nutritional value: the 65% OMD limit is reached 15 days earlier on heavily fertilized meadows than on poor meadows.

References

1. Aufrère J (1982) Annales de Zootechnie (INRA — Versailles — France) 31, 111–130
2. Demarquilly C and Chenost M (1969) Annales de Zootechnie (INRA — Versailles — France) 18, 419–436
3. Fleury Ph, Dorioz J M, Jeannin B (1987) Etudes et Recherches du SAD (INRA — Versailles — France) (in press)
4. Jarrige R (1963) Annales de Biologie Animale (INRA — Versailles — France) 3, 143–190

The effect of regular application of cattle slurry on grassland during the growing season on grass yield and daily milk production

Tj BOXEM and G J REMMELINK
Research Station for Cattle, Sheep and Horse Husbandry, Runderweg 6, 8219 PK Lelystad, The Netherlands

Key words: slurry application, milk yield, grass productivity

Abstract. In two experiments (expt. 1 and 2) the effect of regular applications of 10 t ha^{-1} (expt. 1) or 20 t ha^{-1} (expt. 2) of cattle slurry during the growing season on grass yield and daily milk production was investigated. Grass yield decreased about 5% in expt. 1 and 17% in the first cut after slurry application in expt. 2. There was generally no effect on milk yield. Only in two cases when rainy conditions prevailed was milk yield slightly depressed by slurry application.

Introduction

It is expected that in the near future more slurry will be applied to grassland during spring and summer. This can increase scorching and smothering of grass and lead then to a depression in milk yield and grassland productivity. Present advise is to apply no more than 20 t slurry ha^{-1} per application. The question is whether negative effects can be avoided by the regular application of smaller quantities (8–10 t ha^{-1} per application).

Materials and methods

In experiment 1 on sandy soil, from 1984–1986, applications of 8 t slurry ha^{-1} for each of the first 5 cuts ($5 \times 8 = 40$ t ha^{-1} in total, system A) were compared with a situation where no slurry was applied (system B). Slurry was applied as soon as possible after grazing or cutting. Under windy conditions the distribution of such a small quantity of slurry was uneven. Therefore, in 1986 the distribution system of a vacuum slurry spreader was adjusted to reduce this problem. The grassland utilization was comparable with normal practice in The Netherlands, *viz.* a combination of grazing (day and night) and silage cuts. Yield potential of the cows in systems A and B was comparable. In 1986 on each of 11 plots (paddocks) in system A where

H.G. v.d. Meer, et al. (eds.), Animal Manure on Grassland and Fodder Crops.
 ISBN 978-94-010-8137-5.

slurry was applied, 10 strips were covered with plastic during slurry application. Dry matter (DM) yield in the first cut after application was measured on covered strips and on comparable strips with slurry application.

In experiment 2, on peat soil, in the period May-August 1986 the effect of one application of 20 t slurry ha^{-1} on milk yield and grass production was determined. The slurry was applied to a sward that had been cut. During the summer this experiment was repeated 3 times. At each time 4 plots (paddocks) were treated. On one half of a plot 20 t slurry ha^{-1} was applied (system C), on the other half no slurry was applied (system D). In both cases these 4 plots were grazed after assessing DM yield. To obtain a comparable grass supply per cow in both systems, the number of cows was adjusted by means of a restgroup.

Results and discussion

Data from experiment 1 are presented in Table 1. In 1985 experiment 1 was terminated after a few months because heavy rainfall made grazing impossible. During experiment 1 no significant problems with scorch were noticed. The surplus grass was cut for silage. Total annual DM yield, as cut for silage, was 43.9 t with system A (164% of the area cut for silage) and 53.6 t with system B (191% cut for silage). Half of this difference was obtained in the first cut, for reasons probably not related to slurry application. It was necessary to top the pasture more often in system A than in system B: 26 and 15%, respectively. Dry matter yield in 1986 on the covered strips on system

Table 1. Results of experiment 1. Data are averages of 1984 and 1986

System	Slurry (A)	No slurry (B)
Land area (ha)	11.2	11.2
Number of plots (paddocks)	10	10
Number of cows	30	30
Average duration of the experiments (days per year)*	138	138
Cattle slurry ($t\,ha^{-1}\,yr^{-1}$)	40.1	–
P ($kg\,ha^{-1}\,yr^{-1}$)	34**	21
K ($kg\,ha^{-1}\,yr^{-1}$)	199**	68
N from fertilizer ($kg\,ha^{-1}\,yr^{-1}$)	405	454
N from slurry ($kg\,ha^{-1}\,yr^{-1}$)	188	–
Average sward length at start of grazing (cm)	13.8	14.1
Grazing time per paddock (days per rotation)	3.0	3.0
Concentrates ($kg\,cow^{-1}\,day^{-1}$)	2.5	2.5
Milk production (kg FCM $cow^{-1}\,day^{-1}$)	25.0	24.9

* Start of the grazing experiments: about 10 May; first slurry application: about 1 April.
**From slurry only.

Table 2. Dry matter yield (kg ha^{-1}), milk yield (kg FCM cow^{-1} day^{-1}) and sward height after grazing (cm) on slurried (C) and untreated fields (D). C received 20 t slurry ha^{-1} application^{-1} plus 40 kg fertilizer N ha^{-1}; D received 40 kg fertilizer N ha^{-1} application^{-1}

Average date of application*	Dry matter yield kg ha^{-1}		Milk production kg FCM cow^{-1}		Sward height cm	
	C	D	C	D	C	D
May 25	1245	1357	29.6	30.5	5.9	5.8
June 20	1628	1938	25.0	25.2	6.4	6.0
July 15	1280	1736	26.4	26.4	4.4	4.2

* Each date is an average of 4 plots (paddocks)

A was 91 kg ha^{-1} lower (*ca.* 5%) than on strips with slurry application. Only in 2 cases were differences between covered and non covered strips significant, however ($P < 0.05$).

There was no difference in milk production between systems A and B (Table 1). It is possible that cows became accustomed to the slurry application [1]. It is interesting to note that in an experiment reported by Kleter (Agricultural University of Wageningen, Dpt. of Field Crops and Grassland Science, unpublished) treatment of fouled grass with molasses decreased intake differences from that of clean grass. In the short experimental period in 1985 (with rainy conditions) milk production was about 1 kg cow^{-1} day^{-1} lower in system A. Odour could play a bigger role under wet conditions.

In comparable experiments with zero grazing there was no difference in grass intake and milk yield in 1985 [2]. In 1986 intake of grass from the slurried fields was 0.5 kg DM cow^{-1} day^{-1} lower than that from the untreated plots (12.3 compared to 12.8).

The effect of slurry on grass yield and on milk production in experiment 2 is shown in Table 2. On average, grass yield was 17% higher without slurry. The first application was made in May under rainy conditions. Milk was 0.9 kg cow^{-1} day^{-1} lower with slurry. After the other two applications there were no differences in milk yield.

References

1. Reid G W, Greenhalgh J F D and Aitken J N (1972) The Journal of Agricultural Science (Cambridge) 78, 491–496
2. Meijer A B (1987) Jaarverslag 1986 ROC Cranendonck (in press)

Trials on slurry application techniques for grassland

J MORKEN
Norwegian Institute of Agricultural Engineering, P.O. Box 65, N-1432 Ås-NLH, Norway

Key words: stripe-spreading, slurry application

Abstract. To improve the slurry application techniques for grassland, the Norwegian Institute of Agricultural Engineering has developed an experimental slurry spreader. This spreader gives an improved crosswise distribution of the slurry spread, and the application rate can be easily controlled. Experiments on striped-spreading techniques in 1986 showed that the utilization of the nitrogen in slurry is not better when slurry is spread in stripes than when it is broadcast on the surface. The experiments on these techniques will continue.

Introduction

In Norway most livestock farmers have to use grassland for the spreading of agricultural wastes, because they normally have very little arable land. Cattle and pig slurry form the bulk of these wastes, because faeces and urine are usually collected together, with more or less water added. At our Institute we are working on a project to improve the application techniques. The objects are:
— better control of the application rate;
— more even distribution of the slurry spread;
— better utilization of the nitrogen (N) in the slurry;
— better quality of the crop from the grassland (less contamination).

Materials and methods

To undertake this investigation, we had to design a slurry spreader that could help us attain the objects mentioned above. The resulting spreader has a tank of 540 litres and is attached to the tractor through the normal three-point linkage. A positive displacement pump (lobepump) is driven by an 8 horse-power (hp) engine with a tachometer. This pump delivers slurry in close proportion to the speed of the pump. The application rate is a function of the rotational speed of the pump, the tractor travelling speed, and the spreading width. When the application rate has been decided, the

H.G. v.d. Meer, et al. (eds.), Animal Manure on Grassland and Fodder Crops.
 ISBN 978-94-010-8137-5.

speed of the pump is selected, and the correct travelling speed can then be determined.

To provide an even, crosswise distribution and a constant spreading width (2.5 m), we developed a system based on a distribution chamber with one central inlet and 12 evenly spaced, peripheral outlets. In this chamber there is a driven rotor through which the slurry flows, leaving its periphery through one chamber outlet at a time. The rotor is driven by the 8 hp engine. Each of the 12 outlets is connected to a lip spreader through a 50 mm hose.

The study of various application techniques, *i.e.* broadcast spreading and various striped-spreading systems (slurry placed onto the soil surface below the vegetation cover), resulted in the development of two different spreading booms.

In 1986 we conducted two experiments at different places in Norway. Each experiment consisted of 39 plots of 2.5 m × 8.0 m in three blocks. Each block comprised all the different techniques. The treatments are described in Table 1. Treatments A to F were applied at rates of 20 and 40 $m^3 ha^{-1}$.

Results and discussion

We obtained no significant yield differences ($P = 0.05$) between the various treatments (spreading techniques). However, there were significant differences between sites (Jæren and Ås), between first and second cut, and between application rates. The average N content in the slurry was 3.6 $kg m^{-3}$, and the NH_4^+-N content was 1.8 $kg m^{-3}$. At Jæren the rainfall was high (431 mm in May, June, July and August), and there was a very high yield on the control plots. This may explain the results. The soil was of a moraine type with a high content of organic matter. The soil at Ås was of a sandy clay type, with a low content of organic matter. At Ås the rainfall was low (285 mm in May, June, July and August), and this together with the sandy clay soil may explain the results there.

Table 1. Experimental plan for spreading techniques 1986

Technique	Code
Normal spreading	A
Striped spreading, breaking soil surface, stripe spacing 20 cm	B
Striped spreading, stripe spacing 20 cm	C
Striped spreading, breaking soil surface, stripe spacing 40 cm	D
Striped spreading, stripe spacing 40 cm	E
Slurry injection, injection spacing 50 cm (Kyllingstad)	F
Control	G

Although the experiments indicate that striped spreading does not increase the crop yield, the fact that the wind cannot cause drift when the slurry is placed directly onto the soil surface should normally bring about less pollution of the air and surrounding areas than does broadcasting of the slurry. At this time we can say nothing about the effects that the various application techniques may have on the quality of the grass regarding contamination, palatability, and ensiling processes.

Injector design for minimum grass sward damage

M J HANN, N L WARNER and R J GODWIN
Department of Engineering for Agriculture, Silsoe College, Cranfield Institute of Technology, Silsoe, MK45 4DT, UK

Key words: slurry injector design, grass sward damage

Abstract. The effects of slurry injector leg rake angle have been investigated and show reductions of sward damage with increasing angle. The introduction of press wheels immediately after slurry injection has shown significant increases in grass yields. A study has also been made of the effects of press wheel weight on crop yield.

Introduction

Concern over the environment issues of spreading liquid manure has focused attention on soil injection as an environmentally acceptable technique whilst also reducing the nutrient loss through ammonia volatilization. The injection technique may, however, lead to an increase in sward damage. The preliminary experiments reported here were planned to quantify the effects of injector design on grass sward damage and subsequent yield response.

Materials and methods

Replicated tests in the soil bin and field sites were carried out using four different rake angled legs. Measurements of the soil surface disturbance were made using a profilemeter.

Slurry injection was carried out during a period of high soil moisture deficit on three soil types (clay, clay loam and sandy loam) using press wheels immediately following the injector leg on one half of the replicated plots and running without the press wheels on the remainder. Assessment of crop yield response to the treatments was made by harvesting strips across the soil slot and determining the dry matter yield. Finally, injection was carried out on two soil types, clay loam and sandy loam, using press wheels with differing load conditions. Measurements were taken of the resulting soil disturbance and crop response.

H.G. v.d. Meer, et al. (eds.), Animal Manure on Grassland and Fodder Crops.
 ISBN 978-94-010-8137-5.

Results and discussion

Increasing rake angle reduces the amount of surface disturbance (Fig. 1). This reduction in sward damage and slot heave results in an increase in crop yield as earlier work has shown [1]. There is, however, a draught penalty for increasing the rake angle from 45° to 135° of up to 33% [1]. Introducing press wheels immediately behind the injector leg has a significant effect on crop response (Table 1), due mainly to the action of the press wheel illiminating much of the surface disturbance in the vicinity of the leg slot, the area where most growth reduction occurs. Adding weight has a positive effect on grass sward response [Fig. 2] due to the improved efficiency of the press wheel in reducing soil disturbance [2]. However, there appears to be an optimum weight (at approximately 50 kg) above which no significant improvement was obtained.

Conclusions

Sward damage is reduced by adopting large injector leg rake angles. Rolling immediately after injection with correctly weighted press wheels results in significant increases in grass yields.

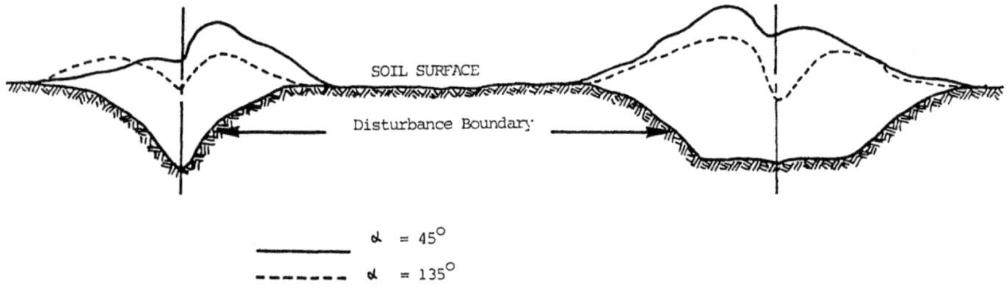

Figure 1. The effect of injector leg rake angle (α) on surface disturbance in a clay soil.

Table 1. The increase in yield resulting from immediate rolling of the sward following slurry injection

Soil type	Dry matter yield ($t\,ha^{-1}$)		Yield increase due to press (%)
	No press	Press	
Clay loam (Mead)	1.7	2.0	18
Sandy loam (Cottenham)	4.3	6.0	40
Clay (Wicken)	1.8	2.6	44

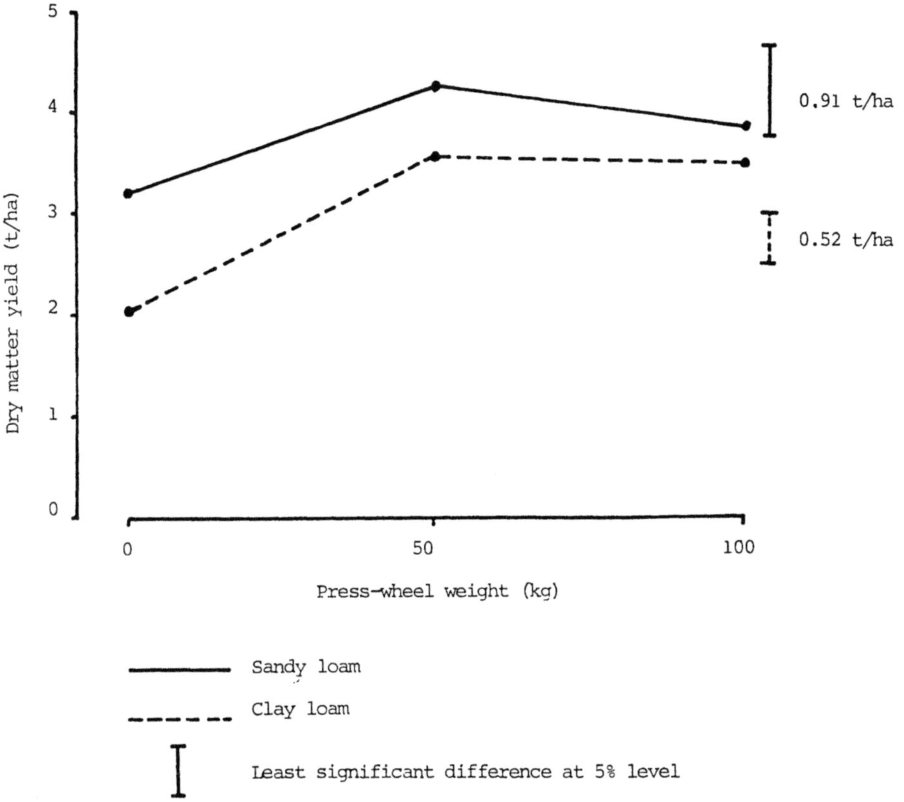

Figure 2. The effect of press-wheel weight on grass sward response

Acknowledgements

The authors wish to acknowledge the financial support of the Science and Engineering Research Council and the Water Research Centre.

References

1. Godwin R J, Warner N L and Hall J E (1985) Proceedings of the 5th international Symposium on Agricultural Wastes. St. Joseph, Michigan: American Society of Agricultural Engineers
2. Kohannejad H and Semple D A (1986) Unpublished M.Sc.Thesis, Silsoe College. Silsoe, UK: Cranfield Institute of Technology

Investigations of the fertilizer value of sheep excrements left on pasture

P SKRIJKA
Department of Meadow and Pasture Cultivation, Agricultural University, Krakow, Poland

Key words: sheep manure, fertilizer value

Abstract. The quantity of manure deposited by sheep on pasture during grazing and when folded was measured. The importance of this manure in the fertilization of the pasture with N, P, K, Ca, Mg, Na, B, Cu, Co, Mn, Mo and Zn is discussed. The effects of the manure on the plants and on botanical changes in the sward are considered.

Introduction

During grazing sheep leave considerable amounts of manure on pastures [1, 2]. About 20% of the pasture is covered in a year [2]. Little is known about the effects of this manure.

Materials and methods

The investigations were undertaken on mountain pasture. The sward was dominated (65%) by *Festuca pratensis*. Milk ewes weighing 40–50 kg were grazed for 10 hours a day. The remaining 14 hours (6 during the day and 8 during the night) were spent in a fold. Observations were made on the frequency of urinating and defecating during grazing. The mass of single depositions was determined, and the results multiplied to give total loadings. These investigations were repeated five times in each grazing month, three days each with eight ewes. To measure the manure voided in the fold, special cages were used to ensure a separate and full collection of urine and faeces. The measurements lasted six days each (5 replications) with three changes of sheep (four in each case) every two days. Chemical analysis was undertaken on cumulative daily samples. Some results given in the present paper were calculated empirically.

H.G. v.d. Meer, et al. (eds.), Animal Manure on Grassland and Fodder Crops.
 ISBN 978-94-010-8137-5.

Results and discussion

During grazing the ewes defecated 4–7 times and urinated 3–5 times daily. The average daily production per ewe amounted to 0.5 kg faeces and 0.6 kg urine. In the fold an ewe left as an average 0.85 kg faeces and 0.95 kg urine. The chemical compositions are shown in Tables 1 and 2. Table 1 gives the amounts of macroelements in faeces and urine, Table 2 the amounts of microelements in faeces; in urine only traces of these elements were found.

During 10 hours grazing 1 ewe leaves on the pasture on average: 12.4 g N, 5.4 g K, 1.6 g Ca, 1.0 g P, 0.4 g Na, 0.3 g Mg as well as 0.9 mg B, 0.1 mg Co, 3.3 mg Cu, 72.2 mg Mn, 0.1 mg Mo, and 22.5 mg Zn. In the fold over the 14 hours stay the quantities amount to: 21.0 g N, 10.0 g K, 3.2 g Ca, 2.1 g P, 0.7 g Na, 0.5 g Mg, and 1.6 mg B, 0.2 mg Co, 5.8 mg Cu, 126.2 mg Mn, 0.1 mg Mo, and 39.3 mg Zn.

The field experiments showed that 67% of N was utilized by the pasture plants. This refers principally to urine. The faecal-N did not show an effect on yield [3]. The rest of the N, *i.e.* about 33%, volatilizes in the form of NH_3 or was retained by the soil. With a grass yield of 8.8 t DM ha^{-1} sheep manure provided about 23% of the N requirement of the sward. Thus the impor-

Table 1. Chemical composition of sheep excrements

	% N	% P	% K	% Ca	% Mg	% Na
Faeces	0.7	0.2	0.3	0.3	0.05	0.05
Urine	1.6	0.01	0.7	0.02	0.01	0.03

Table 2. Content of microelements in faeces of sheep

	B	Co	Cu	Mn	Mo	Zn
mg kg^{-1}:	6.4	0.92	30.3	657.3	0.55	204.9

Table 3. Balance of macroelements

	P	K	Ca	Mg	Na
Uptake, kg ha^{-1}	31.8	213.9	59.2	16.8	1.8
Amount supplied in folded area, kg ha^{-1}	4.3	20.5	6.6	1.0	1.4
Requirement of sward provided at the fold, %	13.5	9.6	11.1	5.9	77.8
Amount supplied at pasture, kg ha^{-1}	2.0	11.1	3.3	0.6	0.8
Requirement of sward provided at pasture, %	6.3	5.2	5.6	3.6	44.4

Table 4. Balance of microelements

	Co	Cu	Mn	Mo	Zn
Uptake, g ha^{-1}	0.7	60.6	1877.6	1.4	602.0
Amount supplied in folded area, g ha^{-1}	0.4	11.9	259.3	0.2	80.9
Requirement of sward provided at the fold, %	52.8	19.7	13.8	16.4	13.4
Amount supplied at pasture, g ha^{-1}	0.2	6.8	148.4	0.1	46.2
Requirement of sward provided at pasture, %	28.6	11.2	7.9	8.6	7.7

tance of the N was considerable Analogous calculations for other nutrients are given in Table 3.

Similarly, it was possible to compute the part played by microelements in the fertilization of a pasture (Table 4).

The manure also affected the botanical structure of the sward [2]. The urine caused scorching of many plants whilst the faeces protected a part of the sward from too frequent defoliation by sheep. The development of *Festuca pratensis, Poa pratensis, Trifolium repens, Alchemilla pastoralis, Achillea millefolium, Plantago lanceolata,* was encouraged as well as other valuable plants.

Conclusions

During 150 days summer grazing sheep leave between 4.3 and 8.6 t ha^{-1} of manure on pasture. The fertilizer value is equal to a dose of 10–20 t ha^{-1} of farmyard manure. Analyses showed that considerable amounts of macro- and microelements entered the system. A considerable proportion of the microelements may originate from the salt given to the sheep. The manure is an important source of nutrients for the pasture plants. The sward utilizes about 70% of the N.

References

1. Skrijka P (1973) Acta Agraria et Silvestria XIII/2, 85–93
2. Skrijka P (1978) Acta Agraria et Silvestria XVIII/1, 117–124
3. Skrijka P (1979) Acta Agraria et Silvestria XVIII/2, 139–146

The characterization of cattle slurry and the response of a range of crop plants to slurry application

M H OPPERMAN, M WOOD, L McBAIN and P J HARRIS
Department of Soil Science, University of Reading, London Road, Reading, Berkshire RG1 5AQ, UK

Key words: cattle slurry, microbial populations, crop responses

Abstract. The microbial and chemical composition of slurry changes during collection and storage, but general characteristics can be measured. Following the application of slurry to soil, yields of barley and wheat increased but red and white clover yields decreased.

Introduction

Changes in methods of livestock rearing have led to the handling of farm wastes as slurry rather than farmyard manure [3], and much of the slurry is disposed of by land-spreading. Cattle slurry is a potentially valuable fertilizer, containing plant nutrients and organic matter; however, little work has been done on the microflora of the slurry. Experiments were carried out to characterize cattle slurry and to examine the response of a range of crop plants to slurry application.

Materials and methods

Whole cattle slurry was collected within 3 hours of deposition and used within 24 hours. Total viable bacteria were determined by plate counts on nutrient agar (Oxoid) at 25 °C, coliforms were also determined by plate counts using violet-red bile agar (Difco) at 35 °C. Protozoa were estimated by the method of Darbyshire *et al.* [2], fungal mycelium by a modified Jones and Mollison technique [5] and nematodes by direct counting in a soil/water matrix. Total N was determined using Kjeldahl apparatus and percentage carbohydrate determined colorimetrically at 625 nm [6]. Inorganic N was determined by extraction in 1 *M* KCl followed by analysis using a Chem-lab autoanalyser.

Cattle slurry was added to 1.5 kg soil of the Rowland series [5] at a rate equivalent to 75 kg N ha^{-1} and covered with 150 g soil. Seeds of maize,

H.G. v.d. Meer, et al. (eds.), Animal Manure on Grassland and Fodder Crops.
 ISBN 978-94-010-8137-5.

barley, wheat, perennial ryegrass (PRG), lucerne, red clover and white clover were sown (5 replicate pots) and maintained in a glasshouse for 10 weeks until harvest.

Results and discussion

Cattle slurry varied depending on temperature, amount of washing water, animal diet and bedding at the source of collection. General characteristics are shown in Table 1. During storage of slurry the microbial population decreased and NO_3^--N remained constant. For 6 days after collection the NH_4^+-N increased, possibly due to urea hydrolysis, before decreasing to a stable level.

Germination of maize and wheat was reduced or inhibited by slurry, whereas all other species were unaffected. Slurry had no effect on yield of maize, ryegrass and lucerne. Shoot dry weight of barley and wheat was increased but that of red and white clover decreased (Table 2). The phytotoxic effects may have been due to toxic volatile organic acids [1] or high NH_4^+-N concentration in the slurry.

Table 1. General characteristics of cattle slurry collected within 3 hours of deposition and sampled within 24 hours of collection. Means of 10 samples (except * where means of 4 samples)

Property	Mean (± S.D.)	Range
Moisture (%)	87.8 (±2.6)	83–92
pH	8.0 (±0.3)	7.4–8.3
NO_3^--N ($\mu g\,g^{-1}$ in fresh weight)	35.4 (±22.8)	13–65
NH_4^+-N ($\mu g\,g^{-1}$ in fresh weight)	1673 (±731)	840–2705
Total N (% of dry weight)	3.4 (±0.27)	3.1–3.7*
Soluble carbohydrate (% of dry weight)	0.7 (±0.04)	0.63–0.78*
Total viable bacteria (Number g^{-1} in dry weight)	1.4×10^9 ($\pm 1.0 \times 10^9$)	7×10^7–6×10^9
Presumptive coliforms (Number g^{-1} in dry weight)	8.2×10^6 ($\pm 7.0 \times 10^6$)	3×10^5–2×10^7
Protozoa (Number g^{-1} in fresh weight)	1.6×10^3 ($\pm 2.1 \times 10^3$)	4×10^2–8×10^3
Nematodes (Number g^{-1} in fresh weight)	< 5.0 –	–
Fungal hyphae ($m\,g^{-1}$ in fresh weight)	9500 (±6000)	750–18000

Table 2. Effect of slurry on shoot dry weight (g per pot) after 10 weeks (means of 5 pots)

Slurry	maize	barley	wheat	PRG	lucerne	red clover	white clover
−	2.43	1.35	1.36	1.13	0.76	2.10	1.67
+	2.45	2.07	1.94	1.42	0.86	0.71	0.77
	NS	**	**	NS	NS	**	**

NS = No significant difference between means
** = Significant at 10% level

Conclusions

Slurry is highly variable and its properties can change during storage. Clovers appeared to be adversely affected by the presence of slurry in soil, whereas barley and wheat showed increases in yield. Germination of some crops was inhibited.

References

1. Cooper P and Cornforth I S (1978) Journal of the Science of Food and Agriculture 29, 19–27
2. Darbyshire J F, Wheatley R F, Greaves M P and Inkson R H E (1974) Revue d'Ecologie et de Biologie du Sol 11, 465–475
3. Flowers T H and Arnold P W (1983) Soil Biology and Biochemistry 15, 329–335
4. Jones P C T and Mollison J E (1948) Journal of General Microbiology 2, 54–69
5. Kay F F (1936) A soil survey of the University Farm, Sonning, Berkshire. University of Reading, Faculty of Agriculture and Horticulture Bulletin XLIX. Reading, Berkshire, UK: Bradley and Son Ltd
6. MAFF (1981) Analysis of Agricultural Materials (2nd Edition) RB427. Method 14: Carbohydrates, soluble, in herbage, pp 36–37. London: HMSO

Scorching of *Lolium perenne* caused by cattle slurry

U KÜNTZEL[a], R KRAUSE[b] and C JONUSCHEIT[ab]
[a] *Institute of Grassland and Forage Research, FAL*
[b] *Institute of Technology, FAL, Bundesallee 50, D-3300 Braunschweig, Federal Republic of Germany*

Key words: scorching, top dressing, liquid manure, cattle slurry, grassland

Abstract. *Lolium perenne* was raised in Kick-Brauckmann containers in a greenhouse. At 20 cm plant height differently treated (untreated, separated, diluted) liquid cattle manure was applied either as a top dressing or to the soil surface. For comparison additional containers were treated with urea ammonium nitrate (UAN). Visible scorching of leaves occurred when liquid manure and UAN were dressed. The degree of damage increased with the rate of application. Increased scorching normally caused decreased yields. Washing plants three hours after application caused a reduction of damage only from the diluted manure and from UAN.

Introduction

To increase the utilization of nutrients, liquid manure should be applied during the growing season. However, on grassland negative effects can be observed like scorching (burning of plants), the suppression of fine grass and non-acceptance by grazing animals. There are only few investigations concerning scorching in relation to the rate and method of manure application. In a greenhouse the scorching of *Lolium perenne* by liquid cattle manure was investigated. Two methods of application were compared as well as the influence of simulated rain to wash the plants.

Materials and methods

In 132 Kick-Brauckmann containers (21.7 cm diameter) *Lolium perenne* (Barenza, tetraploid) was raised from March 27th to May 7th in a non-conditioned greenhouse. The soil was prepared from 50% Tschernosem (mollisol) and 50% sand and fertilized with 100 kg N, 35 kg P, and 100 kg K ha^{-1}. The containers were kept at 60% of their maximum water holding capacity. At 20 cm height the plants were treated with liquid cattle manure and urea ammonium nitrate solution (UAN). The manure was

H.G. v.d. Meer, et al. (eds.), Animal Manure on Grassland and Fodder Crops.
 ISBN 978-94-010-8137-5.

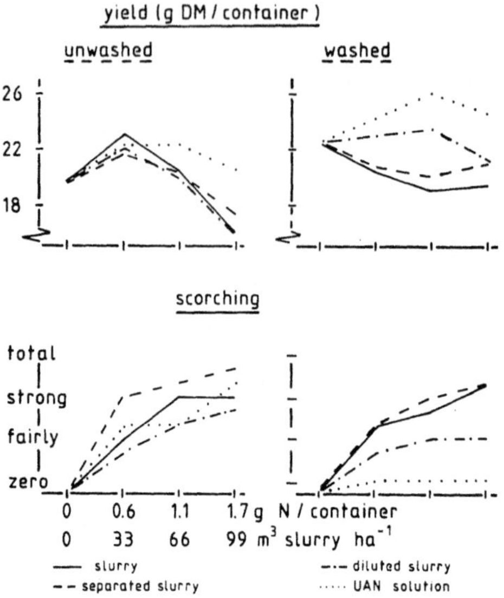

Figure 1. Influence of the amount of cattle slurry and UAN on yield and scorching of *Lolium perenne*.

applied untreated (10.6% DM), separated (8.8% DM) and diluted with water (1:1; 5% DM). A standard test procedure proved the separated slurry to be nearly as or even more adhesive as untreated, while diluted slurry was substantially less adhesive. Rates of slurry equivalent to 0, 150, 300, 450 kg N ha^{-1} were applied. Application was to the plant or on the soil surface. Three hours after top dressing some of the plants were washed by sprinkling them with 10 ml of water per container. Damage by scorching was evaluated on the first and third day after application. Harvesting followed after 14 days.

Results and discussion

Top dressing caused increasing plant damage with increasing rates of application of cattle slurry as well as of UAN (Fig. 1). Washing of plants reduced scorching only in the case of diluted manure and UAN. The untreated and the separated slurry were only partially washed from the plants and caused substantial damage. Scorching thus seems to be a result of adhesion (manure/plant), but additionally, specific manure ingredients like ammonia, benzoates and phenols may be involved. If separation is used to reduce scorching, lowering the DM content to 5% or less seems to be necessary. Yields showed a negative correlation with the degree of scorching, *i.e.* the damage to the plants could not be compensated by the additional nitrogen applied with the slurry. Yield depressions were severest from untreated and separated slurry.

The application of slurry and UAN directly on the soil surface caused no visible scorching and did not decrease yields. Plant damage by the slurry occurred only to leaves and stems but not to the root system.

The effect of cattle slurry on clover in grass/clover swards

R CHAPMAN and S B HEATH
Department of Agriculture, University of Aberdeen, Aberdeen AB9 1UD, UK

Key words: white clover, grass, cattle slurry, nitrogen, potassium

Abstract. The effect of cattle slurry on the maintenance of clover in grass/clover swards was examined. The nutrient value of slurry was also simulated by applying nitrogen (N) and potassium (K) fertilizer. The results support the hypothesis that the damaging effects of N on clover may be offset by the provision of sufficient K. 30 t ha^{-1} of undiluted cattle slurry applied in spring brought a response in clover content in line with the quantity of available N and K supplied. However, a higher slurry rate (60 t ha^{-1}) and summer applications caused a greater than expected reduction in the clover content of the herbage.

Introduction

Low input grass/clover systems depend on maintaining an adequate clover content for efficient N fixation. Clover is a poor competitor for K and consequently responds to a higher K level than does grass [1, 2]. Application of N fertilizers to grass/clover swards reduces the clover proportion of the herbage dry matter (clover %) [3]. Slurry contains N, which is likely to reduce the clover %, but it also has a high K content. The N:K ratio of slurry is 1:2.5 (available nutrients) which is less than the usual 1:0.4 ratio of inorganic fertilizers applied to grassland. As well as a nutrient effect, slurry may have other deleterious secondary effects. The hypothesis that when slurry is applied to grass/clover swards the harmful effect of the slurry N on the clover % is offset by its high K content was investigated.

Materials and methods

A small plot trial was conducted on a perennial ryegrass/white clover (var. Milkanova) sward in 1986. Two silage cuts were taken in early June and mid July. The response of the clover % of the grass/clover sward was investigated using mineral fertilizers at various N:K ratios and N rates (Table 1). These treatments were compared with two rates of undiluted cattle slurry. The 30 S treatment was expected to supply 30 kg available N ha^{-1}, and the

H.G. v.d. Meer, et al. (eds.), Animal Manure on Grassland and Fodder Crops.
 ISBN 978-94-010-8137-5.

Table 1. Experimental treatments applied for each cut

Code		
0 N-0 K	Untreated control	(N:K 0:0)
30 N-MK	30 kg N ha^{-1}, 12.5 kg K ha^{-1}	(N:K 1:0.4)
30 N-HK	30 kg N ha^{-1}, 100 kg K ha^{-1}	(N:K 1:3.3)
60 N-10 K	60 kg N ha^{-1}, 0 kg K ha^{-1}	(N:K 1:0)
60 N-MK	60 kg N ha^{-1}, 25 kg K ha^{-1}	(N:K 1:0.4)
60 N-1 HK	60 kg N ha^{-1}, 240 kg K ha^{-1}	(N:K 1:3.3)
30 S	30 t ha^{-1} undiluted cattle slurry	
60 S	60 t ha^{-1} undiluted cattle slurry	

60 S treatment 60 kg available N ha^{-1}. The treatments were applied in the spring and reapplied after the first cut.

Results and discussion

The results of the two silage cuts are presented in Table 2. The application of NK fertilizer increased total DM yield at both cuts. Compared with the control the application of NK as 1:0.4 fertilizer reduced clover yield and clover %. Application of the high K rate with N (1:3.3) resulted in an improvement in clover yield and clover % at both N levels for both cuts, indicating that available soil K was inadequate. The addition of K increased clover yield to a greater extent than total DM yield.

For cut 1 the 30 S treatment gave similar total DM yield, clover yield and clover % to 30 N-HK. At cut 2 the 30 S treatment gave a reduced total DM yield compared with 30 N-HK, whilst the clover DM yield was reduced to an even greater extent with a consequent decline in clover %. Increasing the slurry rate from 30 S to 60 S did not increase total DM yield but caused a significant reduction in clover DM yield and clover % at both cuts.

This trial confirms previous reports [1, 2] that, where soil K is limiting, the response to K is greater for clover % than for the total DM yield of grass/clover swards. The results support the hypothesis that the damaging effects of inorganic N on clover % may be offset by ensuring the sward has sufficient K. Slurry applied at a low rate in spring results in a clover % in line with the quantity of available N and K supplied. Summer slurry applications and slurry applied at high rates cause a significant reduction in clover DM yield and clover %. It is inferred that this is likely to be caused by secondary effects of slurry, such as coating of clover leaves and stolons.

Table 2. The effect of N, K and cattle slurry on total dry matter (DM) yield (kg ha^{-1}) and clover DM yield (kg ha^{-1}), and on the proportion of clover DM in the herbage (clover %)

	Treatments								
Code	0 N-0 K	30 N-MK	30 N-HK	60 N-0 K	60 N-MK	60 N-HK	30 S	60 S	*LSD
Ratio N:K	0:0	1:0.4	1:3.3	1:0	1:0.4	1:3.3			
Cut 1									
Total DM	2460	3490	3730	4230	4210	4260	4000	4300	610
Clover DM	780	740	860	630	680	940	890	520	330
Clover %	31	21	23	15	16	23	22	12	8
Cut 2									
Total DM	3810	3990	4550	4020	4330	4620	3540	3590	760
Clover DM	2450	1880	2440	1510	2060	2290	1570	430	670
Clover %	64	46	54	38	47	50	45	14	13

* LSD: Least significant difference ($P < 0.05$)

References

1. Brockman J S (1972) In Fisons Grassland Research, pp 23–26. Suffolk: Fisons Ltd
2. Chestnutt D M B (1965) Record of Agricultural Research in Northern Ireland 14, 71–81
3. Reid D (1983) Journal of Agricultural Science (Cambridge) 100, 613–623

Factors controlling vegetation dynamics in hay meadows under high levels of organic fertilizer in the French Northern Alps

J M DORIOZ[a], Ph FLEURY[b] and B JEANNIN[b]
[a] *INRA – Science du Sol, 74200 Thonon, France*
[b] *INRA – SAD, Route de St Cyr, 78000 Versailles, France*

Key words: permanent pasture, organic fertilizing, exploitation pattern, pedoclimate, botanical composition, herbs

Abstract. On permanent mountain pastures cut for hay in the humid French Alps, high levels of organic fertilizers (40 t slurry $ha^{-1} yr^{-1}$) produce an invasion by various herbs which is especially marked in meadows cut late and/or lying on north-facing slopes.

Introduction

Following the recent changes in agriculture, easily accessible and mechanizable hay meadows in mountain areas may sometimes get very high dressings of slurry. An on-going study in the French Northern Alps provides information on the dynamics of plant populations in permanent swards lying between 800 and 1600 m when the first growth cycle is cut for hay. Particular attention was given to the content of herbs in the different swards (all herbs being dicots excepting legumes). This variable, expressed as a relative volume (percent total volume of the vegetation) has a bearing on the utilization value of the sward as it reflects nutritive value, palatability, intake and suitability for mowing. It was investigated because of the trend towards a proliferation of various herbs in hay meadows.

Materials and methods

The work is based on the concurrent study of the physical environment and of management practices at 33 sites. The meadows investigated belong to the same phytosociological grouping (*Triseto–Polygonion bistorti* alliance). They have all been managed similarly over at least the past five years. The sample selected is representative of management practices and of the soils and topoclimates occurring in the region [2].

H.G. v.d. Meer, et al. (eds.), Animal Manure on Grassland and Fodder Crops.
 ISBN 978-94-010-8137-5.

Vegetation analysis was carried out using the handful method, *i.e.* 50 random clips of a handful of herbage were taken on the experimental sites. The relative content of different species is then assessed and the relative volume of the species in the vegetation is calculated, and from this the relative volume of various plant categories (grasses, legumes, herbs) [3]. Each site was further characterized for its management (date of cutting, rate of organic fertilizer, *etc.*) and for the physical environment (topography, soils, *etc.*).

Results

In these pastures managed according to the annual sequence: 1st cut - 2nd cut - grazing or cutting - grazing - grazing, there is a large diversity in species abundances. This contrasts with the fairly homogeneous botanical composition previously reported [2]. High organic fertilizer applications (40 t slurry $ha^{-1} yr^{-1}$) promote species such as *Anthriscus silvestris, Rumex obtusifolius, Ranunculus repens* and *R. acris, Taraxacum officinalis, Poa trivialis,* and induce a considerable increase in the proportion of herbs (Fig. 1). At these high fertilizer levels, good and very good forage grasses (*Dactylis*

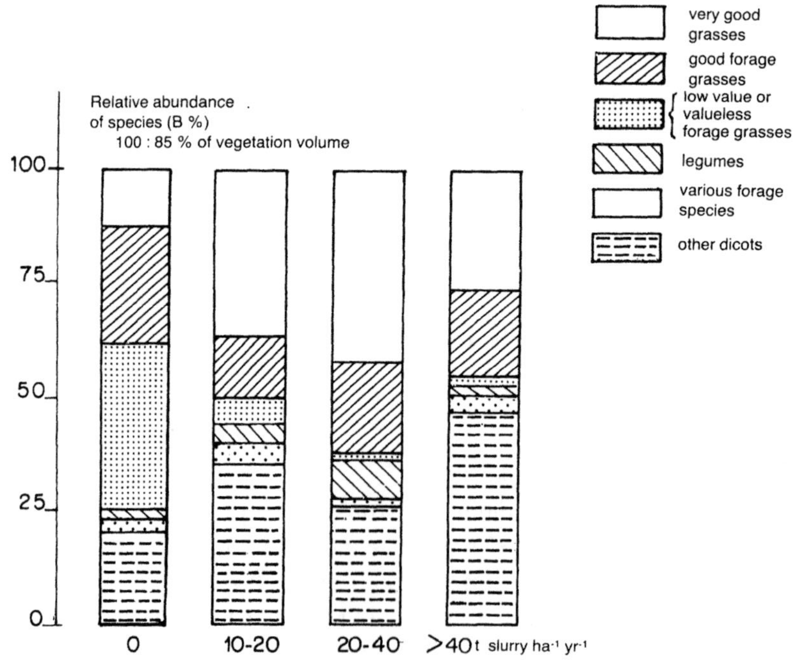

Figure 1. Effects of rate of slurry application on the botanical composition.

glomerata, Festuca pratensis, Trisetum flavescens) regress as a result of competition from the nitrophilous herbs.

Fertilizer does not act alone: the proportion of herbs also depends on the regular cutting date as shown by the positive relation between date of first cutting and herb abundance, as well as on specific physical conditions. These factors were ordered using multiple regression analysis, which underlined the prevalence of the factor "mowing date of the first cycle" [2] over others. This variable accounts for 40% of the variation in herb content; fertilizer levels for 25%; and summer water status of the sites for 10%. Good water availability, and thus nitrogen availability, favours herbs, all other factors being equal. Those meadows which are cut late (during and after seeding of *Dactylis glomerata*) and/or lying in the coolest sites (often facing North) are the most susceptible to invasion by herbs under high levels of organic fertilizer.

Conclusion

The agronomic conclusions are that the mountain environment, and in particular its coolest biotopes, are favourable to herbs, some of which are specific to high areas (*Geranium silvaticum, Chaerophyllum hirsutum*). This degeneration is not inevitable [1]. High fertilizer levels may be applied, provided the cutting date of the first cycle can be controlled.

References

1. Dorioz J M, Fleury Ph and Jeannin B (1987) Séminaire "Agrométéorologie et moyenne montagne", Toulouse, 16–17 avril 1987 (in press)
2. Fleury Ph, Dorioz J M and Jeannin B (1985) Coll "Etudes et Recherches du SAD" Nr 3, 52 pp. Versailles, France: INRA ed
3. Poissonnet P and J (1969) Document Nr 50, CEPE-CNRS, 120 pp. Montpellier, France: CNRS ed

Effect of cattle dung on the spread of couch grass (*Elymus repens* L.) in intensive grassland

W WASSHAUSEN
Grünlandlehranstalt und Marschversuchsstation für Niedersachsen, Infeld, D-2890 Nordenham, Sarver Str. 42, BRD

Key words: cattle dung, couch grass (*Elymus repens* L.), dung patch distribution, slurry

Abstract. Some paddocks of the experimental farm were mapped. The proportion of cover with couch grass (*Elymus repens* L.) and the number of dung patches were determined. A linear multiple regression analysis between number of dung patches and proportion of cover with couch grass gave a correlation coefficient r^2 between 0.23 and 0.32. This result was confirmed in an experimental plot. Dung patches allow the development of couch grass in grazing land. The paddocks were mapped for three years and the plot trial was mapped after four years.

Introduction

Couch grass is a nitrogen (N) loving plant and earlier investigations [2] have shown that N supply is important in the spread of couch grass in grassland. The dung patches produced by cows supply much N. The effect increases when several dung patches occur at the same place, and only species with great competitive power can penetrate the dung cover. Dung patches allow the spread of couch grass provided there are rhizomes in the soil. The influence of dung patches on the spread of couch grass was examined in two experiments.

Materials and methods

A preliminary investigation was carried out by the point-square-method. Each paddock was divided into 156 squares of 10 m × 10 m. With experience of this method couch grass density was estimated visually with an occasional calibration by the point-square-method.

To test the effect of dung patches, couch grass sods of 10 cm radius were planted in grassland plots with 6 replicates as a Latin square in a second experiment. When rooted they were covered with 0, 1 or 2 dung patches per

H.G. v.d. Meer, et al. (eds.), Animal Manure on Grassland and Fodder Crops.
 ISBN 978-94-010-8137-5.

Table 1. Effects of dung and slurry on spread of couch grass (*Elymus repens* L.). Data are % cover (P = 0.05)

Dung patch	Slurry	
	25 m^3 ha^{-1}	50 m^3 ha^{-1}
0	4.3	6.0
1	10.8	8.3
2	14.3	15.8

lsd = 8.9

sod and with slurry at 1 × 25 m^3 or 2 × 25 m^3 ha^{-1}. The second application of slurry and dung was given after an interval of 4 weeks.

Results and discussion

A linear multiple regression gave a significant correlation (P = 0.05) between the number of dung patches and the distribution of couch grass in the sward (r^2 = 0.23 to r^2 = 0.32). At first glance this is a poor correlation but it appears better when it is considered that there are many other factors influencing the spread of couch grass.

One cow produces approximately 20 kg fresh dung daily [1]. We found 12 dung patches cow^{-1} day^{-1} with an average total N supply of 10.25 g (0.5 g ammonium N) per patch on an average area of 1000 cm^2; this is equivalent to an N supply of 1025 kg ha^{-1}.

In the second experiment a near linear correlation was found between the spread of couch grass and dung cover (Table 1).

There is an uneven dung return within a paddock and subsequently an uneven distribution of couch grass. Dung patches allow the spread of couch grass provided there are rhizomes in the soil. The influence of slurry is less but slurry may smother productive grasses causing their death. The effect is the same as that of dung.

Acknowledgement

I thank Mr. F. Taube (Dipl Ing agr), Lehrstuhl für Grünlandwirtschaft, University of Kiel, for help with computing the results.

References

1. Holmes W (1980) in Holmes W, ed. Grass its production and utilization, pp 125–173. Oxford, UK: Blackwell Scientific Publications
2. Wasshausen W (1979) Mitteilungen der Deutschen Landwirtschaftsgesellschaft 94, 200–202

Comparative responses to applied fertilizer phosphorus on manured and unmanured soils

H I GRACEY
Greenmount College of Agriculture and Horticulture, Antrim BT41 4PU, Northern Ireland

Key words: organic manures, fertilizer P, plant uptake, soil P

Abstract. The response to fertilizer phosphorus (P) was investigated in a field experiment. On a manured area the omission of fertilizer P did not reduce herbage dry matter (DM) yield over a 6 year period, and herbage P concentration was not affected by fertilizer P and did not decline in the absence of P fertilizer. On an unmanured area, the omission of P fertilizer for the first 3 years depressed yields by 6%, but this increased to 15% during the subsequent 3 years. Significant linear regressions were obtained between ammonium acetate extractable P (NH_4OAc-P) in the soil and the various fertilizer rates, and the application of 31.7 kg P ha^{-1} yr^{-1} to the unmanured soil would have left NH_4OAc-P at 0–5 cm unchanged.

Introduction

The rate of phosphorus (P) application to Northern Ireland soils has remained fairly static over the past 30 years at around 10,000 t yr^{-1} [4]. However, with 90% of the farm land under grass coupled with intensive pig and poultry enterprises based largely on bought-in feedingstuffs, considerable quantities of organic manures are available for returning to the land [5]. It has been estimated that there is an annual surplus of some 16,000 t P for the country as a whole [1], and the enrichment of P in the soils is reflected in the results of routine soil analysis [2]. In a growth chamber study it was found that the availability of P in organic manures was not as good as that in monoammonium phosphate [3] and the main objective of the experiment reported here was to study the release of soil P from two grassland soils, one soil with a history of heavy organic manuring and the other which had received no organic manure and minimal amounts of inorganic P in recent years.

Materials and methods

A field experiment was set up on a Basaltic soil under permanent pasture. The site had previously been used to study the effect of disposing high rates

H.G. v.d. Meer, et al. (eds.), Animal Manure on Grassland and Fodder Crops.

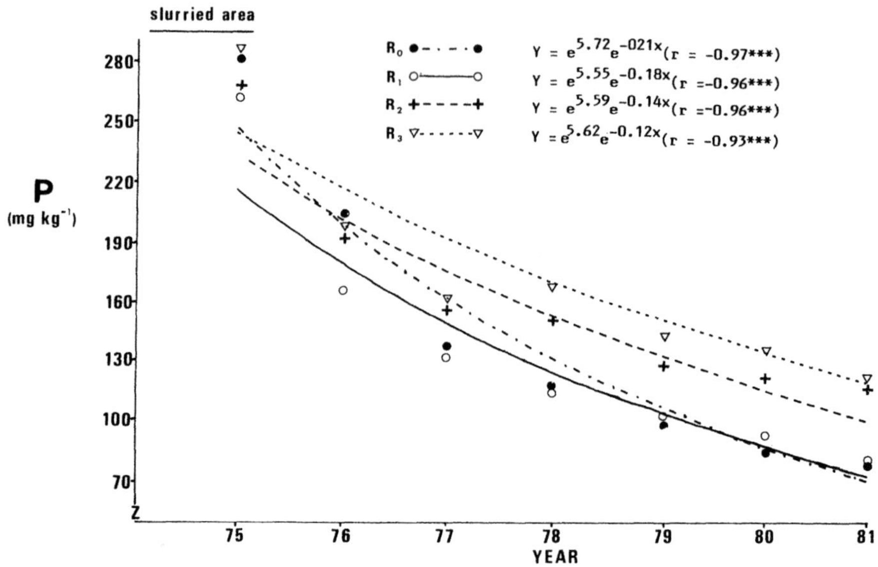

Figure 1. Changes in 0.5 *M* NH_4OAc-P (mg kg^{-1}) in 0–5 cm soil for fertilizer P treatments (mean of one and three seasonal P applications).

of copper-rich pig slurry applied to grassland on the health of grazing sheep [7]. The manured soil had received 1500 kg P ha^{-1} [6] over a 6 year period, while over the same period a control area received no P. Two sets of plots were marked out in the unmanured and manured areas, and P as superphosphate was applied at rates equivalent to 20, 40, 60 kg P $ha^{-1} yr^{-1}$ (R_1, R_2 and R_3, respectively) in either three equal dressings over the year, a single spring application each year or a single spring application every third year. A control area (R_0) received no P and all treatments received nitrogen (N) and potassium (K) equivalent to 300 and 200 kg $ha^{-1} yr^{-1}$, respectively, in three equal dressings. Treatments were harvested three times each year over a 6 year period.

Results and discussion

On the manured area the omission of fertilizer P did not reduce herbage DM yield, and both rate and time of fertilizer application had no effect on yield. Herbage P concentration was not affected by fertilizer P and did not decline in the absence of P fertilizer. Changes in 0.5 *M* NH_4OAc-P in the soil at 0–5 cm are given in Fig. 1, where it can be seen that levels declined exponentially, and the application of P surplus to that removed in the herbage did not halt this decline. On the unmanured area the omission of P fertilizer for the first 3 years depressed yields by 6%, but this increased to 15% during the next 3 years. Similar yields were obtained by applying P for each harvest, as a spring application in each year or as a spring application every third year. The application of 20 kg P $ha^{-1} yr^{-1}$ and 40 kg P $ha^{-1} yr^{-1}$ yielded 94 and 97%, respectively, of the yield given by the 60 kg P $ha^{-1} yr^{-1}$ rate and of P removed in the herbage. Significant linear regressions were obtained between 0.5 M NH_4OAc-P in the soil at 0–5 cm and the various fertilizer rates. The application of 31.7 kg P $ha^{-1} yr^{-1}$ would have left NH_4OAc-P at 0–5 cm unchanged and the soluble P in the soil at which significant responses to applied P were obtained corresponded to 13 mg kg^{-1} NH_4OAc-P at 0–15 cm.

References

1. Adams S N (1975) Paper presented at the Greenmount Association Spring Conference, Antrim, N Ireland, 15 March 1975
2. Gracey H I (1981) PhD Thesis, The Queen's University of Belfast
3. Gracey H I (1984) Agricultural Wastes 11, 133–141
4. Gracey H I (1986) Irish Farmers' Journal, April 1986
5. Gracey H I (1986) British Grassland Society Winter Meeting, 5 December 1985 (in press)
6. Gracey H I & Stewart T A (1975) Record of Agricultural Research 23, 65–66
7. Gracey H I, Stewart T A, Woodside J D & Thompson R H (1976) Journal of Agricultural Science (Cambridge) 87, 617–623

Effect of animal manure on phosphorus and potassium content of herbage

M DURU
INRA – Station d'Agronomie, Unité de Recherche sur les Systèmes Agraires et le Développement, BP 27, 31326 Castanet Tolosan, France

Key words: fertilization, manure, plant nutrition, grassland, mineral diagnosis

Abstract. To diagnose phosphorus (P) and potassium (K) nutrition, the relations between the contents of nutrients in herbage (% P: % N, % K: %N) for two plots of different soil fertility were established. These relations were used to study the cumulative effect of animal manure. A rapid improvement of plant nutrition was shown for the initially deficient plot. This conclusion was checked by comparing nitrogen (N) content and herbage dry matter (DM) yield. The economic benefits of animal manure in disadvantaged areas are discussed.

Introduction

The fertilizer value of organic manure is poorly understood because of losses after spreading and of the variability of nutrient release.

In mountainous areas many grasslands are deficient in mineral nutrients and animal manure is the most common fertilizer [1]. It is important to study whether manure improves the phosphorus (P) and potassium (K) status of different meadows.

Materials and methods

Two permanent meadows located at an altitude of 1250 m in the Pyrenees mountains were studied. The two meadows, B and C, were characterized by adequate and poor soil fertility, respectively. Twenty t ha^{-1} yr^{-1} of manure, containing 140 kg nitrogen (N), 26 kg P, 160 kg K were spread for five years (1981–1985). Each year, 2 or 4 cuts were taken from the same places.

To diagnose P and K nutrition, a comparison of DM production between the plots or with a control plot is inadequate without a check of N nutrition. Knowing only the P and K content of the herbage is inadequate because of their reduction during growth [5]. Linear relations between P and K content and N content have been shown for different meadows [6, 4]. These correlations allow the mineral status to be defined.

H.G. v.d. Meer, et al. (eds.), Animal Manure on Grassland and Fodder Crops.
 ISBN 978-94-010-8137-5.

When P and K nutrition is adequate, the relation for P but not for K is independant of N nutrition [4]. These relations were established for the plots B and C from several samples taken during spring growth, with either restrictive or non restrictive N nutrition.

Results and discussion

For plot B, there was little change compared with the characteristic straight lines established before treatment (Fig. 1). For plot C, the method demonstrated an improvement of P and K nutrition, particularly from the fourth year (Fig. 1). No analysis was carried out the third year.

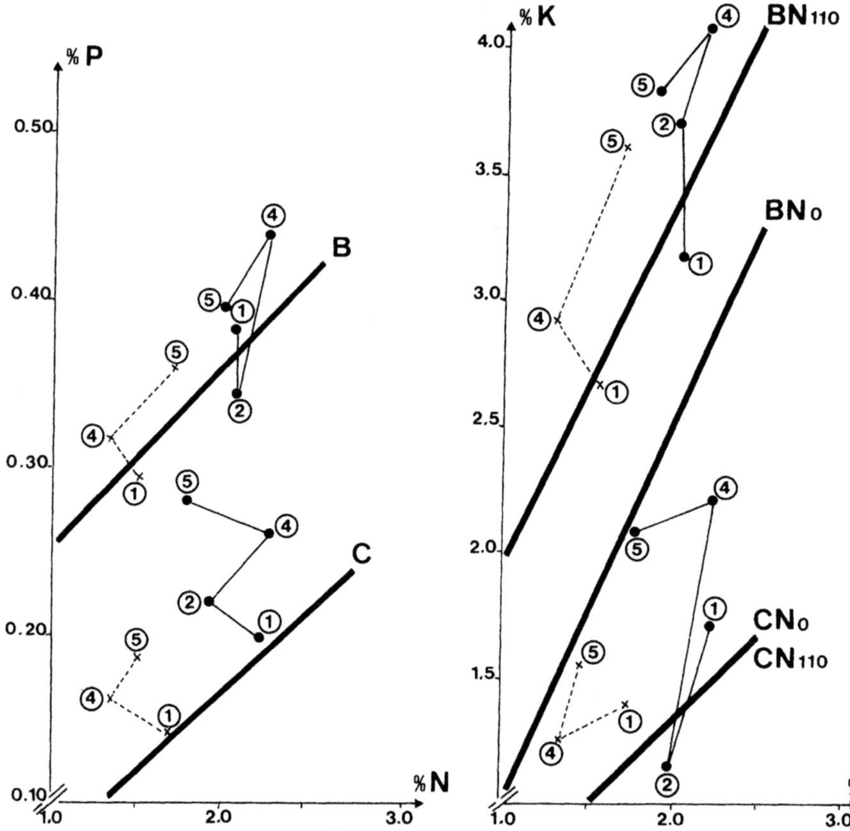

Figure 1. Phosphorus (1a) and potassium (1b) nutrition changes for plots B and C. Cumulative effect of animal manure over 5 years and 2 cutting frequencies (——: 4 cuts year^{-1}, - - - -: 2 cuts year^{-1}) 1–5: experimental year. ——: characteristic initial straight line, showing the relation between % P or % K and % N in the herbage on the plots B and C. N_0: no nitrogen fertilization, N_{110}: 110 kg N ha^{-1}.

To check this interpretation, DM production and N nutrition were compared by a dilution curve method [3], for every plot from 1981 to 1985. The observations were consistent with the P and K nutrition evolution described above. For plot C, P and K nutrition improved quickly in spite of low rates of organic fertilization.

Conclusion

The diagnostic method should be used in more cases. The method allows an easy and quick diagnosis and makes it possible to test the effect of cultural treatments. For permanent meadows, this diagnosis must be complemented by any change in botanical composition, which may be related to an improvement of P and K nutrition.

From an economic point of view, animal manure is especially important for farms with limited finance [2]. If all the animal manure was efficiently used, it would maintain or improve the P and K nutrition of at least 50% of the cut area.

References

1. Balent G and Duru M (1984) Agronomie 4 (2), 113–124
2. Duru M and Gibon A (1985) BTI 399/401, 323–329
3. Lemaire G and Salette J (1984) Agronomie 4 (5), 423–430
4. Salette J (1982) Proc. 12th Congr. Int. Potash Institute (Berne), 117–144
5. Salette J, Lemaire G and Robichet J (1982) Occasional Symposium no. 14, pp 165–168. Hurley, UK: The British Grassland Society
6. Salette J, Dumas Y and Sabesky O (1978) L'agronomie tropicale 28, 833–845

Utilization of potassium excreted by grazing cattle

O JOURDAN
Centre de Recherches SCPA, Aspach le Bas, 68700 Cernay, France

Key words: potassium, excretion, tall fescue, cocksfoot, potassium content

Abstract. A field trial compared zero grazing and grazed pasture with and without potassium (K) fertilization. Half the plots were sown with cocksfoot and the other half with tall fescue. Potassium fertilizer had an immediate effect on dry matter (DM) production. In the first year K content in the grass (especially in cocksfoot) increased from the effects of dung and urine, but five years were required to detect differences of DM production. The recycling ratio of K, variable between species and fertilization, reached fifty per cent.

Introduction

Nitrogen (N), phosphorus (P), and potassium (K) contained in animal manure can be used as fertilizers when manure is spread on grassland. What happens when dung and urine go directly to the grass? Measurements have been made of the coefficient of recovery of K in intensive grassland in order to understand the role of dung and urine in meeting the fertilizer needs of grass [1]. The final estimates were about 50% [2, 3, 4].

Materials and methods

A split-plot design crossed two levels of K fertilization, 0 and 185 $kg\,ha^{-1}\,yr^{-1}$, with two systems of pasture management (i) grazing on 1400 m^2 plots, and (ii) mowing 12 m^2 plots. P fertilization was 40 $kg\,ha^{-1}\,yr^{-1}$ and N fertilization was about 300 $kg\,ha^{-1}\,yr^{-1}$. Four blocks were grown with tall fescue (*Festuca arundinacea*) and four blocks with cocksfoot (*Dactylis glomerata*). Two groups (K = 0 and K = 185) of steers grazed the meadow from April to October. Results are given for 9 years (from 1975 to 1983).

Results and discussion

Grazing and fertilization affected DM yield (Table 1), although grazing effects were small even after 5 years. The effect of grazing is the overall effect

H.G. v.d. Meer, et al. (eds.), Animal Manure on Grassland and Fodder Crops.
 ISBN 978-94-010-8137-5.

Table 1. DM yield (t ha^{-1} yr^{-1}), average of cocksfoot and tall fescue

DM yield t ha^{-1} yr^{-1}	K-0 mown	K-0 grazed	K-185 mown	K-185 grazed
average of 5 years	9.04	9.40	9.88	9.96
average of 9 years	8.42	9.78	9.68	10.64

Table 2. Potassium content (% of DM) in grass. Average of 9 years

K content % of DM	K-0 mown	K-0 grazed	K-185 mown	K-185 grazed
Cocksfoot	1.73	2.77	2.70	3.47
Tall fescue	1.66	2.24	2.52	2.97

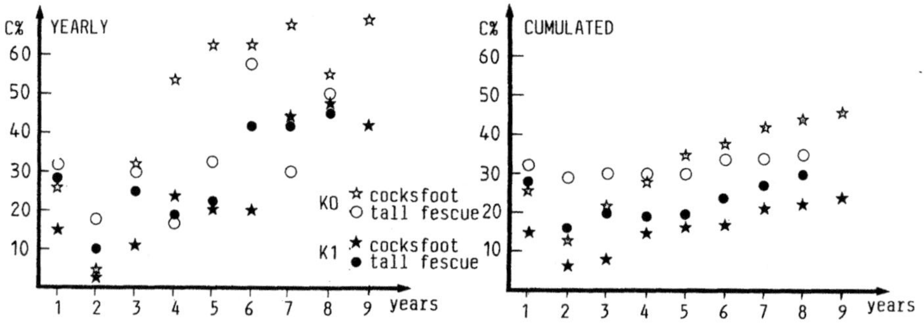

Figure 1. Change with time of the coefficient of recovery of K (C %).

of N, P, and K in manure. The response of the cocksfoot pasture was higher than that of the tall fescue pasture (Table 2).

The coefficient of recovery (C %) of K was calculated as:

$$C\,\% = \frac{\text{K uptake from grazing} - \text{K uptake from mowing}}{\text{K returned with dung and urine}} \times 100$$

97% return of K ingested with grass is assumed [5].

The coefficient of recovery of K improved during the experimental period but it took a very long period to reach equilibrium. Recycling rates differed according to fertilization level and grass species.

References

1. Jourdan O (1986) Dossiers Agronomiques d'Aspach SCPA France
2. Garaudeaux J (1975) C.R. Académie Agriculture France no 10, 571–580
3. Pfitzenmeyer C (1963) Fourrages no 15, 67–81
4. Lombaert V (1983) Thèse Université de Bruxelles
5. Hutton J B (1964) N.Z. Journal of Agric. res. Vol. 7, 583–595

The effect of cattle slurry on the mineral content of pastures and blood serum in dairy cows

I KUCZYNSKA and A LACHOWSKI

Institute for Land Reclamation and Grassland Farming, Al. Ossolińskich 12, 85–093 Bydgoszcz, and Veterinary Institute, Powstańców Wielkopolskich 10, 85–090 Bydgoszcz, Poland

Key words: cattle slurry, pasture, blood serum, mineral content, sodium, potassium

Abstract. Cattle slurry supplemented by mineral fertilizer was applied to pastures, in 1984 and 1985. The contents of potassium (K), sodium (Na), and magnesium (Mg) were determined in herbage and in the blood serum of 78 cows grazed on the fodder under study. High K and low Na contents were found in both herbage and blood serum. The high K content did not affect Mg absorption by the cows.

Introduction

Using slurry as fertilizer is the simplest way of utilizing animal manure. To incorporate slurry into a fertilizer program it is necessary to know the mineral content and to use it in doses to meet the crop needs. The aim of the present studies was to determine the effect of slurry on the yield and quality of hay and on the absorption of Mg and other elements by animals.

Materials and methods

The experiments were carried out on pastures lying on alluvial soils in the Vistula Valley. The experimental area was about 30 ha, 15 ha of which were fertilized with slurry. The investigations were carried out on 78 dairy cows of the Lowland Black and White Breed, divided into two groups, experimental and control. The cattle have been grazed on plots for two years so far. Cattle slurry was applied to the plots as follows: from March 20 to April 11, 1984, 40 m^3 ha^{-1} plus mineral fertilizers to give a total of 276 kg K ha^{-1}, 45 kg P ha^{-1} and 258 kg N ha^{-1}. In 1985, 124 m^3 slurry ha^{-1}, plus mineral fertilizers amounted to 412 kg K ha^{-1}, 80 kg P ha^{-1} and 307 kg N ha^{-1}.

The pasture for the control group received 124 kg K ha^{-1}, 42 kg P ha^{-1} and 144 kg N ha^{-1} every year in the early spring. Samples of plant material were taken the day before letting the cows onto the pasture.

H.G. v.d. Meer, et al. (eds.), Animal Manure on Grassland and Fodder Crops.
 ISBN 978-94-010-8137-5.

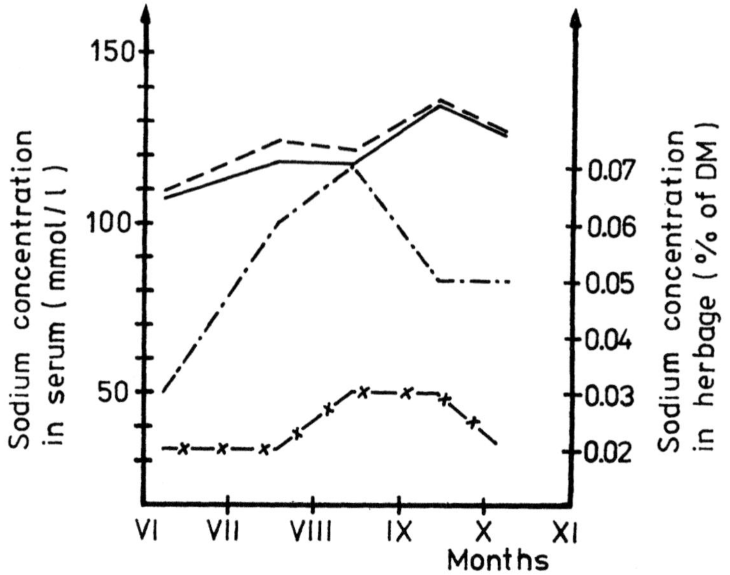

Figure 1. Sodium concentration in herbage and blood serum of cows during the grazing season. Herbage: –×– control; –·– fertilized with slurry. Serum: --- control; — fertilized with slurry.

Results and discussion

Seventy per cent of the pasture plants are grasses: dominant are Cocksfoot and Meadow bluegrass. Other species found are Meadow fescue, Chewings fescue, Meadow foxtail and Perennial ryegrass. The herbage yields obtained in 1984 from the plots fertilized with slurry showed comparatively little variation: from 12.34 to 14.69 t DM ha^{-1}. In 1985 the variations in yield were much greater due to unfavourable weather condition.

The mineral contents of the herbage were relatively high, except for Na which was very low and did not meet the requirements of dairy cows (Fig. 1). In the plots not fertilized with slurry, the Na content of herbage was even lower. The Mg content in the grass was its lowest at the beginning of the growing season. The other components are within standard figures although the K content in the herbage on the control plot was very high (Fig. 2).

Magnesium level in grasses and blood serum throughout the grazing season did not exceed the accepted standards for pasture plants and blood serum. The low Na level in pasture plants reduced the Na content in blood serum (Fig. 1) below the critical levels established by Pinkiewicz and other authors [1, 2]. The high K content in plants (3.0–4.0% K in the DM) caused a corresponding increase in K content in the blood serum (Fig. 2). In both

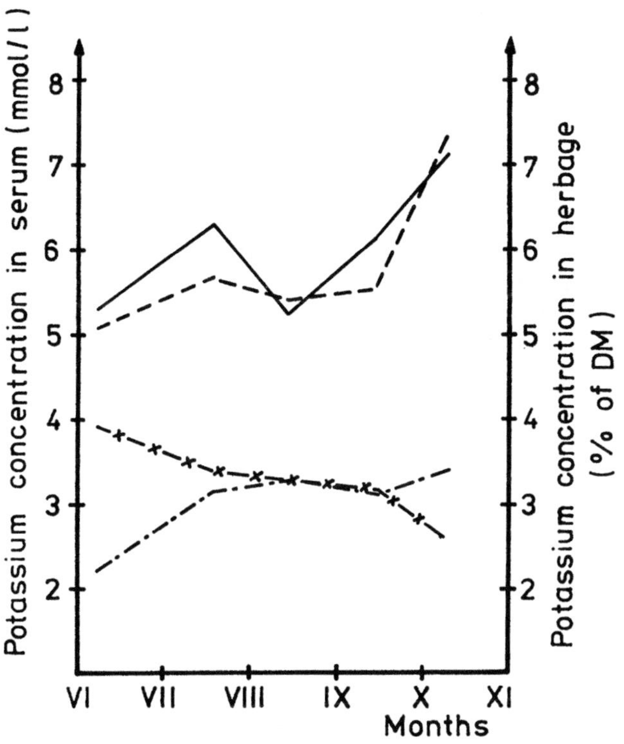

Figure 2. Potassium concentration in herbage and blood serum of cows during the grazing season. Herbage: –×– control; –·– fertilized with slurry. Serum: --- control; — fertilized with slurry.

groups this was maintained in the upper standard zone according to Pinkiewicz, 3.84–5.51 mmol l^{-1} [2]. In July of the second year of the experiment the K content in the experimental group increased to 6.4 mmol l^{-1}, and from September to October it increased again to more than 7 mmol l^{-1}.

The copper content in the blood serum did not increase, and towards the end of the grazing season in 1985 there was even a decrease to 9.5–10.0 μmol l^{-1}, which would indicate a deficiency according to accepted standards.

Clinical examinations were also carried out. Diarrhoea lasting several weeks was observed in the experimental group.

Conclusion

Fertilization of pastures with slurry did not raise the Na content of pasture plants to desirable amounts and caused a moderate deficiency in the blood serum. It was responsible for a high K content in the herbage and blood serum, particularly towards the end of the grazing season. The high K content in herbage did not produce the expected disturbances in Mg absorption by cows.

References

1. Blood D C, Henderson J H and Radostits O M (1979) Veterinary Medicine London, Baillier Tindall, pp 886-887
2. Pinkiewicz E (1971) Podstawowe badania laboratoryjne w chorobach zwierzat, Warszawa PWRiL, pp 59–77

Ammonia volatilization from urine patches in grassland

N VERTREGT and B RUTGERS
Centre for Agrobiological Research (CABO), P.O. Box 14, 6700 AA Wageningen, The Netherlands

Key words: ammonia volatilization, ammonia, nitrogen losses, urine, grazed swards

Abstract. Ammonia (NH_3) volatilization from artificial urine applied to perennial ryegrass (*Lolium perenne* L.) pastures was measured during the grazing season of 1986. The amount of NH_3 volatilized in 10 days after application ranged from 7–12% of nitrogen (N) applied at a urine-N application rate equivalent to 285 kg N ha^{-1} and from 4–18% for an application rate of 600 kg urine-N ha^{-1}. Depending on the experimental conditions, 0–45% of the N applied was not recovered, suggesting high N losses by other loss pathways.

Introduction

In intensively managed grassland with an N input of about 400 kg N $ha^{-1} yr^{-1}$, about 65% of the N ingested by the cattle is excreted in the urine [3]. Rates of NH_3 volatilization mentioned in literature are in the range 6–70% of N voided [1, 4, 5]. Urine patches are considered to be a major source of air-polluting ammonia in The Netherlands. Therefore, the extent of NH_3 volatilization from urine under Dutch conditions was studied.

Materials and methods

Experiments were carried out on a sandy soil (pH 5.9), a clay soil (pH 7.6) and a peat soil (pH 5.0). Artificial urine (6–12 g N l^{-1}), prepared according to Doak [2], was spread evenly over plots of 2 m × 0.55 m at a rate of 5 l m^{-2}. The experimental area was covered by a clear polycarbonate tunnel, 0.6 m high, mounted on an iron frame. Ammonia-free filtered air was blown through the tunnel at a speed of 0.5 or 1.5 m s^{-1}. The outflowing air was sampled through gas washing bottles in which the air was drawn through a 0.3 *M* orthophosphoric acid solution. The flowrates of the air through the tunnel and the sampling device were measured by gasmeters. Volatilized NH_3 was determined using an automated analysis system based on the Berthelot reaction. The detection limit of the system is 0.01 kg N $ha^{-1} h^{-1}$.

H.G. v.d. Meer, et al. (eds.), Animal Manure on Grassland and Fodder Crops.
 ISBN 978-94-010-8137-5.

Ammonium and nitrate contents in the soil were determined at the start and at the end of the experimental period and N uptake in the grass was estimated.

Results and discussion

The results of the experiments carried out on the sandy soil during the grazing period of 1986 are summarized in Table 1. From applications equivalent to 600 kg N ha^{-1}, the amount of NH_3 volatilized during subsequent 10 days ranged from 26 to 104 kg N ha^{-1}, depending on the experimental conditions. The NH_3 volatilization increased slightly in the course of the season, possibly as a result of lower soil moisture and higher temperatures. A simulated heavy rainfall (12 mm) immediately after urine application resulted in a 30% reduction in NH_3 volatilization (exp. 4.1 and 4.2). A large fraction of the urine-N applied was not accounted for as soil inorganic N, as crop N or as volatilized NH_3. The quantity of N unaccounted for was low in the plots treated with the nitrification inhibitor dicyandiamide (exp. 7.1 and 7.2), which may indicate large short-term denitrification losses in urine patches. Results of the experiments on the clay and peat soils were similar to those obtained on the sandy soil.

Conclusions

Average NH_3 volatilization losses from artificial urine patches amounted to 10% of the N applied in the first 10 days after application. Under the

Table 1. Ammonia volatilization, N uptake by the grass and accumulation of inorganic N in the soil profile after application of artificial urine to grassland under different conditions

No. experiment and period, 1986		Urine-N applied	Inorganic N in soil		NH_3 volatilized		N uptake by the grass
			start	end	10 days	10th day	
		(kg N ha^{-1})					
1.1	21/4– 1/5	284	33	196	34	1.6	53
1.2	21/4– 1/5	565	46	313	100	3.0	31
4.1	9/6–19/6	604	47	347	67	0.8	53
4.2	9/6–19/6	604	56	468	47	0.4	81
6.1	7/7–17/7	610	16	295	77	2.9	54
6.2	7/7–17/7	610	128	443	82	2.7	60
7.1	4/8–14/8	607	78	556	76	4.4	15
7.2	4/8–14/8	607	20	570	104	6.0	15

assumption that the 10th day emission level gradually decreases to zero in the next 20 days, an additional quantity of 5% of the N volatilizes during the remaining 20-day period. On intensively managed grassland the excretion of N in urine averaged over the whole field is about 190 kg N ha^{-1} in the grazing season. Under these circumstances NH_3 losses amount to 28 kg N ha^{-1}.

References

1. Ball P R and Ryden J C (1984) Plant and Soil 76, 23–33
2. Doak B W (1952) Journal of Agricultural Science 42, 162–171
3. Meer H G van der (1983) In Corrall A J, ed. Efficient grassland farming, pp 61–68. Proceedings 9th General Meeting of the European Grassland Federation (Reading). Hurley, UK: The British Grassland Society
4. Ryden J C (1986) In Meer H G van der, Ryden J C and Ennik G C, eds. Nitrogen fluxes in intensive grassland systems, pp 59–73. Dordrecht, The Netherlands: Martinus Nijhoff Publishers
5. Sherlock R R and Goh K M (1984) Fertilizer Research 5, 181–195

Limitations for slurry application to peat grassland

R BARTELS and B SCHEFFER
Institute of Soil Technology of the Geological Survey of Lower Saxony, Friedrich-Mißler-Straße 46–50, D-2800 Bremen 1, FRG

Key words: peat soil, slurry, grassland, leaching of phosphorus and nitrogen

Abstract. Manuring high moor grassland with slurry should be limited because of intensive leaching of phosphorus. After applying slurry on low moor peat soils high rates of nitrate are leached from arable land, but not from grassland. It is concluded that slurry application on peat soils should be limited to the equivalent of two cows per ha.

Introduction

In Lower Saxony there is an increased slurry manuring even on peat soils. Slurry with its high concentration of nutrients, and readily decomposable organic matter accelerates the decomposition of high moor peat soil and has an even greater effect on low moor soils. After slurry manuring on organic soils there is an increased risk of leaching of phosphates and nitrates.

Field trials

1. High moor: For five years different forms of manuring were tested on high moor grassland:
1) $40\,m^3\,ha^{-1}\,yr^{-1}$ of pig slurry, applied in spring;
2) $80\,m^3\,ha^{-1}\,yr^{-1}$ of pig slurry, applied in spring;
3) mineral fertilizer, equivalent to 1).
The pig slurry contained 4 kg nitrogen (N) per m^3 and 1.5 kg phosphorus (P) per m^3; the drainage depth was 1 m and the groundwater level was also 1 m. Yield and nutrient uptake of plants and the leaching of phosphates and nitrates in the drainage water were measured.
2. Low moor: Since 1982 the leaching of nutrients from a low moor soil, with and without pig slurry, has been investigated by analysing drainage water from arable land and grassland [2]. Slurry application was 22 to $36\,m^3\,ha^{-1}$ of pig slurry containing 130 to 260 kg N $ha^{-1}\,yr^{-1}$.

H.G. v.d. Meer, et al. (eds.), Animal Manure on Grassland and Fodder Crops.
 ISBN 978-94-010-8137-5.

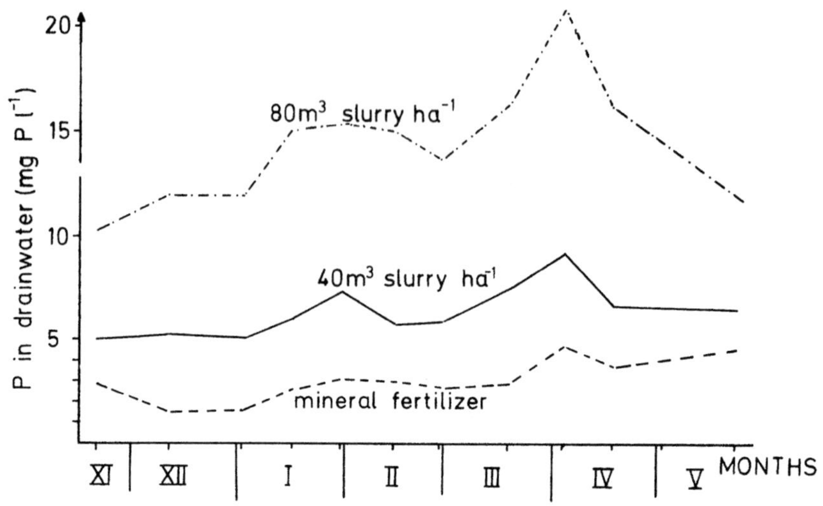

Figure 1. Phosphate-P content of drainage water (1977–1978), high moor peat soil.

Results and discussion

Fig. 1 shows the phosphate content of drainage water after application of slurry compared to fertilizer on high moor grassland. When manuring with $40\,m^3\,ha^{-1}\,yr^{-1}$ in average 9 kg phosphate-P $ha^{-1}\,yr^{-1}$ were leached in drainage water. Under the same conditions nitrate leaching on this acid organic soil was similar to mineral soils (7–8.5 kg N $ha^{-1}\,yr^{-1}$). When more slurry was applied — for instance $80\,m^3\,ha^{-1}\,yr^{-1}$ — there was an increase in leaching of nitrates and phosphates and also an increase of organic matter in the drainage water. It is necessary to limit the application of slurry [3] to such soils depending on the P-content.

When manuring N rich, low moor soils, with slurry on tilled land (continuous corn) we found a significant rise of nitrate content in the drainage water (Fig. 2). When manuring grassland on these soils with slurry applied during the growing season, nitrate leaching is diminished by the uptake of the plants and denitrification in the relatively wet organic soils [1]. The losses of phosphates measured in the drainage water are lower in these soils than in mineral soils. The field trials show that after manuring with slurry, the content of oxidizable organic matter in drainage water increases from 750 mg to 1100 mg organic matter l^{-1}. The influence of slurry application on the transport of organic matter from slurry and on the decomposition of the organic matter in the soil should be analysed. In Lower Saxony it is now

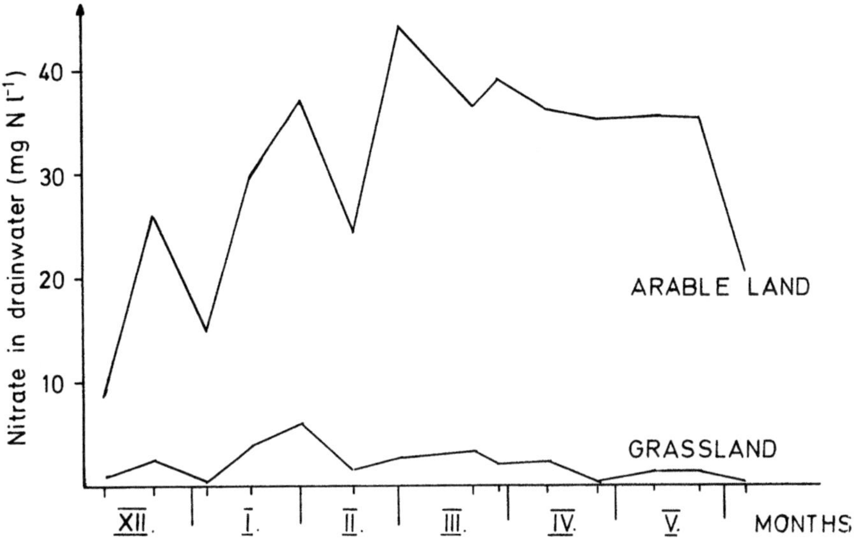

Figure 2. Nitrate-N in drainage water (1982–1983), low moor peat soil.

proposed to limit the application of slurry to peat soils to an equivalent of approximately two cows per ha.

References

1. Richter G, Scheffer B and Kuntze H (1985) Mitteilungen Deutsche Bodenkundliche Gesellschaft 43, 599–604
2. Scheffer B and Foerster P (1987) Jahresbericht Deutscher Verband für Wasserwirtschaft und Kulturbau (in press)
3. Scheffer B, Kuntze H and Bartels R (1982) Landwirtschaftliche Forschung, Sonderheft 38, 288–296

Effect of treated pig slurry on herbage and groundwater quality

W DRAGUN and M SOROKO
Institute for Land Reclamation and Grassland Farming ul. Powstańców Sl. 98 53–333 Wroclaw, Poland

Key words: treated pig slurry, *Dactylis glomerata*, soil percolations, nitrate content

Abstract. Cocksfoot (*Dactylis glomerata*) was irrigated with pig slurry which had received mechanical and mechanical/biological treatment. The effect of the irrigation on the growth of grass and its chemical composition, as well as on the soil and on the amount and composition of soil leachate, was examined.

Introduction

On large pig farms where the liquid manure is applied by means of a sprinkler, mechanical treatment of the slurry is often necessary for technological reasons. Biological treatment can also be applied in the case of high sanitary requirements.

The physical and chemical properties of slurry are changed by the treatment, as is also its fertilizing value.

Materials and methods

For the irrigation of cocksfoot (*Dactylis glomerata*) mechanically treated pig slurry (slurry I) was used. The average composition was: dry matter (DM): 1.1%, N: 0.13%, NH_4^+-N: 0.08%, P: 0.05%, and K: 0.06%. Pig slurry after mechanical and biological treatment (slurry II) was also applied. Its average composition was: DM: 0.23%, N: 0.06%, NH_4^+-N: 0.05%, and K: 0.05%. The amount of N supplied with the liquid manure varied from 300 to 1000 kg ha^{-1} yr^{-1}. The soil leachates were sampled in receivers placed under the plots at a depth of 0.7 m.

H.G. v.d. Meer, et al. (eds.), Animal Manure on Grassland and Fodder Crops.
 ISBN 978-94-010-8137-5.

Results and discussion

An increase in the rate of both pig slurries resulted in an increased NO_3^--N content in grass (Table 1). At the third rate of irrigation with slurry I (about 1000 kg N ha^{-1}) and at the second rate with slurry II (about 700 kg N ha^{-1}) it considerably exceeded the value of 0.20–0.25% in DM, which is considered permissible for grazing livestock. The amount of NH_4^+-N supplied with these applications was similar (about 615 kg N $ha^{-1} yr^{-1}$).

The results obtained are similar to those of experiments reported by other authors, with irrigation of both mineral fertilizers and liquid manure, when similar amounts of inorganic N and K were compared [3, 5, 6].

The volume of leachate in the dry year of 1979 (500 mm rainfall) was between 8 and 22 mm, and in the other years (700 mm rainfall) it was 50 to 100 mm. These values are lower than expected for comparable climatic and soil (sandy loam) conditions. We consider it to be an artefact of our experimental method, which produced consistently lower volume of leachates.

The concentration of NO_3^--N in the soil leachates and the nitrate loads increased with increasing rate of both slurries (Table 2). This increase was moderate at rates up to 600–700 kg N $ha^{-1} yr^{-1}$, due to a considerable uptake of N by grass. A large increase was observed at rates varying up to 1000 to 1200 kg N $ha^{-1} yr^{-1}$. From the results of other experiments it follows that because of the risk of groundwater pollution great caution must be paid when applying doses of 450–500 kg N $ha^{-1} yr^{-1}$. Such doses should not be applied every year to the same field [1, 2, 4, 7].

Table 1. Mean contents of NO_3^--N and K in cocksfoot fertilized with pig slurry after mechanical treatment (I) and mechanical/biological treatment (II) in the years 1979–1981

Kind of fertilizer	Annual rate $m^3 ha^{-1}$	NO_3^--N (% of DM)				K (% of DM)			
		Cuts							
		1	2	3	4	1	2	3	4
Control	0	0.03	0.02	0.03	–	2.51	2.51	2.94	–
Pig slurry I	225	0.03	0.08	0.13	0.04	1.94	2.79	2.89	2.16
	450	0.11	0.28	0.24	0.10	2.49	3.48	3.29	2.47
	675	0.29	0.33	0.40	0.18	2.91	3.35	3.99	2.93
Pig slurry II	525	0.06	0.12	0.18	0.06	2.65	3.17	3.96	2.79
	1050	0.36	0.36	0.41	0.18	4.05	3.99	4.66	3.39
	1575	0.44	0.33	0.44	0.13	4.06	4.08	4.94	3.97
NPK[a] (200 + 100 + 200)	–	0.03	0.04	0.05	0.03	3.05	3.28	3.49	3.23

[a] N applied as ammonium nitrate

Table 2. Contents and loads of NO_3^--N in leachate, from cocksfoot pasture, fertilized with pig slurry after mechanical treatment (I) and mechanical/biological treatment (II) in the years 1979–1981

Kind of fertilizer	N load in liquid manure			Mean contents of NO_3^--N in leachates			Annual loads of NO_3^--N in leachates		
	1979	1980	1981	1979	1980	1981	1979	1980	1981
	(kg N $ha^{-1}yr^{-1}$)			(mg l^{-1})			(kg ha^{-1})		
Pig slurry I	412	277	326	23.4	12.9	9.4	1.9	10.3	5.8
	824	553	652	33.8	25.1	23.0	4.0	19.8	10.8
	1236	829	978	59.4	31.7	88.9	8.1	33.7	71.0
Pig slurry II	433	324	379	39.1	7.8	12.8	7.3	7.0	7.2
	866	648	758	56.4	26.4	27.3	12.3	26.4	20.4
	1299	972	1137	97.8	74.7	60.3	21.7	61.7	27.3
NPK[a] (200 + 100 + 200)	–	–	–	–[b]	6.1	3.5	–[b]	3.7	1.1

[a] Ammonium nitrate
[b] No sampes taken

References

1. Foerster P (1978) Zeitschrift für Acker und Pflanzenbau 146 No 1, 20–32
2. Krüger W and Hirte W (1972) Wasserwirtschaft-Wassertechnik 6, 201–204
3. Maas G and Behnke V (1975) Archiv für Acker und Pflanzenbau und Bodenkunde 19, 827–836
4. Ruszkowska M (1984) Sympozjum naukowe. Falenty 1979 cz. 1, 17–25
5. Stuczyńska J (1973) Pamiętnik puławski Z. 58, 155–180
6. Thun J and Leves D (1975) Archiv für Acker und Pflanzenbau und Bodenkunde 24, 191–199
7. Vetter H and Steffens G (1981) Zeitschrift für Kulturtechnik und Flurbereinigung 22, 159–172

Nitrate in soil water at different depths during the growing season on permanent pasture (Comparison of mineral fertilizer and liquid manure)

M WERMKE
Institute of Grassland and Forage Research, Federal Research Center of Agriculture (FAL), Bundesallee 50, D-3300 Braunschweig, Federal Republic of Germany

Key words: pasture, nitrogen balance, nitrate in soil water

Abstract. In a pasture experiment for fattening cattle, which had been running several years, suction cups were installed on the 19th of April 1985 at different depths in the treatments receiving either mineral fertilizer or liquid manure. Nitrate nitrogen (NO_3^--N) was measured during the vegetation period every 14 days in 8 samples per treatment. On average, the N content in soil water increased with increasing depth from 50 to 200 cm. At 50 cm: in 65% of the cases N values varied between 4 and 11 mg N l^{-1} for both treatments. At 100 cm: higher N values *viz*. 15 to 52 mg N l^{-1} resulted from the mineral fertilizer treatment in 81% and from the liquid manure treatment in 25% of the cases. At 200 cm: lower nitrate contents (17 to 39 mg N l^{-1}) were obtained from the addition of liquid manure than from the mineral fertilizer treatment (46 to 88 mg l^{-1}).

Introduction

Nitrate content in soil water has been reported in some recent publications [1, 2]. The results show that nitrate content is strongly influenced by factors such as climate, soil and pasture management.

Materials and methods

In a pasture experiment on sandy soil running since 1977 [2], suction cups were installed on the 19th of April 1985 at depths of 50, 100 and 200 cm. There were 8 samples per treatment. Treatment A received mineral fertilizer at a rate of 150 kg N ha^{-1} divided into 5 doses of 30 kg N ha^{-1} for the first 5 rotations. Treatment B received liquid manure at a rate of 140 kg N ha^{-1} divided into 2 doses applied on 21.02.1985 at 68 kg N ha^{-1} and on 06.03.1985 at 76 kg N ha^{-1}. Soil water was sampled every 14 days from the 15th of May 1985 to the beginning of the frost period on the 11th of

H.G. v.d. Meer, et al. (eds.), Animal Manure on Grassland and Fodder Crops.
 ISBN 978-94-010-8137-5.

December 1985. The nitrate content in water was analysed with an ion-sensitive electrode.

Results and discussion

The following values of nitrate-N (NO_3^--N) were measured in 1985 in soil water at different depths (Fig. 1). On average, the NO_3^--N content in soil water increased with increasing depth from 18 to 65 mg N l^{-1} on treatment A and from 12 to 24 mg N l^{-1} on treatment **B**.

- At 50 cm: 66% of the values from both treatments were lower than the FAO tolerance limit of 11.3 mg NO_3^--N l^{-1} (4–11 mg N l^{-1}).
- At 100 cm: higher NO_3^--N values (16–52 mg N l^{-1}) were found for liquid

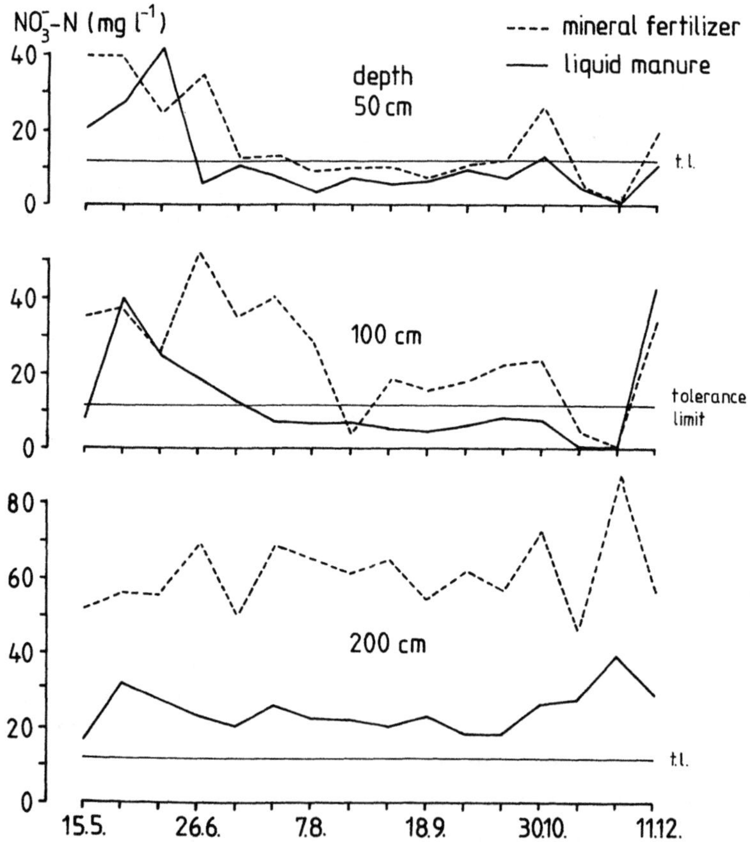

Figure 1. Nitrate-N content in soil water under permanent pasture. Comparison of mineral fertilizer and liquid manure; Völkenrode 1985.

Table 1. N input and N output ($kg\,ha^{-1}\,yr^{-1}$) on permanent pasture for fattening cattle. Comparison of mineral fertilizer and liquid manure

	mineral fertilizer	liquid manure $kg\,N\,ha^{-1}\,yr^{-1}$
N input		
Precipitation: NO_3^--N + NH_4^+-N	15[2]	15[2]
Fertilizer or manure	150[4]	140[4]
N fixation by legumes	12[1,2]	36[1,2]
Total	177	191
N output		
Weight gains of cattle	20[4,3]	20[4,3]
Losses to the air		
– from soil as $N_2O + N_2 + NH_3$	46[2,3]	46[2,3]
– from urine, faeces and liquid manure (laying on the soil)	24[2,3]	45[2,3]
Total	90	111
Difference (N input – N output)	87	80
Fixation in soil		
Topsoil: 0–13 cm	+33[1]	+38[1]
Subsoil: 10–30 cm	+5[1]	+27[1]
Calculated N balance	+49	+15

[1] calculated from analytical data of pasture exp. No. 155, period 1961–1985
[2] calculated from analytical data of a shorter period
[3] estimation from analytical data from literature
[4] analytical data 1985

manure in 25% and for mineral fertilizer in 81% of the cases (16–52 mg N l^{-1}). The difference between the treatments reached 33 mg N l^{-1}.

- At 200 cm: lower NO_3^--N values (17–39 mg N l^{-1}) were obtained after liquid manure than after mineral fertilizer (46–88 mg N l^{-1}).

Calculating the leaching losses in kg N $ha^{-1}\,yr^{-1}$, which was possible on the basis of 1) an estimated amount of leaching water [3] and the nitrate contents on a depth of 200 cm, and 2) the N surplus from the N balance (Table 1) [2] the following results were obtained:

	Mineral fertilizer	*Liquid manure*	
from the soil water	58	23	kg N $ha^{-1}\,yr^{-1}$
from the N balance	49	15	kg N $ha^{-1}\,yr^{-1}$

The difference between the N surplus and the amount of N found in the soil water was small (8–9 kg N $ha^{-1}\,yr^{-1}$). The N losses are higher (34–35 kg N $ha^{-1}\,yr^{-1}$) with mineral fertilizer than with liquid manure.

References

1. Wermke M and Dyckmans A (1983) EEC Workshop: Nitrogen fluxes in intensive grassland systems, Wageningen, 12 – 14.10.1983, pp 50–62
2. Wermke M (1984) Landwirtschaftliche Forschung, Sh. 41, 201–213
3. Wermke M und Laws W (1986) Vorträge auf der Jahrestagung 1986 der AG Grünland und Futterbau vom 4.–6.9.1986 in Nandlstadt, 168–188

The effect of a nitrification inhibitor on the utilization of nitrogen by grass from injected and surface-applied pig slurry

J H GEURINK and J L R M VERBEEK-OVERMEER
Centre for Agrobiological Research (CABO), P.O. Box 14, 6700 AA Wageningen, The Netherlands

Key words: nitrification inhibitor, slurry, nitrogen utilization, grass

Abstract. Addition of the nitrification inhibitor DCD to pig slurry applied by injection or in surface spreading to grassland in winter, considerably improved the utilization of nitrogen (N). In winter and in spring, injection of slurry resulted in an improved utilization of slurry N compared to surface application. In spring, DCD had a small negative effect on the utilization of N. Although the utilization of N from winter-applied slurry was markedly improved by both injection and the addition of DCD, it was less than that from spring applications without DCD, applied either by injection or surface spreading.

Introduction

Owing to a shortage of storage capacity many farmers apply slurry to grassland in winter. This may lead to large losses of nitrogen (N) by volatilization of ammonia, surface run-off, denitrification and leaching of nitrate. The effectiveness of dicyandiamide (DCD) in inhibiting nitrification in soils, and of injection in reducing volatilization of ammonia have been demonstrated [1, 2, 3]. This study was conducted to provide more information on the effects of slurry injection and DCD on the utilization of slurry N applied to grassland in winter.

Materials and methods

In a field experiment on permanent grassland on a sandy soil, the effects of injected: 30 or 60 t ha^{-1} (30 Inj or 60 Inj) and surface-applied: 30 t ha^{-1} (30 SA) pig slurry with and without DCD, on the uptake of N and dry matter (DM) production were studied for two years. The treatments were replicated four times. For the winter treatments, slurry was applied on 17 December 1983 and on 27 November 1984; for the spring treatments on 18 April 1984

H.G. v.d. Meer, et al. (eds.), Animal Manure on Grassland and Fodder Crops.
 ISBN 978-94-010-8137-5.

and 19 March 1985. The average composition of the pig slurry was (% of fresh material): DM: 8.2; total N: 0.63; NH_4^+-N: 0.32. DCD was mixed with the slurry; 30 kg DCD ha^{-1} in winter and 15 kg DCD ha^{-1} in spring. Fertilizer N was not applied. Phosphorus and potassium were maintained at the optimum supply by frequent dressings of inorganic fertilizer to the whole experimental field. The herbage of all treatments was mown on the same dates; 5 or 6 cuts were taken per year. Herbage samples were analysed for DM and N.

Results and discussion

The efficiency of use of N from slurry can be expressed in two ways: (i) as the apparent recovery of N (ANR) which is the increase of the amount of N contained in the harvested herbage expressed as a percentage of that applied in slurry:

$$\text{N recovery (ANR)} = \frac{\text{N uptake treated plot} - \text{N uptake untreated plot}}{\text{slurry N applied}} \times 100$$

and (ii) as the apparent efficiency of N (ANE) which is the increase of the herbage DM per kg N applied in slurry:

$$\text{N efficiency (ANE)} = \frac{\text{DM yield treated plot} - \text{DM yield untreated plot}}{\text{slurry N applied}}$$

Fig. 1 shows that addition of DCD to surface-spread and injected slurry applied in winter improved both ANR and ANE. This was very evident on the 30 Inj treatment. No explanation can be given for the relatively small effect of DCD on the 60 Inj treatment in winter. For winter applications, ANR and ANE from injected slurry were higher than from surface-spread slurry. In the spring treatments DCD had a small negative effect on both ANR and ANE. ANR and ANE of the spring treatments were considerably higher than those of comparable winter treatments; the difference in ANE between winter and spring treatments was smaller than that in ANR. In spring there was only a small difference in the growth responses between injected and surface-applied slurry. This is in contrast with results from

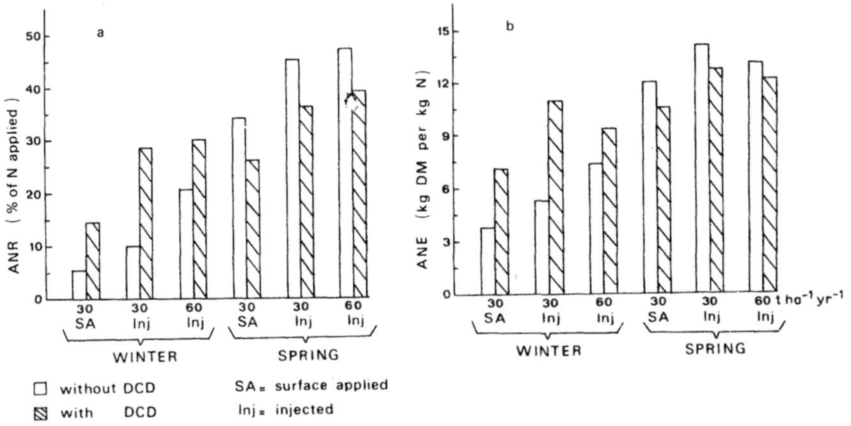

Figure 1. Average apparent recovery of N (ANR) (a) and apparent efficiency of N (ANE) (b) from surface-applied or injected pig slurry with and without nitrification inhibitor (DCD).

experiments with cattle slurry on grassland, where ANR and ANE from injected slurry were considerably higher than ANR and ANE from surface-applied slurry [2]. Possibly, the losses of N by volatilization of ammonia from surface-applied pig slurry are smaller than from surface-applied cattle slurry. Although the utilization of N from winter-applied slurry was greatly improved by both injection and the addition of DCD, it was less than that from spring applications, without DCD, applied either by injection or surface spreading.

References

1. Amberger A (1983) Symposium Nitrifikationshemmstoffe, pp 22–47. Weihenstephan, Federal Republic of Germany: Verband Deutscher Landwirtschaftlicher Untersuchungs- und Forschungsanstalten
2. Meer H G van der, Thompson R B, Snijders P J M and Geurink J H (1987) This volume
3. Solanski S (1981) Symposium on SKW-Didin, Nitrogen stabilizer for liquid manure, pp 87–96. Augsburg, Federal Republic of Germany: SKW Trostberg AG

The influence of ground superphosphate on the pH value and the ammonia release from slurry

G JÜRGENS
BASF Agricultural Research Station, 6703 Limburgerhof, Federal Republic of Germany

Key words: cattle slurry, pig slurry, superphosphate, pH value, ammonia volatilization

Abstract. In pot trials and in a trial under practical conditions, ground superphosphate was found to lower the pH value of pig and cattle slurry and to reduce the ammonia (NH_3) content in the air space above the slurry. In comparison with other slurry additives, such as dicyandiamide, bentonite with organic substances or a 50% cyanamide solution, ground superphosphate proved to be superior in lowering the pH value of the slurry and the NH_3 content in the air.

Introduction

The use of ground superphosphate in livestock housing for reducing the odour nuisance and improving the homogeneity and flowability of slurry has already been described [1, 2]. In the years from 1982 to 1984, small-scale pot trials and one trial under practical conditions (cattle shed with 70 cows) were conducted in order to investigate the influence on the pH value and the release of ammonia (NH_3) from pig and cattle slurry.

Materials and methods

In pot trials, cattle slurry containing 0.3–0.4% total N, was diluted with water in the ratio of 1:1. Ground superphosphate, as the commercial product Stall-Super containing 7.9% P, was used. The pH value of the slurry was measured by means of a pH meter, the NH_3 content in the air space above the slurry was measured with Dräger tubes (Ammoniak 5/a, Ref. N. 4044106).

Results and discussion

The effects can be summarized as follows:
1. Ground superphosphate at an application rate of 1% reduced the pH

H.G. v.d. Meer, et al. (eds.), Animal Manure on Grassland and Fodder Crops.
 ISBN 978-94-010-8137-5.

Table 1. The influence of ground superphosphate on the pH of slurry and on the NH_3 content in the air space above the slurry. (Values are means of daily measurements over the period indicated)

	Pig slurry 1983–1984 3 pot trials 3–4 weeks		Cattle slurry 1982 1 pot trial 3 weeks		Cattle slurry 1982 1 trial under practical conditions (cattle shed) 7 weeks	
	pH value	NH_3 content ppm	pH value	NH_3 content ppm	pH value	NH_3 content ppm
control	7.6	17.5	7.9	19.6	7.3	8.6
ground superphosphate	7.2	11.8	7.2	12.5	6.9	4.9
LSD (5%)	0.1	2.9	0.1	4.9	0.1	1.1

Table 2. The effect of various additives on pig slurry

	Application rate kg m^{-3}	pH value of the slurry	NH_3 content* ppm
control	–	8.25	199
dicyandiamide	3	8.30	239
bentonite	3	8.27	200
cyanamide solution (50%)	3	8.37	224
ground superphosphate	3	7.90	163
ground superphosphate	6	7.70	150
* in the air space above the slurry	LSD (5%)	0.49	33

value in all trials with pig and cattle slurry. On average, the pH value was lowered in three pot trials with pig slurry from 7.6 to 7.2 and in one pot trial with cattle slurry from 7.9 to 7.2 (Table 1).

2. The NH_3 release, which is one of the causes of the unpleasant odour of slurry and which was measured as the NH_3 content in the air space above the slurry, was reduced in the three pot trials with pig slurry from 17.5 ppm to 11.8 ppm and in the pot trial with cattle slurry from 19.6 to 12.5 ppm.
3. The positive influence of ground superphosphate on the reduction of the pH value and NH_3 release was confirmed in a 7-week trial under practical conditions (with cattle slurry from 70 cows, each group of 35 cows). The pH value of the slurry was lowered from 7.3 to 6.9 and the NH_3 content in the air space above the slurry from 8.6 to 4.9 ppm. In addition, a significant improvement of the air in the cattle shed and of the flowability of the slurry was observed during the trial.
4. A comparison with other slurry additives, such as dicyandiamide, bentonite with organic substances or a 50% cyanamide solution showed that after 9 days ground superphosphate was superior to these products with regard to the decrease of pH value and the reduction of NH_3 content in the air above the slurry (Table 2).

With the addition of ground superphosphate to cattle slurry at a daily rate of 0.6 kg per head of cattle and to pig slurry at a rate of 0.06 kg m^{-2} housing area, the slurry was enriched by about 0.7–0.8 kg P m^{-3}. This can be taken into account in a fertilizer programme. The net costs for the purchase of the slurry additive are thereby reduced compared with other materials.

References

1. Cramer N (1983) Gülleversuche in Schleswig-Holstein 1977–1983, pp 1–30. Kiel, FRG: Landwirtschaftskammer Schleswig-Holstein
2. Kosmat H (1974) Bericht über die 6. Arbeitstagung "Fragen der Güllerei", pp 1–47. Gumpenstein, Österreich: Bundesversuchsanstalt für alpenländische Landwirtschaft

Separation of pig slurry by sedimentation

S PICCININI, L CORTELLINI and G BONAZZI
Centro Ricerche Produzioni Animali, Via Crispi 3, Reggio Emilia, Italy

Key words: landspreading, pig slurry, solid separation, nutrient removal

Abstract. A pig slurry treatment system of five lagoons for the separation of a clarified fraction and a thickened fraction was monitored. Compared to the composition of fresh slurry, the thickened fraction contained 69% of the total solids (TS), 80% of the volatile solids (VS), 94% of the total suspended solids (TSS), 91% of the volatile suspended solids (VSS), 86% of the chemical oxigen demand (COD), 30% of the Kjeldahl total nitrogen (NTK), 14% of the ammonium nitrogen (N-NH_4^+) and 86% of the total phosphorus (P). This treatment can be used on farms in areas of high livestock density to reduce costs: the high volume clarified fraction can be irrigated on nearby land, the thickened fraction (6–10% TS; 7% of total volume) can be transported off the farm for application to arable land.

Introduction

In some areas of northern Italy there is a high density of pig farms together with cattle or poultry farms. For use in agriculture, the law states that the maximum quantity of slurry to be applied to one hectare should be equal to the effluent produced by 4 t of liveweight (LW). Local rules may reduce this quantity and prohibit the disposal in certain areas. Hence, many pig farms have to find more land for their effluent disposal. One way of reducing transport costs may be an easy straightforward system whereby two fractions are separated: a low volume fraction with a high amount of solids and nutrients for transport off the farm and a clarified fraction for spreading on surrounding land.

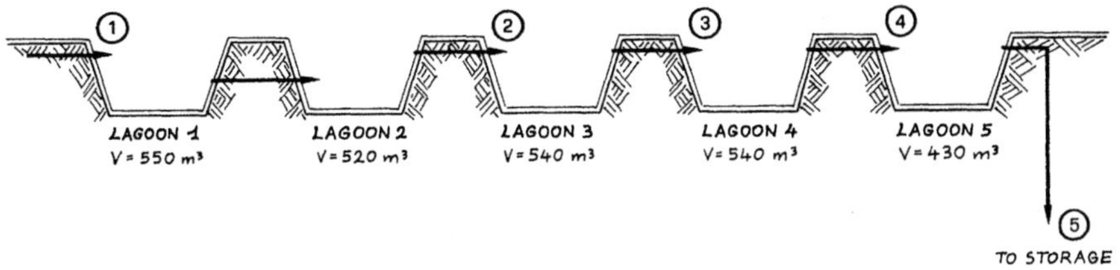

Figure 1. The treatment system: 1 – 5 sampling points.

H.G. v.d. Meer, et al. (eds.), Animal Manure on Grassland and Fodder Crops.
 ISBN 978-94-010-8137-5.

Table 1. Chemical characteristics of slurries at the different sampling points (1 – 5 in Fig. 1)

Parameter	Sampling points				
	1	2	3	4	5
	mean ± SD	mean ± SD	mean ± SD	mean ± SD	mean ± SD
TS ($g l^{-1}$)	10.00 ± 3.01	3.77 ± 0.60	3.34 ± 0.37	3.08 ± 0.28	3.07 ± 0.24
VS ($g l^{-1}$)	6.36 ± 2.16	1.80 ± 036	1.44 ± 0.22	1.26 ± 0.19	1.24 ± 0.16
TSS ($g l^{-1}$)	6.12 ± 2.04	1.04 ± 0.61	0.67 ± 0.57	0.50 ± 0.30	0.34 ± 0.15
COD ($mg l^{-1}$)	11730 ± 3220	2580 ± 1230	2020 ± 1010	1720 ± 830	1680 ± 780
COD_s ($mg l^{-1}$)	3810 ± 1490	1660 ± 590	1200 ± 480	1150 ± 500	1090 ± 495
NTK ($mg l^{-1}$)	1246 ± 340	968 ± 93	921 ± 143	859 ± 112	821 ± 141
N-NH_4^+ ($mg l^{-1}$)	810 ± 167	804 ± 67	752 ± 155	721 ± 119	671 ± 134
P ($mg l^{-1}$)	272 ± 130	91 ± 35	69 ± 25	45 ± 11	37 ± 10

Table 2. Removal efficiencies (%). S1: chemical characteristics of raw waste; S2–S5: chemical characteristics of clarified effluent leaving lagoons 1–5

Parameter	$\frac{S1 - S2}{S1}$	$\frac{S1 - S3}{S1}$	$\frac{S1 - S4}{S1}$	$\frac{S1 - S5}{S1}$
TS	62.3	66.6	69.2	69.3
VS	71.7	77.4	80.2	80.5
TSS	83.0	89.0	91.8	94.4
COD	78.0	82.8	85.3	85.7
NTK	22.3	26.1	31.1	34.1
$N\text{-}NH_4^+$	0.7	7.2	11.0	17.2
P	66.5	74.6	83.4	86.4

Materials and methods

A lagoon system adjoining a pig farm of 200 t LW producing 60 m^3 of slurry per day was monitored for about one year. The system as shown in Fig. 1 is composed of five sedimentation lagoons, 2.5 m deep with a total volume of 2600 m^3; the clarified effluent passes to a storage lagoon. Each year the lagoons are emptied by removing thickened slurries in July or August and clarified effluent in the irrigation season. From September 1985 to July 1986 the raw slurry and the effluent in lagoons 1–5 were sampled once or twice a month. The samples were analysed for TS, VS, TSS, COD, soluble COD (COD_s), NTK, $N\text{-}NH_4^+$ and total P. At the end of the cycle, the settled sludge from the first four lagoons was analysed for TS, VS, NTK, $N\text{-}NH_4^+$, and P after homogenization.

Results and discussion

Tables 1 and 2 show the average concentration and the efficiency of removal of the monitored parameters, respectively. The highest removal efficiency was found in lagoon 1 and 2: over 65% TS, over 70% COD, VSS, TSS, and total P, achieved both by sedimentation and by anaerobic digestion as can

Table 3. Characteristics of settled sludge in lagoons 1, 2, 3, and 4

	Volume m^3	TS (%)	VS (% of TS)	NTK (% of TS)	$N\text{-}NH_4^+$ (% of NTK)	P (% of TS)
lagoon 1	550	8.2	52.8	4.4	41.2	5.2
lagoon 2	520	9.6	49.4	5.6	42.5	4.1
lagoon 3	540	6.3	42.9	6.6	50.0	4.5
lagoon 4	540	0.7	46.7	14.1	87.6	6.1

be deduced from the COD_s data. Table 3 shows the features of sludge in lagoons 1, 2, 3, and 4; the suspended solids are almost totally removed in the first three lagoons; the well mineralized sludge has a volume of 1600 m^3 which is about 7% of the raw waste. Average TS content in the sludge is 8%. NTK and total P content is respectively 5.5% and 4.6% of TS.

Conclusions

The results indicate that 3–4 lagoons may be used to separate out 10% in volume of the raw waste as a thickened sludge. This is achieved by 2–3 sedimentation lagoons with a total detention time of 30 days, and a much larger one for receiving and storing the clarified fraction.

GPSR Compliance

The European Union's (EU) General Product Safety Regulation (GPSR) is a set of rules that requires consumer products to be safe and our obligations to ensure this.

If you have any concerns about our products, you can contact us on ProductSafety@springernature.com

In case Publisher is established outside the EU, the EU authorized representative is:

Springer Nature Customer Service Center GmbH
Europaplatz 3
69115 Heidelberg, Germany

Batch number: 10815974

Printed by Printforce, the Netherlands